IMPULSIVE SYSTEMS AND CONTROL: THEORY AND APPLICATIONS

IMPULSIVE SYSTEMS AND CONTROL: THEORY AND APPLICATIONS

TAO YANG

Nova Science Publishers, Inc.
Huntington, New York

Editorial Production: Susan Boriotti and Donna Dennis
Office Manager: Annette Hellinger
Graphics: Wanda Serrano
Information Editor: Tatiana Shohov
Book Production: Jonathan Rose, Jennifer Vogt, Matthew Kozlowski,,
and Lynette Van Helden
Circulation: Cathy DeGregory Ave Maria Gonzalez, Ron Hedges, and Andre Tillman

Library of Congress Cataloging-in-Publication Data
Available Upon Request

ISBN 1-59033-058-7.

227 Main Street, Suite 100
Huntington, New York 11743
Tele. 631-424-6682 Fax 631-425-5933
E Mail Novascil@aol.com; Novascience@earthlink.net

Printed in the United States of America

To my thirties

Preface

Although there are more than five books on the mathematical aspects of "impulsive differential equations", so far there is no book on the engineering applications of "impulsive differential equations". Within the last decade, there are many developments in control systems with impulsive effects or control systems via impulsive control laws. As the other control paradigms, an impulsive control system consists of a *plant*, which is usually a continuous-time dynamical system, and a *control law*. Unlike the other control paradigms, in an impulsive control system, *impulsive effects* that can change the state variables of the plant in a very short period are concerned. The impulsive effects can happen in the dynamics of the plant or can be introduced by the control laws. Thus, there are mainly two types of impulsive control systems; namely,

(1) controlling impulsive systems;
(2) impulsively control dynamical systems (plants).

This book will study both of them.

In the first type of impulsive control, the plant itself is modeled by impulsive differential equations and thus the control law can be either continuous or impulsive. The engineering examples of this type of impulsive control system came from the emerging field of nanoelectronics where the devices are fabricated in such a small scale that quantum effects dominate device behaviors. A typical example is a device called *single electron tunneling junction* (SETJ) where an electron can tunnel through the barrier in a time period of the order of 10^{-15}s and causes the voltage of the junction capacitor a jump in the order of 10^{-3}V. Since the junction capacitor voltage is a state variable of the SETJ model, it is practical to model SETJ using impulsive differential equations. The impulsive control becomes the most important method to design and program nanodevices which will be the building block for the next generation of computers.

In the second type of impulsive control, the plant itself has no impulsive effects but the control laws introduce impulsive effects to the state variables of the plant. The engineering examples come from communication systems with nonlinear carriers. For a communication system with nonlinear carriers, the critical technology is called *impulsive synchronization* which uses an impulsive control scheme to synchronize transmitter and receiver. In this case, impulsive effects are introduced by using strong current sources to charge capacitors such that the capacitor voltages can be changed in a very short time. In this case, the entire controlled systems can be modeled as impulsive differential equations.

This book will benefit three parties:

(1) give mathematicians the real applications of impulsive differential equations and evoke new activities in pure and applied mathematical researches on impulsive differential equations;
(2) provide engineers with a tool box and a well organized mathematical theory for impulsive control problems.
(3) provide physicists and engineers with a new framework of modeling many impulsive effects caused by quantum effects of nanodevices.

This book will provide real engineering applications and experimental results of impulsive control to demonstrate the importance and the unique ability of impulsive control paradigm. In many engineer applications, the impulsive control is the most beautiful and clearest solutions if it is not the only one.

This book is intended to serve as a graduate textbook for a class in impulsive control, as well as a useful monograph for the control engineers and physicists. It can also serve as a useful reference for a class in impulsive differential equation, nonlinear control theory, nanoelectronics, and nonlinear communication systems.

Berkeley, California
December 2000

Tao Yang

Contents

Chapter 1
Introduction

Although the impulsive effects in mechanical systems had been found and modeled many years ago and there are at least 5 English books[LBS89; BS89; BC94; SP95; BS93] and one journal[Jou] devoted to the theory of impulsive differential equations, it was until recent years engineers found industrial applications of impulsive differential equations and transferred it from mathematics to engineering tools. Three kinds of engineering applications had contributed to this transient.

First, the advances of digital control technology introduced many hybrid control structures in which the state variables can be suddenly changed because of the nature of digital controllers. Although many control engineers were confronted with this kind of phenomenon, they might treat them with standard digital control strategy rather than methods based on theory of impulsive differential equations; namely, impulsive control. This is because the mathematical theory of impulsive differential equations is still out of the reach of most control engineers.

Second, the recent develop of communication systems based on nonlinear carriers. In this kind of communication system, the technology call "impulsive synchronization" is in fact an "impulsive control" problem. The application of impulsive control can benefit this kind of communication system in many aspects such as high channel efficiency, multiple access and high security. Many experimental results performed in this area can be easily analyzed by impulsive control theory.

Finally, since the development of nanoelectronics in recent years, the basic devices which will replace CMOS transistors become so small that one digital bit is represented by a single electron. On of this kind of device is called "single electron tunneling junction" in which an electron can tunnel through the so called Coulomb blockade and introduce an impulsive effect to the junction voltage. Impulsive control can be used to design different strategies for implementing digital logic operations based on nanodevices. Since nanoelectronics is growing rapidly and since it will be the jointed research between physicists and electrical engineers, impulsive control theory can benefit this emerging field very much.

This book will cover the structures and designing principles of different types of impulsive controllers. The stability of impulsive control will be systematically studied. Many real-life applications and real experimental results will be presented to show mathematicians what kind of engineering problem should be further studied and show engineers and physicists how to use impulsive differential equations to solve their problems in a precise and simple way.

1.1 Basic knowledge of control systems

A *control system* is a paradigm which consists physical variables of a system called a *plant*. It behaves in a prescribed way under the influence of uncertainties and disturbances. The plant is usually modeled as a continuous-time, discrete-time or hybrid dynamical system. A general automatic control system is shown in Fig. 1.1. Observe that the automatic control system consists of a plant and a set of control laws. The output $y(t)$ of the plant can be observed through a measuring channel where some noise $n(t)$ is added. Then the control laws use the noisy measurement $y(t) + n(t)$ and a reference signal $r(t)$ to generate a control signal $u(t)$. $u(t)$ is then fed into the input port of the plant. The control objective is to make the system output $y(t)$ related to reference signal $r(t)$ in a prescribed manner. We usually use $r(t)$ to determine the positions of equilibrium points of the control systems while $u(t)$ is used to make these equilibrium points stable.

The control system in Fig. 1.1 is very general in the sense that it covers both feedforward and feedback control paradigms. And the control laws can be very general and include adaptive control laws, fuzzy control laws, neural network control laws, optimal control laws, robustness control laws and computational verb control laws[Yan00],etc. For example, a computational verb feedback control system is shown in Fig. 1.2 as a special case of this general control paradigm. Observe that in a verb feedback control system,

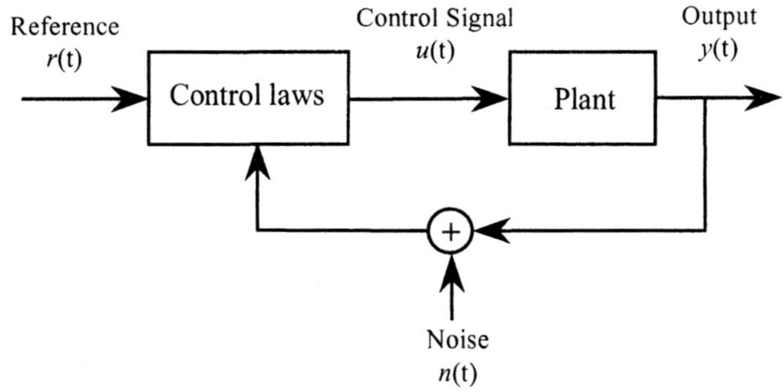

Fig. 1.1 The block diagram of a general automatic control system.

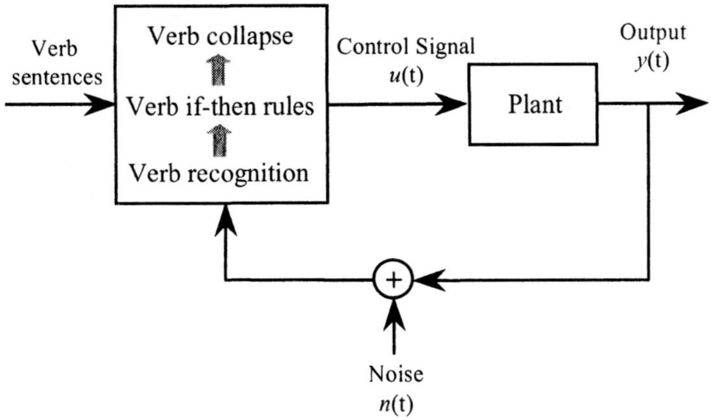

Fig. 1.2 The block diagram of a computational verb feedback control system.

the reference signal is given in the form of verb sentences. The control laws are replaced by a series of verb computations including verb recognition, verb if-then rules and verb collapse.

1.2 What is impulsive control ?

Impulsive control is a control framework based on impulsive differential equations. A plant in an impulsive control system should have at least one *changeable* state variable or have impulsive effects in at least one state variable.

Definition 1.1 *Impulsive control*
Given an *nonimpulsive plant* $\mathcal{P}$ whose state variable is denoted by $X \in \mathbb{R}^n$, a set of *control instants* $T = \{\tau_k\}$, $\tau_k \in \mathbb{R}$, $\tau_k < \tau_{k+1}$, $k = 1, 2, \cdots$, and *control laws* $U(k, X) \in \mathbb{R}^n$, $k = 1, 2, \cdots$. At each τ_k, X is changed impulsively by $X(\tau_k^+) = X(\tau_k) + U(k, X)$ such that the output $Y = f(X)$, $f : \mathbb{R}^n \to \mathbb{R}^m$, $Y \in \mathbb{R}^m$, approaches a goal $Y^* \in \mathbb{R}^m$ as $k \to \infty$. If the plant $\mathcal{P}$ is impulsive, then the control law can be nonimpulsive.

Remarks:

(1) At least one state-variable in a plant $\mathcal{P}$ can be changed instantaneously to any value which is given by a control law. In this sense, not all physical systems can be controlled by impulsive control schemes.
(2) We only need to change the *changeable state variables* at discrete instants called *control instants.* Control instants are not necessary to be equi-distant. Control instants can also be chosen whenever some conditions are satisfied.
(3) The control law $U(k, X)$ is something related to the concept of "input" in other control methods. However, in an impulsive control, $U(k, X)$ gives a sudden change of X at instant τ_k.
(4) The purpose of invention of impulsive control is not to compete with the other control schemes, on the contrary, impulsive control provides a new viewpoint when the plant has at least one changeable state-variable or when the plant has impulsive effects.

To address the necessary of impulsive control we should first give some examples of plants whose state-variables can be changed instantaneously.

Example 1.1 If we want to control the population of a kind of insect by using its natural enemies, we can model this problem by a system whose state-variables are numbers of insects and their natural enemies. We can cultivate the natural enemies in a laboratory and then lease them to the nature at some proper instants. In this sense, some state variables of this system can be changed instantaneously.

Example 1.2 In a chemical reactor, the quantities of different chemicals service as state variables. To control the process of reaction, we can add chemicals instantaneously to change state variables. In this sense, we can use impulsive control here.

Example 1.3 In a financial system, we suppose that one state variable is the amount of money in a market and the other state-variables are savings rates of a central bank. As it always occurs, the former can be controlled to a desired value by changing the latter instantaneously. In fact, this kind of impulsive control has been used for a long time to control the financial market by different governments.

The necessary of inventing impulsive control is that in some cases impulsive control may give an efficient way to deal with plants which cannot endure continuous control inputs. Or, it is impossible to provide continuous control input. Let us consider the following examples.

(1) In some cases, the plant can not be controlled by using continuous control. For example, a government can not change savings rates of its central bank everyday.
(2) In some cases, impulsive control is more efficient. For example, suppose that the population of a kind of bacterium and the density of a bactericide are two state variables of a system. We can control the population by instantaneously changing the density of the bactericide without enhance the drug resistance of bacteria which may be caused by continuous control.
(3) For plants that are impulsive in nature, impulsive control is the best choice if not the only choice. The nanodevices are among this kind of plant.

1.3 Different types of impulsive control schemes

Impulsive systems can be classified into the following types based on the characteristics of plants and control laws.

The type-I impulsive control system is given by

$$\begin{aligned} \dot{\mathbf{x}} &= \mathbf{f}(t, \mathbf{x}), \quad t \neq \tau_k(\mathbf{x}), \\ \Delta\mathbf{x} &= U(k, \mathbf{y}), \quad t = \tau_k(\mathbf{x}), \\ \mathbf{y} &= \mathbf{g}(\mathbf{x}). \end{aligned} \tag{1.1}$$

In this kind of system, the control input is implemented by the "sudden jumps" of some state variables.

The type-II impulsive control system is given by

$$\begin{aligned} \dot{\mathbf{x}} &= \mathbf{f}(t,\mathbf{x}) + \tilde{\mathbf{h}}(\tilde{\mathbf{u}}), \quad t \neq \tau_k(\mathbf{x}), \\ \Delta\mathbf{x} &= U(k,\mathbf{y}), \quad t = \tau_k(\mathbf{x}), \\ \mathbf{y} &= \mathbf{g}(\mathbf{x},\tilde{\mathbf{u}}). \end{aligned} \tag{1.2}$$

In this system, there are two kinds of control inputs. The first kind is continuous control input and the second one is impulsive control input.

The type-III impulsive control system is given by

$$\begin{aligned} \dot{\mathbf{x}} &= \mathbf{f}(t,\mathbf{x}) + \tilde{\mathbf{h}}(\tilde{\mathbf{u}}), \quad t \neq \tau_k(\mathbf{x}), \\ \Delta\mathbf{x} &= \mathbf{j}_k(\mathbf{x}), \quad t = \tau_k(\mathbf{x}), \\ \mathbf{y} &= \mathbf{g}(\mathbf{x},\tilde{\mathbf{u}}). \end{aligned} \tag{1.3}$$

In this kind of systems, the plant itself is impulsive and the control is continuous.

There are many other types of impulsive control systems such as impulsive plants with impulsive laws are less important in engineering applications, we do not discuss them in this book. However, those types can be easily understood by using the same methods presented in this book.

Chapter 2

Basic Knowledge of Impulsive Differential Equations

Many books on impulsive differential equations have been published[LBS89; BS89; BC94; SP95; BS93]. In this chapter, we present only some relevant results of impulsive differential equations to make this book self-contained.

2.1 General notations and definitions

In this book we use the following notations:

$\triangleq$		defined as
$\mathbb{N}$	$\triangleq$	set of all positive integers
$\mathbb{Z}$	$\triangleq$	set of all integers
$\mathbb{R}$	$\triangleq$	set of all real numbers
$\mathbb{R}_+$	$\triangleq$	$[0, +\infty)$
$\mathbb{C}$	$\triangleq$	set of all complex numbers
$\mathbb{R}^n$	$\triangleq$	real n-dimensional Euclidean space
$\mathbb{C}^n$	$\triangleq$	complex n-dimensional Euclidean space
$\mathrm{Re}(z)$	$\triangleq$	real part of the complex number z
z^*	$\triangleq$	complex conjugate of the complex number z

$$
\begin{aligned}
\mathbf{x} &\triangleq \text{col}(x_1, x_2, \cdots, x_n);\ \text{namely, element of } \mathbb{R}^n \text{ or } \mathbb{C}^n \\
\|\mathbf{x}\| &\triangleq \text{norm of the element } \mathbf{x} \in \mathbb{R} \text{ or } \mathbf{x} \in \mathbb{C} \\
\|\mathbf{x}\|_2 &\triangleq \sqrt{\sum_{i=1}^{n} x_i x_i^*}
\end{aligned}
$$

$$
\begin{aligned}
I &\triangleq n \times n \text{ identity matrix} \\
I_m, I_{m\times m} &\triangleq m \times m \text{ identity matrix}
\end{aligned}
$$

$$
\begin{aligned}
A = A(a_{ij}) &\triangleq n \times n \text{ matrix with entries } a_{ij} \\
A = A(a_{ij})_{m\times m} &\triangleq m \times m \text{ matrix with entries } a_{ij} \\
A^{-1} &\triangleq \text{inverse of the matrix } A \\
A^{\dagger} &\triangleq \text{pseudo-inverse of the matrix } A \\
A^{\top} = A(a_{ji}) &\triangleq \text{transposed matrix of the matrix } A = (a_{ij}) \\
\det A &\triangleq \text{determinant of the matrix } A \\
\text{diag}(A_1, A_2, \cdots, A_m) &\triangleq \text{block-diagonal matrix with blocks } A_1, A_2, \cdots, A_m \\
S^{\perp} &\triangleq \text{the orthogonal complement of } S
\end{aligned}
$$

$$
\begin{aligned}
\mathcal{H} &\triangleq \{h \in C[\mathbb{R}_+ \times \mathbb{R}^n, \mathbb{R}_+]|\ \inf_{\mathbf{x}\in\mathbb{R}^n} h(t, \mathbf{x}) = 0 \text{ for each } t \in \mathbb{R}_+\}, \\
\mathcal{K} &\triangleq \{\kappa \in C[\mathbb{R}_+, \mathbb{R}_+]|\ \kappa(t) \text{ is monotone strictly increasing and } \kappa(0) = 0\}, \\
\mathcal{KR} &\triangleq \{\kappa \in \mathcal{K}|\ \lim_{t\to+\infty} \kappa(t) = +\infty\}, \\
\mathcal{CK} &\triangleq \{\kappa \in C[\mathbb{R}_+ \times \mathbb{R}_+, \mathbb{R}_+]|\ \kappa(t, s) \in \mathcal{K} \text{ for each } t \in \mathbb{R}_+\}.
\end{aligned}
$$

$\mathcal{PC}(D, F)$: set of functions $\psi : D \to F$ which are continuous for $t \in D, t \neq \tau_k$, have discontinuities of first kind at the points τ_k and are left continuous.

$\mathfrak{J}^+(t_0, \mathbf{x}_0)$: the maximal interval of the form (t_0, ω) in which the solution $\mathbf{x}(t, t_0, \mathbf{x}_0)$ is defined.

$\mathfrak{J}^-(t_0, \mathbf{x}_0)$: the maximal interval of the form (α, t_0) in which the solution $\mathbf{x}(t, t_0, \mathbf{x}_0)$ is defined.

$$\mathfrak{S}_\rho \triangleq \{\mathbf{x} \in \mathbb{R}^n | \ \|\mathbf{x}\| < \rho\}.$$

2.2 Mathematical model of systems with impulsive effects

The mathematical model of an evolving process with impulsive effects is given by

(1) A differential equation

$$\frac{d\mathbf{x}}{dt} = \mathbf{f}(t, \mathbf{x}), \tag{2.1}$$

where $\mathbf{x} \in \Omega$ and $t \in \mathbb{R}$.

(2) A *switching set* $\Sigma_t \in \Omega \times \mathbb{R}$.

(3) A *jumping operator* $\mathcal{J}_t$ defined on the switching set Σ_t such that $\mathcal{J}_t \circ \Sigma_t \subset \Omega \times \mathbb{R}$.

The set Ω is called the *phase space*; namely, the set of all possible states of this evolving process. The *state variable* $\mathbf{x}(t)$ represents the state of the evolving process at time t. If we need only n parameters to describe $\mathbf{x}(t)$ at t and n is a finite number, then this evolving process is *finite dimensional.* In this case, $\mathbf{x}(t)$ is an n-dimensional vector of the Euclidean space $\mathbb{R}^n$ and Ω is a subset of $\mathbb{R}^n$. We call $\Omega \times \mathbb{R}$ the *extended phase space.*

With an *initial state,* $\mathbf{x}(t_0^+) = \mathbf{x}_0$, Eq. (2.1) defines a *dynamical system,* the solution to Eq. (2.1) is often written as $\phi_t(\mathbf{x}_0)$, or $\mathbf{x}(t) = \mathbf{x}(t, t_0, \mathbf{x}_0)$. The one-parameter family of mappings $\phi_t : \mathbb{R} \to \mathbb{R}$ satisfies

$$\phi_{t+s} = \phi_t \circ \phi_s$$

and

$$\phi_0(\mathbf{x}) = \mathbf{x},$$

and is called the *flow.* The set of points $\{\phi_t(\mathbf{x}_0) | -\infty < t < \infty\}$ is called a *trajectory* of (2.1) through $\mathbf{x}_0$. If an impulsive event happens at $t = \tau_i$, then the trajectory *hits* the switching set Σ_t at $t = \tau_i$; namely, $\mathbf{x}(\tau_i) \in \Sigma_t$. As soon as $\mathbf{x}(t)$ hits the switching set Σ_t at $t = \tau_i$, it "jumps" immediately to a point defined by the jumping operator $\mathcal{J}_t$; namely, $\mathbf{x}(\tau^+) = \mathcal{J}_t \circ \mathbf{x}(\tau)$.

Thus the solution of the impulsive system $\phi_t(\mathbf{x}_0)$, satisfies Eq. (2.1) outside the switching set Σ_t and has discontinuities of the first kind at the points where it hits the switching set Σ_t with the jumps

$$\Delta\mathbf{x}(t) = \phi_{t^+}(\mathbf{x}_0) - \phi_t(\mathbf{x}_0) = \mathcal{J}_t \circ \phi_t(\mathbf{x}_0) - \phi_t(\mathbf{x}_0). \tag{2.2}$$

We then have a concise form of impulsive differential equation as

$$\begin{aligned} \frac{d\mathbf{x}}{dt} &= \mathbf{f}(t,\mathbf{x}), \qquad (t,\mathbf{x}) \notin \Sigma_t, \\ \Delta\mathbf{x} &= \mathcal{J}_t \circ \mathbf{x} - \mathbf{x}, \qquad (t,\mathbf{x}) \in \Sigma_t. \end{aligned} \tag{2.3}$$

Equation (2.3) can have the following three kinds of solutions:

(1) $\phi_t(\mathbf{x}_0)$ does not hit Σ_t or the hitting points are fixed points of $\mathcal{J}_t$. In this case, there is no impulsive event.
(2) $\phi_t(\mathbf{x}_0)$ hits Σ_t at a finite number of points that are not fixed points of $\mathcal{J}_t$. In this case, there are finite number of impulsive events.
(3) $\phi_t(\mathbf{x}_0)$ hits Σ_t in a countable number of points that are not fixed points of $\mathcal{J}_t$. In this case, there are countable number of impulsive events.

Based on the different characteristics of impulsive events, we usually encounter three primary types of impulsive systems

(1) in which impulses occur at fixed time;
(2) in which impulses occur when the trajectory hits a hypersurface in the extended phase space;
(3) which are discontinuous dynamical systems.

2.2.1 *Impulsive events at fixed time*

If the impulsive events occur in a finite or infinite sequence of time $\{\tau_k\}$, then system (2.3) can be written as

$$\begin{aligned} \frac{d\mathbf{x}}{dt} &= \mathbf{f}(t,\mathbf{x}), \qquad t \neq \tau_k, \\ \Delta\mathbf{x} &= J_k(\mathbf{x}), \qquad t = \tau_k. \end{aligned} \tag{2.4}$$

The solution of Eq. (2.4) is a piecewise continuous function $\mathbf{x} = \phi(t)$ that has discontinuities of the first kind at $t = \tau_k$. $\frac{d\phi(t)}{dt} = \mathbf{f}(t, \phi(t))$ is satisfied for all $t \neq \tau_k$. At $t = \tau_k$, $\phi(t)$ satisfies the following jumping condition:

$$\Delta\phi|_{t=\tau_k} = \phi(\tau_k^+) - \phi(\tau_k) = J_k(\phi(\tau_k)). \tag{2.5}$$

2.2.2 *Impulsive events at variable time*

In this kind of system, the impulsive events occurs when trajectories hit a hypersurface $\aleph(t, \mathbf{x}) = 0$. This kind of impulsive differential equation can be written as

$$\begin{aligned} \frac{d\mathbf{x}}{dt} &= \mathbf{f}(t, \mathbf{x}), \quad \aleph(t, \mathbf{x}) \neq 0, \\ \Delta\mathbf{x} &= J(t, \mathbf{x}), \quad \aleph(t, \mathbf{x}) = 0. \end{aligned} \tag{2.6}$$

In this case, the switching set is given by

$$\Sigma_t = \{(t, \mathbf{x}) | \ \aleph(t, \mathbf{x}) = 0\}, \tag{2.7}$$

and the jumping operator $\mathcal{J}_t$ is given by

$$\mathcal{J}_t : (t, \mathbf{x}) \to (t, \mathbf{x} + J(t, \mathbf{x})). \tag{2.8}$$

If the equation $\aleph(t, \mathbf{x}) = 0$ has a countable number of solutions with respect to t, we denote these solutions by $t = \tau_k(\mathbf{x})$ and index then by the set of integers (or a subset of integers) such that $\tau_k(\mathbf{x}) \to \infty$ as $k \to \infty$ and $\tau_k(\mathbf{x}) \to -\infty$ as $k \to -\infty$. In this case, the jumping operator is given by

$$\mathcal{J}_{\tau_k(\mathbf{x})} : \mathbf{x} \to \mathbf{x} + J(\tau_k(\mathbf{x}), \mathbf{x}), \tag{2.9}$$

which can be simplified as

$$\mathcal{J}_{\tau_k(\mathbf{x})} : \mathbf{x} \to \mathbf{x} + J_k(\mathbf{x}). \tag{2.10}$$

Then Eq. (2.6) can be rewritten as

$$\begin{aligned} \frac{d\mathbf{x}}{dt} &= \mathbf{f}(t, \mathbf{x}), \quad t \neq \tau_k(\mathbf{x}), \\ \Delta\mathbf{x} &= J_k(\mathbf{x}), \quad t = \tau_k(\mathbf{x}). \end{aligned} \tag{2.11}$$

System (2.6) is more difficult to study than system (2.4). The solutions of system (2.6) starting at different initial conditions have different points of discontinuity. A solution of system (2.6) may hit the same hypersurface $\aleph(t, \mathbf{x}) = 0$ many times and cause so called "*beating phenomenon*" or "*pulse phenomenon*". Different solutions of system (2.6) may coincide after some time and behave like a single solution thereafter and thus cause "*confluence*".

2.2.3 *Discontinuous dynamical systems*

When the differential equation in Eq. (2.1) does not explicitly depend on t and the jumping operator $\mathcal{J}_t = \mathcal{J}$ for all $t \in \mathbb{R}$, let $\mathcal{J} : \Sigma \to \Sigma_0$ be a mapping between the switching set Σ and a target set Σ_0, then a discountinuous dynamical system is given by

$$\begin{aligned} \frac{d\mathbf{x}}{dt} &= \mathbf{f}(\mathbf{x}), \quad \mathbf{x} \notin \Sigma, \\ \Delta\mathbf{x} &= \mathcal{J}\mathbf{x} - \mathbf{x} = J(\mathbf{x}), \quad \mathbf{x} \in \Sigma. \end{aligned} \tag{2.12}$$

Many electrical circuit models containing switching capacitors belong to this kind of impulsive system. To get interesting phenomena from this kind of impulsive system, we usually assume that the switching set Σ is a compact or locally compact manifold in the phase space with dimension $(n-1)$. Or else, there is a good chance that a large part of the motions will not have impulsive effects.

2.3 Existence and continuation of solutions

Consider the *initial value problem* for the following impulsive differential equation:

$$\begin{aligned} \frac{d\mathbf{x}}{dt} &= \mathbf{f}(t,\mathbf{x}), \quad t \neq \tau_k(\mathbf{x}), \\ \Delta\mathbf{x} &= J_k(\mathbf{x}), \quad t = \tau_k(\mathbf{x}), \\ \mathbf{x}(t_0^+) &= \mathbf{x}_0, \quad t_0 \geq 0 \end{aligned} \tag{2.13}$$

where $\mathbf{f} : D \to \mathbb{R}^n$ and $J_k : \Omega \to \mathbb{R}^n$. Here we let $\Omega \subset \mathbb{R}^n$ be an open set and $D = \mathbb{R}_+ \times \Omega$. We assume that $\tau_k \in C[\Omega, \mathbb{R}_+]$, $\tau_k(\mathbf{x}) < \tau_{k+1}(\mathbf{x})$ and $\lim_{k\to\infty} \tau_k(\mathbf{x}) = \infty$ for $k = 1, 2, \cdots$, and $\mathbf{x} \in \Omega$.

Definition 2.1 A function $\mathbf{x} : (t_0, t_0 + \varepsilon) \to \mathbb{R}^n$, $t_0 \geq 0$, $\varepsilon > 0$, is a solution of Eq. (2.13) if

(1) $\mathbf{x}(t_0^+) = \mathbf{x}_0$ and $(t, \mathbf{x}(t)) \in D$ for $t \in [t_0, t_0 + \varepsilon)$,

(2) $\mathbf{x}(t)$ is continuously differentiable and satisfies $\dot{\mathbf{x}}(t) = \mathbf{f}(t, \mathbf{x}(t))$ for $t \in [t_0, t_0 + \varepsilon)$ and $t \neq \tau_k(\mathbf{x}(t))$,

(3) if $t \in [t_0, t_0 + \varepsilon)$ and $t = \tau_k(\mathbf{x}(t))$, then $\mathbf{x}(t^+) = \mathbf{x}(t) + J_k(\mathbf{x})$, and at such t's we always assume that $\mathbf{x}(t)$ is left continuous and $\alpha \neq \tau_i(\mathbf{x}(\alpha))$ for any $i = 1, 2, \cdots$, $t < \alpha < \delta$, for some $\delta > 0$.

Remark: Observe that instead of the initial condition, $\mathbf{x}(t_0) = \mathbf{x}_0$, used in ordinary differential equations, the limiting condition $\mathbf{x}(t_0^+) = \mathbf{x}_0$ is used. This is because $(t_0, \mathbf{x}_0)$ may be such that $t_0 = \tau_k(\mathbf{x}_0)$ for some k. Whenever $t_0 \neq \tau_k(\mathbf{x}_0)$ for any k, then $\mathbf{x}(t_0^+) = \mathbf{x}_0$ is the same as $\mathbf{x}(t_0) = \mathbf{x}_0$. Unlike ordinary differential equations, Eq. (2.13) may have no solution event if $\mathbf{f}$ is continuous or continuously differentiable since its only solution of the initial value problem may entirely lie on a surface $\Sigma_k : t = \tau_k(\mathbf{x})$. Thus we need some extra condition on τ_k and/or $\mathbf{f}$ besides continuity in order to establish existence theory for Eq. (2.13).

Theorem 2.1 *Assume that*

(1) $\mathbf{f} : D \to \mathbb{R}^n$ *is continuous at* $t \neq \tau_k(\mathbf{x})$, $k = 1, 2, \cdots$, *and for each* $(t, \mathbf{x}) \in D$ *there is an* $l \in \mathcal{L}^1_{\text{loc}}$ *such that for any* $(\alpha, \mathbf{y})$ *in a small neighborhood of* $(t, \mathbf{x})$ *we have*

$$\|\mathbf{f}(\alpha, \mathbf{y})\| \leq l(\alpha); \tag{2.14}$$

(2) for any k, $t_1 = \tau_k(\mathbf{x}_1)$ *implies the existence of a* $\delta > 0$ *such that*

$$t \neq \tau_k(\mathbf{x}) \tag{2.15}$$

for any $t - t_1 \in (0, \delta)$ *and* $\|\mathbf{x} - \mathbf{x}_1\| < \delta$.

Then, for each $(t_0, \mathbf{x}_0) \in D$, *there exists a solution* $\mathbf{x} : [t_0, t_0 + \gamma) \to \mathbb{R}^n$ *of the initial value problem (2.13) for some* $\gamma > 0$.

Proof. If $t_0 \neq \tau_k(\mathbf{x}_0)$ for any k then the conclusion is true. If $t_0 = \tau_k(\mathbf{x}_0)$ for some k, from Eq. (2.14) and since $\mathbf{f}$ is continuous, we know that there exists a local solution $\mathbf{x}(t)$ of $\dot{\mathbf{x}} = \mathbf{f}(t, \mathbf{x})$ and $\mathbf{x}(t_0^+) = \mathbf{x}_0$. Since $\tau_i(\mathbf{x}) < \tau_j(\mathbf{x})$ for $i < j$ and $t_0 = \tau_k(\mathbf{x}_0)$, we know $t \neq \tau_j(\mathbf{x})$ for $j \neq k$ and t sufficiently close to t_0. Condition (2.15) does not permit $t = \tau_k(\mathbf{x}(t))$ in a sufficient small right neighborhood of t_0. Hence $\mathbf{x}(t)$ is a local solution of the system (2.13). □

Theorem 2.2 *Assume that*

(1) $\mathbf{f} : D \to \mathbb{R}^n$ *is continuous;*
(2) $\tau_k : \Omega \to (0, \infty)$ *are diferentiable;*
(3) if $t_1 = \tau_k(\mathbf{x}_1)$ *for some* $(t_1, \mathbf{x}_1) \in D$ *and* $k \geq 1$, *then there is a* $\delta > 0$ *such that*

$$\frac{\partial \tau_k(\mathbf{x})}{\partial \mathbf{x}} \mathbf{f}(t, \mathbf{x}) \neq 1 \tag{2.16}$$

for $(t, \mathbf{x}) \in D$ *with* $t - t_1 \in (0, \delta)$ *and* $|\mathbf{x} - \mathbf{x}_1| < \delta$.

Then, for each $(t_0, \mathbf{x}_0) \in D$, *there exists a solution* $\mathbf{x} : [t_0, t_0 + \alpha) \to \mathbb{R}^n$ *of the initial value problem (2.13) for some* $\alpha > 0$.

Proof. For the cases of $t_0 \neq \tau_k(\mathbf{x}_0)$ for any k the proof is straightforward. If $t_0 = \tau_k(\mathbf{x}_0)$ for some $k \geq 1$ and $\mathbf{x}(t)$ is a solution of $\dot{\mathbf{x}} = \mathbf{f}(t, \mathbf{x})$ and $\mathbf{x}(t_0^+) = \mathbf{x}_0$, set $\gamma(t) = t - \tau_k(\mathbf{x}(t))$. Then $\gamma(t_0) = 0$ and

$$\dot{\gamma}(t) = 1 - \frac{\partial \tau_k(\mathbf{x}(t))}{\partial \mathbf{x}} \mathbf{f}(t, \mathbf{x}(t)). \tag{2.17}$$

Followed Eq. (2.16) we know that $\dot{\gamma}(t) \neq 0$ in a small right neighborhood of t_0. Hence $\gamma(t)$ is either strictly increasing or decreasing in that neighborhood and we have $t \neq \tau_k(\mathbf{x}(t))$ for $t - t_0 \in (0, \delta)$ for some $\delta > 0$. The rest of proof is similar to that of Theorem 2.1. □

We then consider the continuation of solutions of (2.13) to the right. Given a solution $\mathbf{x}(t)$ of (2.13) defined on $[t_0, t_0 + \alpha)$ with $\alpha > 0$, we say that a solution $\mathbf{x}^*(t)$ of (2.13) is a *proper* continuation to the right of $\mathbf{x}(t)$ if $\mathbf{x}^*(t)$ is defined on $[t_0, t_0 + \beta)$ for some $\beta > \alpha$ and $\mathbf{x}(t) = \mathbf{x}^*(t)$ for $t \in [t_0, t_0 + \alpha)$. The interval $[t_0, t_0 + \alpha)$ is called the *maximal interval of existence* of a solution $\mathbf{x}(t)$ of (2.13), if $\mathbf{x}(t)$ is well defined on $[t_0, t_0 + \alpha)$ and it does not have any proper continuation to the right. We then have the following theorem.

Theorem 2.3 *Assume that* $\Omega = \mathbb{R}^n$ *and*

(1) $\mathbf{f} : D \to \mathbb{R}^n$ *is continuous;*
(2) $J_k \in C[\Omega, \mathbb{R}^n]$, $\tau_k \in C[\Omega, \mathbb{R}_+]$ *for all* $k > 1$.

Then, for any solution $\mathbf{x}(t)$ *of (2.13) with a finite* $[t_0, \beta)$ *as its maximal interval of existence, we have*

$$\lim_{t \to \beta^-} \|\mathbf{x}(t)\| = \infty, \tag{2.18}$$

if one of the following three conditions is satisfied:

(1) for any $k \geq 1$, $t_1 = \tau_k(\mathbf{x}_1)$ *implies the existence of a* $\delta > 0$ *such that* $t \neq \tau_k(\mathbf{x})$ *for all* $(t, \mathbf{x})$ *with* $t - t_1 \in (0, \delta)$ *and* $\|\mathbf{x} - \mathbf{x}_1\| < \delta$*;*

(2) for all $k \geq 1$, $t_1 = \tau_k(\mathbf{x}_1)$ *implies that* $t_1 \neq \tau_j(\mathbf{x}_1 + J_k(\mathbf{x}_1))$ *for all* $j \geq 1$*;*

(3) $\tau_k \in C^1[\Omega, \mathbb{R}_+]$ *for all* $k \geq 1$ *and* $t_1 = \tau_k(\mathbf{x}_1)$ *implies that* $t_1 = \tau_j(\mathbf{x}_1 + J_k(\mathbf{x}_1))$ *for some* $j \geq 1$ *and*

$$\frac{\partial \tau_j(\mathbf{x}_1^+)}{\partial \mathbf{x}} \mathbf{f}(t_1, \mathbf{x}_1^+) \neq 1, \tag{2.19}$$

where $\mathbf{x}_1^+ = \mathbf{x}_1 + J_k(\mathbf{x}_1)$.

Proof. See pages 17 to 20 of [LBS89]. □

Some other results on the continuation of solutions can be found in Chapter 3 of [BS89]. We do not repeat them here.

2.4 Beating phenomena

The solutions of an impulsive differential equation may hit the same switching surface finite or infinite number of times causing *beating phenomena*. Consider the following impulsive differential equation:

$$\begin{aligned}
\frac{d\mathbf{x}}{dt} &= \mathbf{f}(t, \mathbf{x}), \quad t \neq \tau_k(\mathbf{x}), \\
\Delta \mathbf{x} &= J_k(\mathbf{x}), \quad t = \tau_k(\mathbf{x}), \\
\mathbf{x}(t_0^+) &= \mathbf{x}_0, \quad t_0 \geq 0
\end{aligned} \tag{2.20}$$

where $\mathbf{f} \in C[\mathbb{R}_+ \times \Omega, \mathbb{R}^n]$ and $\Omega \subset \mathbb{R}^n$ is an open set.

Theorem 2.4 *Assume that*

(1) $\mathbf{f} \in C[\mathbb{R}_+ \times \Omega, \mathbb{R}^n]$, $\tau_k \in C^1[\Omega, \mathbb{R}_+]$, $\tau_k(\mathbf{x}) < \tau_{k+1}(\mathbf{x})$ *for every* k, $\lim_{k\to\infty} \tau_k(\mathbf{x}) = \infty$ *uniformly in* $\mathbf{x} \in \Omega$ *and* $J_k \in [\Omega, \mathbb{R}^n]$*;*

(2) (a) $\frac{\partial \tau_k(\mathbf{x})}{\partial \mathbf{x}} \mathbf{f}(t, \mathbf{x}) \leq 0$ *for* $(t, \mathbf{x}) \in \mathbb{R}_+ \times \Omega$*, and*

(b) $\mathbf{x} + J_k(\mathbf{x}) \in \Omega$ *for* $\mathbf{x} \in \Omega$ *and*

$$\left(\frac{\partial \tau_k(\mathbf{x} + \epsilon J_k(\mathbf{x}))}{\partial \mathbf{x}}\right) J_k(\mathbf{x}) \leq 0,$$

$\epsilon \in [0, 1]$, *for every* k.

Then, each solution of Eq. (2.20) hits any given switching surface $\Sigma_i : t = \tau_i(\mathbf{x})$ at most once.

Proof. See pages 22 and 23 of [LBS89]. □

Theorem 2.5 *Assume that*

(1) $\mathbf{f} \in C[\mathbb{R}_+ \times \Omega, \mathbb{R}^n]$, $\tau_k \in C^1[\Omega, \mathbb{R}_+]$, $\tau_k(\mathbf{x}) < \tau_{k+1}(\mathbf{x})$ *for every* k, $\lim_{k\to\infty} \tau_k(\mathbf{x}) = \infty$ *uniformly in* $\mathbf{x} \in \Omega$ *and* $J_k \in [\Omega, \mathbb{R}^n]$*;*

(2) *(a)* $\frac{\partial \tau_k(\mathbf{x})}{\partial \mathbf{x}} \mathbf{f}(t, \mathbf{x}) \leq \delta$, $0 \leq \delta \leq 1$, *for* $(t, \mathbf{x}) \in \mathbb{R}_+ \times \Omega$*;*

(b) $\mathbf{x} + J_k(\mathbf{x}) \in \Omega$ *for* $\mathbf{x} \in \Omega$ *and*

$$\left(\frac{\partial \tau_k(\mathbf{x} + \epsilon J_k(\mathbf{x}))}{\partial \mathbf{x}}\right) J_k(\mathbf{x}) < 0,$$

$\epsilon \in [0, 1]$, *for every* k.

Then, each solution of Eq. (2.20) hits any given switching surface $\Sigma_i : t = \tau_i(\mathbf{x})$ at most once.

Proof. Similar to that of Theorem 2.4. □

Theorem 2.6 *Assume that*

(1) $\mathbf{f} \in C[\mathbb{R}_+ \times \Omega, \mathbb{R}^n]$, $J_k \in C[\Omega, \mathbb{R}^n]$, $\tau_k \in C^1[\Omega, \mathbb{R}_+]$, $\tau_k(\mathbf{x})$ *is bounded and* $\tau_k(\mathbf{x}) < \tau_{k+1}(\mathbf{x})$ *for every* k*;*

(2) *(a)* $\frac{\partial \tau_k(\mathbf{x})}{\partial \mathbf{x}} \mathbf{f}(t, \mathbf{x}) \leq 1$, *for* $(t, \mathbf{x}) \in \mathbb{R}_+ \times \Omega$*;*

(b)

$$\left(\frac{\partial \tau_k(\mathbf{x} + \epsilon J_k(\mathbf{x}))}{\partial \mathbf{x}}\right) J_k(\mathbf{x}) < 0;$$

(c)

$$\left(\frac{\partial \tau_k(\mathbf{x} + \epsilon J_{k-1}(\mathbf{x}))}{\partial \mathbf{x}}\right) J_{k-1}(\mathbf{x}) \geq 0$$

where $\mathbf{x} + J_k(\mathbf{x}) \in \Omega$ *for* $\mathbf{x} \in \Omega$ *and* $\epsilon \in [0, 1]$, *for every* k.

Then, each solution $\mathbf{x}(t, t_0, \mathbf{x}_0)$ of Eq. (2.20) with $0 \leq t_0 < \tau_1(\mathbf{x}_0)$ hits each switching surface $\Sigma_i : t = \tau_i(\mathbf{x})$ exactly once.

Proof. See pages 24 and 25 of [LBS89]. □

2.5 Solutions of impulsive differential equations

Let us consider the solution of the following linear impulsive differential equation:

$$\begin{cases} \dot{\mathbf{x}} = A(t)\mathbf{x}, & t \neq \tau_k, \\ \Delta\mathbf{x} = B_k\mathbf{x}, & t = \tau_k, \quad k = 1, 2, \cdots. \end{cases} \tag{2.21}$$

Theorem 2.7 *Assume that*

(1)

$$0 = \tau_0 < \tau_1 < \cdots, \quad \lim_{k\to\infty} \tau_k = \infty; \tag{2.22}$$

(2)

$$A \in \mathcal{PC}(\mathbb{R}^+, \mathbb{C}^{n\times n}), \quad B_k \in \mathbb{C}^{n\times n}, \quad k = 1, 2, \cdots. \tag{2.23}$$

Then for any $(t_0, \mathbf{x}_0) \in \mathbb{R}_+ \times \mathbb{R}^n$ *there exists a unique solution* $\phi(t)$ *of system (2.21) which is defined for* $t > t_0$ *with* $\phi(t_0^+) = \mathbf{x}_0$*. Furthermore, if* $\det(I + B_k) \neq 0$ *for every* k*, then this solution is defined for all* $t \in \mathbb{R}_+$*.*

Proof. See the proof of Theorem 3.6 of [BS89]. □

Let the functions $\mathbf{x}_1(t), \mathbf{x}_2(t), \cdots, \mathbf{x}_n(t)$ be solutions of system (2.21) defined in the interval (t_0, ∞). Let $X(t) = [\mathbf{x}_1(t), \mathbf{x}_2(t), \cdots, \mathbf{x}_n(t)]$ be a matrix-valued function. Then $\mathbf{x}_1(t), \mathbf{x}_2(t), \cdots, \mathbf{x}_n(t)$ are linearly independent in the interval (t_0, ∞) if and only if $\det X(t_0^+) \neq 0$. Then we call $X(t)$ a *fundamental matrix* of solutions of system (2.21) in the interval (t_0, ∞).

Let the *state transition matrix* of the linear system

$$\dot{\mathbf{x}} = A(t)\mathbf{x}, \quad t \in (\tau_{k-1}, \tau_k] \tag{2.24}$$

be $\Phi_k(t, s)$ $(t, s \in (\tau_{k-1}, \tau_k])$ and conditions (2.22) and (2.23) hold, then the solution of the system (2.21) is given by

$$\mathbf{x}(t; t_0, \mathbf{x}_0) = \Psi(t, t_0^+)\mathbf{x}_0, \tag{2.25}$$

where $\Psi(t,s)$ is called *state transition matrix* of system (2.21) and given by

$$\Psi(t,s)=\begin{cases}\Phi_k(t,s), \\ \qquad\qquad \text{for } t,s\in(\tau_{k-1},\tau_k], \\ \Phi_{k+1}(t,\tau_k^+)(I+B_k)\Phi_k(\tau_k,s), \\ \qquad\qquad \text{for } \tau_{k-1}<s\le\tau_k<t\le\tau_{k+1}, \\ \Phi_k(t,\tau_k)(I+B_k)^{-1}\Phi_{k+1}(\tau_k^+,s), \\ \qquad\qquad \text{for } \tau_{k-1}<t\le\tau_k<s\le\tau_{k+1}, \\ \Phi_{k+1}(t,\tau_k^+)\left(\prod_{j=k}^{i+1}(I+B_j)\Phi_j(\tau_j,\tau_{j-1}^+)\right) \\ \quad (I+B_i)\Phi_i(\tau_i,s), \\ \qquad\qquad \text{for } \tau_{i-1}<s\le\tau_i<\tau_k<t\le\tau_{k+1}, \\ \Phi_i(t,\tau_i)\left(\prod_{j=i}^{k-1}(I+B_j)^{-1}\Phi_{j+1}(\tau_j^+,\tau_{j+1})\right) \\ \quad (I+B_k)^{-1}\Phi_{k+1}(\tau_k^+,s), \\ \qquad\qquad \text{for } \tau_{i-1}<t\le\tau_i<\tau_k<s\le\tau_{k+1}.\end{cases} \tag{2.26}$$

$\Psi(t,s)$ has the following properties:

$$\begin{aligned}\Psi(t,t) &= I, \\ \Psi(\tau_k^-,\tau_k) &= \Psi(\tau_k,\tau_k^-)=I, \\ \Psi(\tau_k^+,s) &= (I+B_k)\Psi(\tau_k,s), \\ \Psi(s,\tau_k^+) &= (I+B_k)^{-1}\Psi(s,\tau_k), \qquad s\ne\tau_k^+, \\ \frac{\partial\Psi}{\partial t}(t,s) &= A(t)\Psi(t,s), \qquad t\ne\tau_k.\end{aligned} \tag{2.27}$$

We then consider the solution of the following linear impulsive nonautonomous system:

$$\begin{cases}\dot{\mathbf{x}}=A(t)\mathbf{x}+\mathbf{g}(t), & t\ne\tau_k, \\ \Delta\mathbf{x}=B_k\mathbf{x}+\mathbf{j}_k, & t=\tau_k, \quad k=1,2,\cdots,\end{cases} \tag{2.28}$$

where $\mathbf{g}\in\mathcal{PC}(\mathbb{R}_+,\mathbb{C}^n)$, $\mathbf{j}_k\in\mathbb{C}^n$. Then the solution of system (2.28), $\mathbf{x}(t)\triangleq\mathbf{x}(t,t_0,\mathbf{x}_0)$, is given by

$$\mathbf{x}(t)=\begin{cases}\Psi(t,t_0^+)\mathbf{x}(t_0^+)+\int_{t_0}^t\Psi(t,s)\mathbf{g}(s)ds \\ \quad +\sum_{t_0<\tau_k<t}\Psi(t,\tau_k^+)\mathbf{j}_k, & t>t_0, \\ \Psi(t,t_0^+)\mathbf{x}(t_0^+)+\int_{t_0}^t\Psi(t,s)\mathbf{g}(s)ds \\ \quad -\sum_{t\le\tau_k\le t_0}\Psi(t,\tau_k^+)\mathbf{j}_k, & t\le t_0.\end{cases} \tag{2.29}$$

We then consider the solution of the following impulsive differential equations:

$$\begin{cases} \dot{\mathbf{x}} = A(t)\mathbf{x} + \mathbf{g}(t,\mathbf{x}), & t \neq \tau_k, \\ \Delta\mathbf{x} = B_k\mathbf{x} + \mathbf{j}_k(\mathbf{x}), & t = \tau_k, \quad k = 1, 2, \cdots, \end{cases} \tag{2.30}$$

where $A(\cdot) \in \mathcal{PC}(\mathbb{R}_+, \mathbb{R}^{n\times n})$, $B_k \in \mathbb{R}^{n\times n}$, $\mathbf{g} \in \mathbb{R}_+ \times \Omega \to \mathbb{R}^n$, and $\mathbf{j}_k \in \Omega \to \mathbb{R}^n$ where $\Omega \subset \mathbb{R}^n$. Then the solution of system (2.30), $\mathbf{x}(t) \triangleq \mathbf{x}(t, t_0, \mathbf{x}_0)$, is given by

$$\mathbf{x}(t) = \begin{cases} \Psi(t,t_0^+)\mathbf{x}(t_0^+) + \int_{t_0}^t \Psi(t,s)\mathbf{g}(s,\mathbf{x}(s))ds & \\ \quad + \sum_{t_0<\tau_k<t} \Psi(t,\tau_k^+)\mathbf{j}_k(\mathbf{x}(\tau_k)), & t \in \mathfrak{J}^+(t_0,\mathbf{x}_0), \\ \Psi(t,t_0^+)\mathbf{x}(t_0^+) + \int_{t_0}^t \Psi(t,s)\mathbf{g}(s,\mathbf{x}(s))ds & \\ \quad - \sum_{t\leq\tau_k\leq t_0} \Psi(t,\tau_k^+)\mathbf{j}_k(\mathbf{x}(\tau_k)), & t \in \mathfrak{J}^-(t_0,\mathbf{x}_0). \end{cases} \tag{2.31}$$

2.6 Lyapunov's second method

Let us consider the stability of the zero solution of the following impulsive differential equation:

$$\begin{aligned} \dot{\mathbf{x}} &= \mathbf{f}(t,\mathbf{x}), & t \neq \tau_k(\mathbf{x}), \\ \Delta\mathbf{x} &= \mathbf{j}_k(\mathbf{x}), & t = \tau_k(\mathbf{x}), \quad k = 1, 2, \cdots, \end{aligned} \tag{2.32}$$

where $t \in \mathbb{R}_+$, $\mathbf{x} \in \mathfrak{S}_\rho$, $\rho > 0$.

We use the following notation of set:

$$\begin{aligned} \mathfrak{G}_i &\triangleq \{(t,\mathbf{x}(t)) \in \mathbb{R}_+ \times \mathbb{R}^n \mid \tau_{i-1}(\mathbf{x}) < t < \tau_i(\mathbf{x})\}, \\ & \quad i = 1, 2, \cdots, \text{ and } \mathfrak{G} = \bigcup_{i=1}^{\infty} \mathfrak{G}_i, \end{aligned} \tag{2.33}$$

where $\tau_0(\mathbf{x}) = 0$ is assumed.

Definition 2.2 Let $V : \mathbb{R}_+ \times \mathbb{R}^n \to \mathbb{R}_+$, then V is said to belong to class $\mathcal{V}_0$ if

(1) V is continuous in $(\tau_{k-1}, \tau_k] \times \mathbb{R}^n$ and for each $\mathbf{x} \in \mathbb{R}^n$, $k = 1, 2, \cdots$,

$$\lim_{(t,\mathbf{y})\to(\tau_k^+,\mathbf{x})} V(t,\mathbf{y}) = V(\tau_k^+, \mathbf{x}) \tag{2.34}$$

exists;

(2) V is locally Lipschitzian in $\mathbf{x}$.

Definition 2.3 Let $V : \mathbb{R}_+ \times \mathbb{R}^n \to \mathbb{R}_+$, then V is said to belong to class $\mathcal{V}_1$ if $V \in \mathcal{V}_0$ and it is continuously differentiable in $\mathfrak{G}$.

Definition 2.4 For $(t, \mathbf{x}) \in \mathfrak{G}$, we define the *derivative* of the function $V \in \mathcal{V}_1$ with respect to system (2.32) as

$$\dot{V}(t, \mathbf{x}) \triangleq \frac{\partial V(t, \mathbf{x})}{\partial t} + \frac{\partial V(t, \mathbf{x})}{\partial \mathbf{x}} \mathbf{f}(t, \mathbf{x}). \tag{2.35}$$

Definition 2.5 For $(t, \mathbf{x}) \in \mathfrak{G}$, we define the *upper right derivative* of the function $V \in \mathcal{V}_0$ with respect to system (2.32) as

$$D^+V(t, \mathbf{x}) \triangleq \lim_{h \to 0} \sup \frac{1}{h}[V(t + h, \mathbf{x} + h\mathbf{f}(t, \mathbf{x})) - V(t, \mathbf{x})]. \tag{2.36}$$

Theorem 2.8 *Assume that*

(1) the function $\mathbf{f} : \mathbb{R}_+ \times \Omega \to \mathbb{R}^n$ *is continuous in* $\mathbb{R}_+ \times \Omega$, $\mathbf{f}(t, 0) = 0$ *for* $t \in \mathbb{R}_+$ *and there is a constant* $L > 0$ *such that* $\|\mathbf{f}(t, \mathbf{x}) - \mathbf{f}(t, \mathbf{y})\| \leq L\|\mathbf{x} - \mathbf{y}\|$ *for* $t \in \mathbb{R}_+$, $\mathbf{x}, \mathbf{y} \in \Omega$;

(2) the functions $\mathbf{j}_k : \Omega \to \mathbb{R}^n$ *are continuous and* $\mathbf{j}_k(0) = 0$;

(3) There is a constant $\rho \in (0, H)$ *such that if* $\mathbf{x} \in \mathfrak{S}_\rho$, *then* $\mathbf{x} + \mathbf{j}_k(\mathbf{x}) \in \Omega$;

(4) the functions $\tau_k : \Omega \to \mathbb{R}_+$ *are continuous in* Ω *and*

$$0 = \tau_0(\mathbf{x}) < \tau_1(\mathbf{x}) < \tau_2(\mathbf{x}) < \cdots, \quad \lim_{k \to \infty} \tau_k(\mathbf{x}) = \infty, \quad \mathbf{x} \in \Omega;$$

(5) there is no beating phenomenon;

(6) there are $V \in \mathcal{V}_1$ *and* $\alpha \in \mathcal{K}$ *such that*

$$\alpha(\|\mathbf{x}\|) \leq V(t, \mathbf{x}), \quad \text{for } (t, \mathbf{x}) \in \mathbb{R}_+ \times \Omega, \tag{2.37}$$

$$\dot{V}(t, \mathbf{x}) \leq 0, \quad \text{for } (t, \mathbf{x}) \in \mathfrak{G}, \tag{2.38}$$

$$V(t^+, \mathbf{x} + \mathbf{j}_k(\mathbf{x})) \leq V(t, \mathbf{x}), \text{ for } (t, \mathbf{x}) \in \Sigma_k \cap (\mathbb{R}_+ \times \mathfrak{S}_\rho). \tag{2.39}$$

Then the solution $\mathbf{x} = 0$ *of system (2.32) is stable. Furthermore, if for some* $\beta \in \mathcal{K}$ *we have*

$$V(t, \mathbf{x}) \leq \beta(\|\mathbf{x}\|), \quad \text{for } (t, \mathbf{x}) \in \mathbb{R}_+ \times \Omega \tag{2.40}$$

then the solution $\mathbf{x} = 0$ *of system (2.32) is uniformly stable.*

Proof. See that of Theorem 13.1 of [BS89]. □

Let the set

$$\Xi(t_0) = \{\mathbf{x}_0 \in \Omega |\ \mathbf{x}(t, t_0, \mathbf{x}_0) \to 0 \text{ as } t \to \infty\}$$

be the *basin of attraction of the origin* at the instant t_0. And for $t \in \mathbb{R}_+$, $\rho \in \mathbb{R}_+$, $V \in \mathcal{V}_0$ and $\alpha \in \mathcal{K}$ we define the sets

$$\Theta_\rho(t, \alpha) = \{\mathbf{x} \in \Omega |\ V(t^+, \mathbf{x}) < \alpha(\rho)\}.$$

Theorem 2.9 *Let the assumptions (1)-(5) of Theorem 2.8 hold and function $V \in \mathcal{V}_1$ and there are $\alpha, \beta, \varrho \in \mathcal{K}$ such that*

$$\alpha(\|\mathbf{x}\|) \le V(t, \mathbf{x}) \le \beta(\|\mathbf{x}\|), \qquad \textit{for } (t, \mathbf{x}) \in \mathbb{R}_+ \times \Omega, \tag{2.41}$$

$$\dot{V}(t, \mathbf{x}) \le -\varrho(\|\mathbf{x}\|), \qquad \textit{for } (t, \mathbf{x}) \in \mathfrak{G}, \tag{2.42}$$

$$V(t^+, \mathbf{x} + \mathbf{j}_k(\mathbf{x})) \le V(t, \mathbf{x}), \ \textit{for } (t, \mathbf{x}) \in \Sigma_k \cap (\mathbb{R}_+ \times \mathfrak{S}_\rho). \tag{2.43}$$

Then we have the following conclusions:

(1) If $0 < a < \rho$, then $\lim_{t\to\infty} \|\mathbf{x}(t, t_0, \mathbf{x}_0)\| = 0$ uniformly with respect to $(t_0, \mathbf{x}_0) \in \mathbb{R}_+ \times \Theta_a(t, \alpha)$;
(2) The solution $\mathbf{x} = 0$ of system (2.32) is uniformly asymptotically stable.

Proof. See that of Theorem 13.2 of [BS89]. □

Theorem 2.10 *Let the assumptions (1)-(5) of Theorem 2.8 hold and functions $V, V_1 \in \mathcal{V}_1$ and there are $\alpha, \beta, \varrho \in \mathcal{K}$ such that*

$$\alpha(\|\mathbf{x}\|) \le V(t, \mathbf{x}), \ \ \beta(\|\mathbf{x}\|) \le V_1(t, \mathbf{x}), \qquad \textit{for } (t, \mathbf{x}) \in \mathbb{R}_+ \times \Omega, \tag{2.44}$$

$$\dot{V}(t, \mathbf{x}) \le -\varrho(V_1(t, \mathbf{x})), \qquad \textit{for } (t, \mathbf{x}) \in \mathfrak{G}, \tag{2.45}$$

$$V(t^+, \mathbf{x} + \mathbf{j}_k(\mathbf{x})) \le V(t, \mathbf{x}), \ \textit{for } (t, \mathbf{x}) \in \Sigma_k \cap (\mathbb{R}_+ \times \mathfrak{S}_\rho), \tag{2.46}$$

and $\dot{V}_1(t, \mathbf{x})$ is bounded above (resp., below) in $\mathfrak{G}$ and

$$V_1(t^+, \mathbf{x} + \mathbf{j}_k(\mathbf{x})) \le V_1(t, \mathbf{x}) \ (\textit{resp., } V_1(t^+, \mathbf{x} + \mathbf{j}_k(\mathbf{x})) \ge V_1(t, \mathbf{x})), \\ \textit{for } (t, \mathbf{x}) \in \Sigma_k \cap (\mathbb{R}_+ \times \mathfrak{S}_\rho). \tag{2.47}$$

Then we have the following conclusions:

(1) If $0 < a < \rho$ and $t_0 \in \mathbb{R}_+$, then $\Xi(t_0) \supset \Theta_a(t, \alpha)$;

(2) The solution $\mathbf{x} = 0$ of system (2.32) is asymptotically stable.

Proof. See that of Theorem 13.3 of [BS89]. □

From Theorem 2.10 we have the following corollary.

Corollary 2.1 *Let the assumptions (1)-(5) of Theorem 2.8 hold and a function $V \in \mathcal{V}_1$ and there are $\alpha, \varrho \in \mathcal{K}$ such that*

$$\alpha(||\mathbf{x}||) \leq V(t, \mathbf{x}), \quad \textit{for } (t, \mathbf{x}) \in \mathbb{R}_+ \times \Omega, \tag{2.48}$$

$$\dot{V}(t, \mathbf{x}) \leq -\varrho(V(t, \mathbf{x})), \quad \textit{for } (t, \mathbf{x}) \in \mathfrak{G}, \tag{2.49}$$

$$V(t^+, \mathbf{x} + \mathbf{j}_k(\mathbf{x})) \leq V(t, \mathbf{x}), \ \textit{for } (t, \mathbf{x}) \in \Sigma_k \cap (\mathbb{R}_+ \times \mathfrak{S}_\rho). \tag{2.50}$$

Then we have the following conclusions:

(1) If $0 < a < \rho$ and $t_0 \in \mathbb{R}_+$, then $\Xi(t_0) \supset \Theta_a(t, \alpha)$;
(2) The solution $\mathbf{x} = 0$ of system (2.32) is asymptotically stable.

Proof. Let $V_1 = V$ and $\beta = \alpha$ then the proof is immediately followed from Theorem 2.10. □

Theorem 2.11 *Let the assumptions (1)-(5) of Theorem 2.8 hold and a function $V \in \mathcal{V}_1$ and there are positive constants a, b, c and a positive integer p such that*

$$a||\mathbf{x}||^p \leq V(t, \mathbf{x}) \leq b||\mathbf{x}||^p, \quad \textit{for } (t, \mathbf{x}) \in \mathbb{R}_+ \times \Omega, \tag{2.51}$$

$$\dot{V}(t, \mathbf{x}) \leq -c||\mathbf{x}||^p, \quad \textit{for } (t, \mathbf{x}) \in \mathfrak{G}, \tag{2.52}$$

$$V(t^+, \mathbf{x} + \mathbf{j}_k(\mathbf{x})) \leq V(t, \mathbf{x}), \ \textit{for } (t, \mathbf{x}) \in \Sigma_k \cap (\mathbb{R}_+ \times \mathfrak{S}_\rho). \tag{2.53}$$

Then the solution $\mathbf{x} = 0$ of system (2.32) is exponentially stable.

Proof. See that of Theorem 13.5 of [BS89]. □

2.7 Stability in terms of two measures

Stability in terms of two measures[LL93; LL89] provides a much more flexible framework of studying the stability of impulsive control systems. For example, to study the stability of a stepping motor or the stability of phase states of a driven single electron tunneling junction, we need to study stability of periodic solutions instead of equilibrium points. In this kind of impulsive control problem, we need the methods presented in this section.

Let us consider the following impulsive differential equation:

$$\begin{aligned} \dot{\mathbf{x}} &= \mathbf{f}(t,\mathbf{x}), \quad t \neq \tau_k(\mathbf{x}), \\ \Delta\mathbf{x} &= \mathbf{j}_k(\mathbf{x}), \quad t = \tau_k(\mathbf{x}), \end{aligned} \tag{2.54}$$

where $\mathbf{f} \in C[\mathbb{R}_+ \times \mathbb{R}^n, \mathbb{R}^n]$, $\tau_k \in C[\mathbb{R}^n, \mathbb{R}]$ and $\mathbf{j}_k \in C[\mathbb{R}^n, \mathbb{R}^n]$.

Definition 2.6 Let $h_0, h \in \mathcal{H}$, then h_0 is *finer* than h if there exists a $\sigma > 0$ and a function $\alpha \in \mathcal{K}$ such that $h_0(t,\mathbf{x}) < \sigma$ implies $h(t,\mathbf{x}) \leq \alpha(h_0(t,\mathbf{x}))$.

Definition 2.7 The impulsive system (2.54) is (h_0, h)-*stable* if, given $\epsilon > 0$ and $t_0 \in \mathbb{R}_+$, there exists a $\delta = \delta(t_0, \epsilon)$ such that $h_0(t_0, \mathbf{x}_0) < \delta$ implies $h(t, \mathbf{x}(t)) < \epsilon$, $t \geq t_0$ for any solution $\mathbf{x}(t) = \mathbf{x}(t, t_0, \mathbf{x}_0)$ of system (2.54).

$$\mathfrak{S}_\rho(h) \triangleq \{(t,\mathbf{x}) \in \mathbb{R}_+ \times \mathbb{R}^n \mid h(t,\mathbf{x}) < \rho, \quad h \in \mathcal{H}\}. \tag{2.55}$$

Definition 2.8 Let $V \in \mathcal{V}_0$ and $h_0, h \in \mathcal{H}$, then $V(t,\mathbf{x})$ is said to be

(1) h-*positive definite* if there exists a $\rho > 0$ and a function $\beta \in \mathcal{K}$ such that $h(t,\mathbf{x}) < \rho$ implies $\beta(h(t,\mathbf{x})) \leq V(t,\mathbf{x})$;
(2) h_0-*decrescent* if there exists a $\delta > 0$ and a function $\alpha \in \mathcal{K}$ such that $h_0(t,\mathbf{x}) < \delta$ implies $V(t,\mathbf{x}) \leq \alpha(h_0(t,\mathbf{x}))$;
(3) *weakly* h_0-*decrescent* if there exists a $\delta > 0$ and a function $\alpha \in C\mathcal{K}$ such that $h_0(t,\mathbf{x}) < \delta$ implies $V(t,\mathbf{x}) \leq \alpha(t, h_0(t,\mathbf{x}))$.

Theorem 2.12 *Assume that*

(1) $h_0, h \in \mathcal{H}$ and h_0 is finer than h;
(2) there exists a $V \in \mathcal{V}_0$ such that $V(t,\mathbf{x})$ is locally Lipschitzian in $\mathbf{x}$ on each $\mathfrak{G}_i$, and h-positive definite on $\mathfrak{S}_\rho(h)$, and $D^+V(t,\mathbf{x}) \leq 0$ on $G \cap \mathfrak{S}_\rho(h)$;

(3) $V(\tau_k^+, \mathbf{x} + \mathbf{j}_k(\mathbf{x})) \leq V(t, \mathbf{x})$ *on* $\Sigma_k \cap \mathfrak{S}_\rho(h)$.

Then system (2.54) is

(1) (h_0, h)*-stable if* $V(t, \mathbf{x})$ *is weakly* h_0*-decrescent;*
(2) (h_0, h)*-uniformly stable if* $V(t, \mathbf{x})$ *is* h_0*-decrescent.*

Proof. If $V(t, \mathbf{x})$ is weakly h_0-decrescent, then there exists a $\delta_0 > 0$ and a function $\alpha \in C\mathcal{K}$ such that if $h_0(t, \mathbf{x}) < \delta_0$ we have

$$V(t, \mathbf{x}) \leq \alpha(t, h_0(t, \mathbf{x})). \tag{2.56}$$

Since $V(t, \mathbf{x})$ is h-positive definite on $\mathfrak{S}_\rho(h)$, there exists a function $\beta \in \mathcal{K}$ such that

$$\beta(h(t, \mathbf{x})) \leq V(t, \mathbf{x}), \quad (t, \mathbf{x}) \in \mathfrak{S}_\rho(h). \tag{2.57}$$

Followed assumption (1), we can find a $\delta_1 > 0$ and a function $\phi \in \mathcal{K}$ such that if $h_0(t, \mathbf{x}) < \delta_1$ we have

$$h(t, \mathbf{x}) \leq \phi(h_0(t, \mathbf{x})). \tag{2.58}$$

Given $\epsilon > 0$ and $t_0 \in \mathbb{R}_+$, there exist $0 < \delta_2 < \delta_0$ and $0 < \delta_3 < \delta_1$ such that

$$\alpha(t_0, \delta_2) < \beta(\epsilon), \text{ and } \phi(\delta_3) < \rho. \tag{2.59}$$

Let us set $\delta = \min(\delta_2, \delta_3)$ and $m(t) = V(t, \mathbf{x})$ provided $h_0(t_0, \mathbf{x}_0) < \delta$. It follows from assumption (2) that $D^+m(t) \leq 0$ for $t \neq \tau_k$. From assumption (3) we know that $m(\tau_k^+) \leq m(\tau_k)$. Therefore, $m(t)$ is nonincreasing. From Eqs. (2.56)-(2.59) we know that for $t \geq t_0$

$$\beta(h(t, \mathbf{x}(t))) \leq m(t) \leq m(t_0^+) \leq \alpha(t_0, h_0(t_0, \mathbf{x}_0)) < \beta(\epsilon), \tag{2.60}$$

which yields for $t \geq t_0$ and $h_0(t_0, \mathbf{x}_0)) < \delta$,

$$h(t, \mathbf{x}(t)) < \epsilon. \tag{2.61}$$

We than have conclusion (1).

If $V(t, \mathbf{x})$ is h_0-decrescent, we have Eqs. (2.56) and (2.59) with α independent of t. Thus, δ can be chosen independent of t_0 such that Eq. (2.61) holds provided $h_0(t_0, \mathbf{x}_0)) < \delta$. This leads to conclusion (2). □

Definition 2.9 Let $\lambda : \mathbb{R}_+ \to \mathbb{R}_+$ be measurable. Then $\lambda(t)$ is *integrally positive* if

$$\int_{\wp} \lambda(s)ds = \infty \quad \text{whenever} \quad \wp = \bigcup_{i=1}^{\infty} [a_i, b_i],$$

$$a_i < b_i < a_{i+1}, \quad \text{and} \quad b_i - a_i \geq \delta > 0.$$

Theorem 2.13 *Assume that*

(1) $h_0, h \in \mathcal{H}$ and h_0 is finer than h;

(2) there exists a $V \in \mathcal{V}_0$ such that $V(t, \mathbf{x})$ is locally Lipschitzian in $\mathbf{x}$ on each $\mathfrak{G}_i$, h-positive definite on $\mathfrak{S}_\rho(h)$, and weakly h_0-decrescent, and $D^+V(t, \mathbf{x}) \leq -\lambda(t)\alpha(V_1(t, \mathbf{x}))$ on $G \cap \mathfrak{S}_\rho(h)$, where $\alpha \in \mathcal{K}$, $V_1 \in \mathcal{V}_0$, and $\lambda(t)$ is integrally positive;

(3) $V(\tau_k^+, \mathbf{x} + \mathbf{j}_k(\mathbf{x}(\tau_k))) \leq V(\tau_k, \mathbf{x}(\tau_k))$ on $\Sigma_k \cap \mathfrak{S}_\rho(h)$;

(4) $V_1(t, \mathbf{x})$ is locally Lipschitzian in $\mathbf{x}$ on each $\mathfrak{G}_i$ and h-positive definite on $\mathfrak{S}_\rho(h)$, and for every piecewise continuous function $\mathbf{y}(t)$ with discontinuities at $\tau_k = \tau_k(\mathbf{y}(\tau_k))$, $k = 1, 2, \cdots$, such that $(t, \mathbf{y}(t)) \in \mathfrak{S}_\rho(h)$, the function

$$\int_0^t [D^+V_1(s, \mathbf{y}(s))]_+ ds \quad (\text{ resp. } \int_0^t [D^+V_1(s, \mathbf{y}(s))]_- ds)$$

is uniformly continuous on $\mathbb{R}_+$, where $[\cdot]_+$ (resp. $[\cdot]_-$) denotes that the positive (resp. negative) part is considered for all $s \in \mathbb{R}_+$, and $V_1(\tau_k^+, \mathbf{x} + \mathbf{j}_k(\mathbf{x})) \leq V_1(t, \mathbf{x})$ (resp. $V_1(\tau_k^+, \mathbf{x} + \mathbf{j}_k(\mathbf{x})) \geq V_1(t, \mathbf{x})$) on $\Sigma_k \cap \mathfrak{S}_\rho(h)$.

Then system (2.54) is (h_0, h)-asymptotically stable.

Proof. From Theorem 2.12 and the assumptions (1)-(3) we know that the control system (2.54) is (h_0, h)-stable. Therefore for $\rho > 0$ and $t_0 \in \mathbb{R}_+$, there is a $\delta_0 = \delta_0(t_0, \rho) > 0$ such that $h_0(t_0, \mathbf{x}_0) < \delta_0$ implies

$$h(t, \mathbf{x}(t)) < \rho, \qquad t \geq t_0, \tag{2.62}$$

where $\mathbf{x}(t) = \mathbf{x}(t, t_0, \mathbf{x}_0)$ is any solution of system (2.54). We then need to prove that for every solution $\mathbf{x}(t) = \mathbf{x}(t, t_0, \mathbf{x}_0)$ of system (2.54) satisfying Eq. (2.62), the following statement is true:

$$\liminf_{t \to \infty} V_1(t, \mathbf{x}(t)) = 0. \tag{2.63}$$

If it is not the case, then there is a $\xi > 0$ and some $T > 0$ such that

$$V_1(t, \mathbf{x}(t)) \geq \xi, \quad t \geq t_0 + T. \tag{2.64}$$

We can choose a sequence

$$t_0 + T < a_1 < b_1 < \cdots < a_i < b_i < \cdots$$

such that $b_i - a_i \geq \xi$ for $i = 1, 2, \cdots$. Therefore, from Eq. (2.64) and assumption (2) we have

$$\begin{aligned} \lim_{t\to\infty} V(t, \mathbf{x}(t)) &\leq V(t_0, \mathbf{x}_0) - \int_{t_0}^{\infty} \lambda(s)\alpha(V_1(s, \mathbf{x}(s)))ds \\ &\leq V(t_0, \mathbf{x}_0) - \alpha(\xi) \int_{\bigcup_{i=1}^{\infty}[a_i, b_i]} \lambda(s)ds \\ &= -\infty, \end{aligned} \tag{2.65}$$

which leads to a contradiction.

Let us suppose that

$$\lim_{t\to\infty} \sup V_1(t, \mathbf{x}(t)) > 0,$$

then there is an $\eta > 0$ such that

$$\lim_{t\to\infty} \sup V_1(t, \mathbf{x}(t)) > 2\eta.$$

For definiteness, suppose that assumption (4) holds with respect to $[\cdot]_+$. Since Eq. (2.63) holds, we can find a sequence

$$t_0 < t_1^{[1]} < t_1^{[2]} < \cdots < t_i^{[1]} < t_i^{[2]} < \cdots$$

such that, for $i = 1, 2, \cdots$,

$$V_1(t_i^{[1]}, \mathbf{x}(t_i^{[1]})) = \eta, \quad V_1(t_i^{[2]}, \mathbf{x}(t_i^{[2]})) = 2\eta, \tag{2.66}$$

from which and since

$$V_1(\tau_k^+, \mathbf{x} + \mathbf{j}_k(\mathbf{x})) \leq V_1(\tau_k, \mathbf{x}(\tau_k)),$$

we can find a sequence $t_0 < a_1 < b_1 < \cdots < a_i < b_i < \cdots$ such that, for $i = 1, 2, \cdots$, $t_i^{[1]} \leq a_i < b_i \leq t_i^{[2]}$ and

$$\begin{aligned} &V_1(a_i, \mathbf{x}(a_i)) = \eta, \quad V_1(b_i, \mathbf{x}(b_i)) = 2\eta, \\ &V_1(t, \mathbf{x}(t)) \in [\eta, 2\eta] \text{ for } t \in [a_i, b_i]. \end{aligned} \tag{2.67}$$

We then have

$$\begin{aligned} 0<\eta &= V_1(b_i,\mathbf{x}(b_i))-V_1(a_i,\mathbf{x}(a_i)) \le \int_{a_i}^{b_i}[D^+V_1(s,\mathbf{x}(s))]_+ds, \\ & i=1,2,\cdots, \end{aligned} \tag{2.68}$$

from which and assumption (4), we have for some constant $c>0$

$$b_i-a_i\ge c,\quad i=1,2,\cdots. \tag{2.69}$$

Therefore, from assumption (2) and Eqs. (2.67) and (2.69) we have

$$\begin{aligned} \lim_{t\to\infty}V(t,\mathbf{x}(t)) &\le V(t_0,\mathbf{x}_0)-\int_{t_0}^{\infty}\lambda(s)\alpha(V_1(s,\mathbf{x}(s)))ds \\ &\le V(t_0,\mathbf{x}_0)-\alpha(\eta)\int_{\bigcup_{i=1}^{\infty}[a_i,b_i]}\lambda(s)ds=-\infty. \end{aligned} \tag{2.70}$$

This contradiction, together with Eq. (2.63), implies that

$$\lim_{t\to\infty}V_1(t,\mathbf{x}(t))=0,$$

and since $V_1(t,\mathbf{x}(t))$ is h-positive definite, we then have

$$\lim_{t\to\infty}h(t,\mathbf{x}(t))=0.$$

□

From Theorem 2.13 we have the following corollaries.

Corollary 2.2 *Assume that*

(1) $h_0,h\in\mathcal{H}$ and h_0 is finer than h;
(2) there exists a $V\in\mathcal{V}_0$ such that $V(t,\mathbf{x})$ is locally Lipschitzian in $\mathbf{x}$ on each $\mathfrak{G}_i$, h-positive definite on $\mathfrak{S}_\rho(h)$, and weakly h_0-decrescent, and $D^+V(t,\mathbf{x})\le-\lambda(t)\alpha(V_1(t,\mathbf{x}))$ on $G\cap\mathfrak{S}_\rho(h)$, where $\alpha\in\mathcal{K}$, $V_1\in\mathcal{V}_0$, and $\lambda(t)$ is integrally positive;
(3) $V(\tau_k^+,\mathbf{x}+\mathbf{j}_k(\mathbf{x}(\tau_k)))\le V(\tau_k,\mathbf{x}(\tau_k))$ on $\Sigma_k\cap\mathfrak{S}_\rho(h)$;
(4) $V_1(t,\mathbf{x})$ is locally Lipschitzian in $\mathbf{x}$ on each $\mathfrak{G}_i$ and h-positive definite on $\mathfrak{S}_\rho(h)$. $D^+V_1(t,\mathbf{x})$ is bounded from above(resp. from below) on $G\cap\mathfrak{S}_\rho(h)$ and $V_1(\tau_k^+,\mathbf{x}+\mathbf{j}_k(\mathbf{x}))\le V_1(\tau_k,\mathbf{x}(\tau_k))$(resp. $V_1(\tau_k^+,\mathbf{x}+\mathbf{j}_k(\mathbf{x}))\ge V_1(\tau_k,\mathbf{x}(\tau_k))$) on $\Sigma_k\cap\mathfrak{S}_\rho(h)$.

Then system (2.54) is (h_0,h)-asymptotically stable.

Corollary 2.3 *Assume that*

(1) $h_0, h \in \mathcal{H}$ *and* h_0 *is finer than* h*;*

(2) there exists a $V \in \mathcal{V}_0$ *such that* $V(t, \mathbf{x})$ *is locally Lipschitzian in* $\mathbf{x}$ *on each* $\mathfrak{G}_i$*,* h*-positive definite on* $\mathfrak{S}_\rho(h)$*, and weakly* h_0*-decrescent, and* $D^+V(t, \mathbf{x}) \leq -\lambda(t)\alpha(V(t, \mathbf{x}))$ *on* $G \cap \mathfrak{S}_\rho(h)$*, where* $\alpha \in \mathcal{K}$ *and* $\lambda(t)$ *is integrally positive;*

(3) $V(\tau_k^+, \mathbf{x} + \mathbf{j}_k(\mathbf{x}(\tau_k))) \leq V(\tau_k, \mathbf{x}(\tau_k))$ *on* $\Sigma_k \cap \mathfrak{S}_\rho(h)$*.*

Then system (2.54) is (h_0, h)*-asymptotically stable.*

Theorem 2.14 *Assume that*

(1) $h_0, h \in \mathcal{H}$ *and* h_0 *is finer than* h*;*

(2) there exists a $V \in \mathcal{V}_0$ *such that* $V(t, \mathbf{x})$ *is locally Lipschitzian in* $\mathbf{x}$ *on each* $\mathfrak{G}_i$*,* h*-positive definite on* $\mathfrak{S}_\rho(h)$*, and* h_0*-decrescent, and* $D^+V(t, \mathbf{x}) \leq -\lambda(t)\alpha(h_0(t, \mathbf{x}))$ *on* $G \cap \mathfrak{S}_\rho(h)$*, where* $\alpha \in \mathcal{K}$ *and* $\lambda(t)$ *is integrally positive;*

(3) $V(\tau_k^+, \mathbf{x} + \mathbf{j}_k(\mathbf{x}(\tau_k))) \leq V(\tau_k, \mathbf{x}(\tau_k))$ *on* $\Sigma_k \cap \mathfrak{S}_\rho(h)$*.*

Then system (2.54) is (h_0, h)*-uniformly asymptotically stable.*

Proof. From Theorem 2.12 we know that the control system (2.54) is (h_0, h)-uniformly stable. Therefore for $\rho > 0$ there is a $\delta_0 = \delta_0(\rho) > 0$ such that $h_0(t_0, \mathbf{x}_0) < \delta_0$ implies

$$h(t, \mathbf{x}(t)) < \rho, \qquad t \geq t_0, \tag{2.71}$$

where $\mathbf{x}(t) = \mathbf{x}(t, t_0, \mathbf{x}_0)$ is any solution of system (2.54).

Given $0 < \eta < \rho$, let $\delta = \delta(\eta) > 0$, then we need to prove that there is a $T = T(\eta) > 0$ such that, for some $t^* \in [t_0, t_0 + T]$, we have

$$h_0(t^*, \mathbf{x}(t^*)) < \delta. \tag{2.72}$$

If it is not the case, then for any $T > 0$ there is a solution $\mathbf{x}(t) = \mathbf{x}(t, t_0, \mathbf{x}_0)$ of system (2.54) satisfying Eq. (2.71) such that

$$h_0(t, \mathbf{x}(t)) \geq \delta, \qquad t \in [t_0, t_0 + T]. \tag{2.73}$$

Since $V(t, \mathbf{x}(t))$ is nonincreasing, from assumption (2) we know that there is a $\beta \in \mathcal{K}$ such that

$$\int_{t_0}^{\infty} \lambda(s)\alpha(h_0(s, \mathbf{x}(s)))ds \leq \beta(\delta_0) \tag{2.74}$$

for each solution $\mathbf{x}(t) = \mathbf{x}(t, t_0, \mathbf{x}_0)$ of system (2.54) satisfying Eq. (2.71). From the assumption on $\lambda(t)$, there is a $T > 0$ such that

$$\int_{t_0}^{t_0+T} \lambda(s)ds > \frac{\beta(\delta_0)+1}{\alpha(\delta)}. \tag{2.75}$$

Let $\mathbf{x}(t) = \mathbf{x}(t, t_0, \mathbf{x}_0)$ be the solution of system (2.54) satisfying Eq. (2.73) with T given in Eq. (2.75), then from Eqs. (2.74) and (2.75) we have

$$\beta(\delta_0) \geq \int_{t_0}^{\infty} \lambda(s)\alpha(h_0(s, \mathbf{x}(s)))ds > \alpha(\delta) \int_{t_0}^{t_0+T} \lambda(s)ds > \beta(\delta_0)+1,$$

which is a contradiction. □

2.8 Stability based on linear approximation

Consider the following impulsive differential system

$$\begin{cases} \dot{\mathbf{x}} = A(t)\mathbf{x} + \mathbf{g}(t, \mathbf{x}), & t \neq \tau_k, \\ \Delta \mathbf{x} = B_k \mathbf{x} + J_k(\mathbf{x}), & t = \tau_k, \quad k = 1, 2, \cdots, \end{cases} \tag{2.76}$$

where $\mathbf{x} \in \mathbb{R}^n$, $A(t) \in \mathbb{R}^{n\times n}$ and $\mathbf{g} : \mathbb{R}_+ \times \mathfrak{S}_\rho \to \mathbb{R}^n$, $\rho > 0$, $B_k \in \mathbb{R}^{n\times n}$, $J_k(\mathbf{x}) : \mathfrak{S}_\rho \to \mathbb{R}^n$. Let $\Phi(t, s)$ be the fundamental matrix of the following ordinary differential equation:

$$\dot{\mathbf{x}} = A(t)\mathbf{x}. \tag{2.77}$$

Then the stability of system (2.76) based on linear approximation is given by the following theorem.

Theorem 2.15 *Let the following conditions hold:*

(1) $A(t) \in \mathcal{PC}(\mathbb{R}_+, \mathbb{R}^{n\times n})$;
(2) $\|\Phi(t, s)\| \leq \phi(t)\psi(s)$ *for* $\tau_{k-1} < s \leq t \leq \tau_k, k \in \mathbb{N}$, *where* $\phi(\cdot), \psi(\cdot) \in \mathcal{PC}(\mathbb{R}_+, \mathbb{R})$ *and* $\phi(t) > 0$, $\psi(t) > 0$, $\phi(\tau_k^+) > 0$, $\psi(\tau_k^+) > 0$ *for* $t \in \mathbb{R}_+$, $k \in \mathbb{N}$;
(3) the function $\mathbf{g}(t, \mathbf{x})$ *is continuous and locally Lipschitz continuous with respect to* $\mathbf{x}$ *in the sets* $(\tau_{k-1}, \tau_k] \times \mathfrak{S}_\rho$, $k \in \mathbb{N}$, *and for each* k *and* $\mathbf{y} \in \mathfrak{S}_\rho$ *the limit of* $\mathbf{g}(t, \mathbf{x})$ *exists as* $(t, \mathbf{x}) \to (\tau_k, \mathbf{y})$, $t > \tau_k$;
(4) in the domain $\mathbb{R}_+ \times \mathfrak{S}_\rho$, *the following inequality holds:* $\|\mathbf{g}(t, \mathbf{x})\| \leq \alpha(t)\|\mathbf{x}\|^m$, *where* $m \geq 1$ *is a constant and the function* $\alpha(\cdot) \in \mathcal{PC}(\mathbb{R}_+, \mathbb{R})$ *is nonnegative;*

(5) $\|I + B_k\| \le \gamma_k$, $\gamma_k \ge 0$ *is a constant and* I *is the identity matrix;*
(6) the functions $J_k(\mathbf{x})$ *are continuous in* $\mathfrak{S}_\rho$ *and*

$$\|J_k(\mathbf{x})\| \le \rho_k \|\mathbf{x}\|, \qquad \mathbf{x} \in \mathfrak{S}_\rho, \quad k \in \mathbb{N}$$

where $\rho_k \ge 0$ *are constants.*
(7) for any $t_0 \ge 0$ *there is a* $N = N(t_0)$ *such that*

$$\phi(t) \prod_{t_0<\tau_k<t} r_k < N, \quad \textit{for } t > t_0,$$

where $r_k = (\gamma_k + \rho_k)\phi(\tau_k)\psi(\tau_k^+)$
(8)

$$D(0,\infty) \triangleq \int_0^\infty \left(\prod_{0<\tau_k<s} r_k \right)^{m-1} \phi^m(s)\psi(s)\alpha(s)ds < \infty; \quad (2.78)$$

(9) there is a $N > 0$ *such that*

$$\phi(t)\psi(t_0^+) \prod_{t_0<\tau_k<t} r_k < N, \qquad \textit{for } 0 \le t_0 \le t < \infty,$$

$$\psi^{m-1}(t_0^+)D(t_0,\infty) < N, \qquad \textit{for } t_0 \ge 0;$$

(10) for any $\epsilon > 0$ *and* $t_0 \in \mathbb{R}_+$ *there is a* $\sigma = \sigma(\epsilon, t_0)$ *such that*

$$\phi(t) \prod_{0<\tau_k<t} r_k < \epsilon, \textit{ for } \ t \ge t_0 + \sigma; \qquad (2.79)$$

(11) for any $\epsilon > 0$ *there is a* $\sigma = \sigma(\epsilon)$ *such that for each* $t_0 \in \mathbb{R}_+$ *we have*

$$\phi(t)\psi(t_0^+) \prod_{0<\tau_k<t} r_k < \epsilon, \textit{ for } \ t \ge t_0 + \sigma.$$

Let us assume that $m > 1$*, then the solution* $\mathbf{x} = 0$ *of system in Eq.(2.76) is*

(1) stable if assumptions (1)-(8) hold;
(2) uniformly stable if assumptions (1)-(6) and (9) hold;
(3) asymptotically stable if assumptions (1)-(6), (8) and (10) hold;
(4) uniformly asymptotically stable if assumptions (1)-(6), (9) and (11) hold.

Proof. See the proofs of Theorems 8.1-8.4 of [BS89]. □

Theorem 2.16 *Given the following function:*

$$M(t_0,t) = \phi(t)\psi(t_0^+)\left[\prod_{0<\tau_k<t}(\gamma_k+\rho_k)\phi(\tau_k)\psi(\tau_k^+)\right]\exp\left(\int_0^t \phi(s)\psi(s)ds\right)$$

and assume that the assumptions (1)-(6) of Theorem 2.15 hold and $m = 1$, *then the solution* $\mathbf{x} = 0$ *of system in Eq.(2.76) is*

(1) *stable if for any* $t_0 \in \mathbb{R}_+$ *there is a* $N > 0$ *such that* $M(t_0,t) \leq N$ *for* $t \geq t_0 \geq 0$;
(2) *asymptotically stable if for any* $t_0 \in \mathbb{R}_+$ *and* $\epsilon > 0$ *there is a* $\sigma > 0$ *such that* $M(t_0,t) \leq \epsilon$ *for* $t \geq t_0 + \sigma$;
(3) *uniformly stable if there is a* $N > 0$ *such that* $M(t_0,t) \leq N$ *for* $t \geq t_0 \geq 0$;
(4) *uniformly asymptotically stable if there is a* $N > 0$ *such that* $M(t_0,t) \leq N$ *for* $t \geq t_0 \geq 0$ *and for any* $\epsilon > 0$ *there is a* $\sigma > 0$ *such that* $M(t_0,t) \leq \epsilon$ *for any* $t_0 \in \mathbb{R}_+$ *and* $t \geq t_0 + \sigma$.

Proof. See the proof of Theorems 8.5 of [BS89]. □

Let us then consider the following special case:

$$\begin{cases} \dot{\mathbf{x}} = A\mathbf{x} + P(t)\mathbf{x}, & t \neq \tau_k, \\ \Delta\mathbf{x} = B\mathbf{x} + J_k\mathbf{x}, & t = \tau_k, \quad k = 1, 2, \cdots, \end{cases} \tag{2.80}$$

where $A \in \mathbb{R}^{n\times n}$, $P(t) \in \mathbb{R}^{n\times n}$, $B \in \mathbb{R}^{n\times n}$, $J_k \in \mathbb{R}^{n\times n}$, $k \in \mathbb{N}$. Let us also consider the following reference system

$$\begin{cases} \dot{\mathbf{x}} = A\mathbf{x}, & t \neq \tau_k, \\ \Delta\mathbf{x} = B\mathbf{x}, & t = \tau_k, \quad k = 1, 2, \cdots. \end{cases} \tag{2.81}$$

Lemma 2.1 *For* $t \geq t_0$ *let a nonnegative piecewise continuous function* $u(t)$ *satisfy*

$$u(t) \leq c + \int_{t_0}^t v(s)u(s)ds + \sum_{t_0<\tau_i<t} b_i u(\tau_i) \tag{2.82}$$

where $c \geq 0$, $b_i \geq 0$, $v(s) > 0$. $u(t)$ *has discontinuous points of the first kind at* τ_i. *Then we have*

$$u(t) \leq c \prod_{t_0<\tau_i<t}(1+b_i)\exp\left\{\int_{t_0}^t v(s)ds\right\}. \tag{2.83}$$

Proof. Let $\tau_0 = t_0$ and when $t \in [t_0, \tau_1]$ we have from Eq. (2.82)

$$u(t) \le c + \int_{t_0}^{t} v(s)u(s)ds \tag{2.84}$$

from which and based on Cronwell-Bellman inequality we have

$$u(t) \le c\exp\left\{\int_{t_0}^{t} v(s)ds\right\}, \qquad t \in [t_0, \tau_1]. \tag{2.85}$$

Thus, for $t \in [t_0, \tau_1]$ Eq. (2.83) holds. Let us suppose that Eq. (2.83) holds for $t \in (\tau_i, \tau_{i+1}]$, $i = 1, 2, \cdots, k-1$. Then for $t \in (\tau_k, \tau_{k+1}]$ we have

$$\begin{aligned} u(t) \le\ & c + \sum_{i=1}^{k} b_i c \prod_{j=1}^{i-1}(1+b_j)\exp\left\{\int_{t_0}^{\tau_i} v(s)ds\right\} \\ & + \sum_{i=1}^{k}\int_{\tau_{i-1}}^{\tau_i} v(p)c\prod_{j=1}^{i-1}(1+b_j)\exp\left\{\int_{t_0}^{p} v(s)ds\right\}dp + \int_{\tau_k}^{t} v(s)u(s)ds \\ =\ & c\left(1 + \sum_{i=1}^{k}\prod_{j=1}^{i-1}(1+b_j)\exp\left\{\int_{t_0}^{\tau_i} v(s)ds\right\}(1+b_i)\right. \\ & \left. - \sum_{i=1}^{k}\prod_{j=1}^{i-1}(1+b_j)\exp\left\{\int_{t_0}^{\tau_{i-1}} v(s)ds\right\}\right) + \int_{\tau_k}^{t} v(s)u(s)ds \\ =\ & c\prod_{j=1}^{k}(1+b_j)\exp\left\{\int_{t_0}^{\tau_k} v(s)ds\right\} + \int_{\tau_k}^{t} v(s)u(s)ds, \end{aligned} \tag{2.86}$$

from which and Cronwell-Bellman inequality we know for $t \in (\tau_k, \tau_{k+1}]$

$$u(t) \le \left(c\prod_{j=1}^{k}(1+b_j)\exp\left\{\int_{t_0}^{\tau_k} v(s)ds\right\}\right)\exp\left\{\int_{\tau_k}^{t} v(s)ds\right\} \tag{2.87}$$

therefore we have

$$u(t) \le c\prod_{t_0<\tau_i<t}(1+b_i)\exp\left\{\int_{t_0}^{t} v(s)ds\right\}, \qquad t \in (\tau_k, \tau_{k+1}]. \tag{2.88}$$

The proof is completed by induction. □

We then have the following theorem.

Theorem 2.17 *If the solutions of system (2.81) are stable, $I+B$ is nonsingular and*

$$\int_{t_0}^{\infty} \|P(t)\|dt < \infty \text{ and } \prod_{\tau_i > t_0} (1 + \|J_i\|) < \infty \tag{2.89}$$

then the solutions of system (2.80) are also stable.

Proof. The state transition matrix of system (2.81) is given by

$$\Psi(t, t_0) = e^{A(t-\tau_i)} \prod_{t_0 < \tau_j < \tau_i} (I+B)e^{A(\tau_j - \tau_{j-1})}, \quad \tau_0 = t_0, \quad t \in (\tau_i, \tau_{i+1}]. \tag{2.90}$$

Since the matrix $I+B$ is nonsingular, $\Psi(t, t_0)$ is also nonsingular and we have

$$\Psi(t, t_0)\Psi^{-1}(s, t_0) = e^{A(t-\tau_i)} \prod_{t_0 < \tau_j < \tau_i} (I+B)e^{A(\tau_j - \tau_{j-1})}(I+B)e^{A(\tau_{k+1} - s)}, \quad \tau_i < t \le \tau_{i+1}, \quad \tau_k < s < \tau_{k+1}, \quad k < i. \tag{2.91}$$

Since the solutions of system (2.81) are stable, it follows from (2.91) that there is a number $K > 0$ such that

$$\|\Psi(t, t_0)\| \le K, \quad \|\Psi(t, t_0)\Psi^{-1}(s, t_0)\| \le K, \quad t_0 \le s \le t. \tag{2.92}$$

With the initial condition $\mathbf{x}(t_0, \mathbf{x}_0) = \mathbf{x}_0$, any solution of system (2.80) is given by

$$\mathbf{x}(t, \mathbf{x}_0) = \Psi(t, t_0)\mathbf{x}_0 + \int_{t_0}^{t} \Psi(t, s)P(s)\mathbf{x}(s, \mathbf{x}_0)ds + \sum_{t_0 < \tau_i < t} \Psi(t, \tau_i)J_i\mathbf{x}(\tau_i, \mathbf{x}_0). \tag{2.93}$$

Let $\mathbf{x}(t, \mathbf{x}_0)$ and $\mathbf{x}(t, \mathbf{y}_0)$ be any two solutions of (2.80) and from Eq. (2.92) we have

$$\begin{aligned} \|\mathbf{x}(t, \mathbf{x}_0) - \mathbf{x}(t, \mathbf{y}_0)\| \le\ & K\|\mathbf{x}_0 - \mathbf{y}_0\| \\ & + \int_{t_0}^{t} K\|P(s)\|\|\mathbf{x}(s, \mathbf{x}_0) - \mathbf{x}(s, \mathbf{y}_0)\|ds \\ & + \sum_{t_0 < \tau_i < t} K\|J_i\|\|\mathbf{x}(\tau_i, \mathbf{x}_0) - \mathbf{x}(\tau_i, \mathbf{y}_0)\|. \end{aligned}$$

From Lemma 2.1 we have

$$\begin{aligned} & \|\mathbf{x}(t, \mathbf{x}_0) - \mathbf{x}(t, \mathbf{y}_0)\| \\ \leq \ & K \prod_{t_0<\tau_i<t} (1 + K\|J_i\|) \exp\left(\int_{t_0}^{t} K\|P(s)\|ds \right) \|\mathbf{x}_0 - \mathbf{y}_0\|, \qquad t \geq t_0. \end{aligned}$$

Since $1+K\|J_i\| \leq (1+\|J_i\|)^K$ and from Eq. (2.89) we know that $\prod_{t_0<\tau_i<t}(1+K\|J_i\|)$ converges. Thus we know that the solutions of system (2.80) is stable. $\square$

Chapter 3

Linear Impulsive Control

In a linear impulsive control system, the plant and the impulsive control laws are linear.

3.1 Linear impulsive control system with constant parameters

If the parameters in the plant of is constant, then a linear impulsive control system is given by

$$\begin{cases} \dot{\mathbf{x}} = A\mathbf{x}, & t \neq \tau_k, \\ \mathbf{y} = C\mathbf{x}, & \\ \Delta\mathbf{x} = P\mathbf{y} = PC\mathbf{x}, & t = \tau_k, \quad k \in \mathbb{N}. \end{cases} \tag{3.1}$$

The solution of Eq.(3.1) with initial condition $\mathbf{x}(t_0) = \mathbf{x}_0$ and $t_0 \neq \tau_1$ is given by:

$$\begin{aligned} \mathbf{x}(t_0, \mathbf{x}_0) &= \mathbf{x}_0 e^{A(t-\tau_k)} \prod_{j=1}^{k} (I + PC) e^{A(\tau_j - \tau_{j-1})}, \\ & \quad \tau_0 = t_0, \quad t \in (\tau_k, \tau_{k+1}]. \end{aligned} \tag{3.2}$$

Since the impulsive effect at moment $\{\tau_k\}$, the system in Eq. (3.1) is not an autonomous system any more. And the solutions of Eq. (3.1) is no longer invariant with respect to shifts. Given arbitrary matrices A and PC, it is hard to say something on the property of the solutions of Eq. (3.1). Therefore even the linear impulsive control system is more difficult to study than the corresponding continuous control systems. However, if the matrices A and PC are commute then we can get a simpler form of the solution in Eq.(3.2) as

$$\mathbf{x}(t_0, \mathbf{x}_0) = e^{A(t-t_0)}(I + PC)^{n(t,t_0)}\mathbf{x}_0 \tag{3.3}$$

where $n(t, t_0)$ denotes the impulse number in time interval $[t_0, t)$. From Eq.(3.3) one can find that the property of solutions depends on the eigenvalues of matrix A and PC and the $\{\tau_k\}$ sequence. If we use an equidistant impulse moment generating law, i.e., $\tau_k = \tau_1 + (k-1)\delta$, where $\delta > 0$ is the impulse interval, and $(I + PC)$ is nonsingular, then the solutions in Eq.(3.3) can be further simplified into:

$$\mathbf{x}(t_0, \mathbf{x}_0) = e^{A(\tau_1-t_0)}(I + PC)e^{-\alpha \ln(I+PC)}e^{[A+\frac{1}{\delta}\ln(I+PC)](t-\tau_1)}\mathbf{x}_0 \tag{3.4}$$

where $\alpha = \frac{t-\tau_1}{\delta} - \lfloor \frac{t-\tau_1}{\delta} \rfloor$ and $\lfloor x \rfloor$ denotes the integer part of $x \geq 0$. Then we have the following theorem.

Theorem 3.1 *Assume A and PC are commute, $\{\tau_k\}$ is equidistant with interval δ such that $0 < \delta < \infty$, then the origin of the impulsive controlled system in Eq. (3.1) is*

(1) asymptotically stable if all eigenvalues of the matrix $[A + \frac{1}{\delta}\ln(I + PC)]$ are negative;

(2) unstable if at least one eigenvalue of the matrix $[A + \frac{1}{\delta}\ln(I + PC)]$ is in the open left-hand side of s-plane.

Remark: Compare with the corresponding linear conventional control system

$$\begin{cases} \dot{\mathbf{x}} = A\mathbf{x} + \mathbf{u} \\ \mathbf{y} = C\mathbf{x} \\ \mathbf{u} = P\mathbf{y} \end{cases} \tag{3.5}$$

one can see that the property of solutions are determined by matrix $(A + PC)$. From Theorem 3.1 we know for a linear impulsive system the property of solutions is determined by A, PC and δ. On the one hand, given the plant is unstable, by choosing proper PC and δ, we can stabilize it "impulsively". On the other hand, we can also distabilize a stable plant "impulsively".

If A and PC are not commute then we have the following theorem:

Theorem 3.2 *Assume that $\{\tau_k\}$ is equidistant with interval δ such that $0 < \delta < \infty$, then the origin of the impulsive controlled system in Eq. (3.1) is* *asymptotically stable if and only if all eigenvalues of the matrix* $(I+PC)e^{A\delta}$, λ_i, *satisfy* $|\lambda_i| < 1$.

Proof. From the assumption we know the spectral radius of $(I+PC)e^{A\delta}$, $\rho((I+PC)e^{A\delta})$, satisfies $\rho((I+PC)e^{A\delta}) < 1$. From Theorem 5.3.4 on page 198 of [Ort87] we know

$$\lim_{k\to\infty}[(I+PC)e^{A\delta}]^k = 0.$$

From Eq. (3.2) we have

$$\begin{aligned} \mathbf{x}(t_0,\mathbf{x}_0) &= \mathbf{x}_0 e^{A(t-\tau_k)}\prod_{j=1}^{k}(I+PC)e^{A(\tau_j-\tau_{j-1})}, \\ &= \mathbf{x}_0 e^{A(t-\tau_k)}[(I+PC)e^{A\delta}]^k, \\ & \quad \tau_0 = t_0, \quad t\in(\tau_k,\tau_{k+1}], \end{aligned} \tag{3.6}$$

from which we know

$$\lim_{t\to\infty}\mathbf{x}(t_0,\mathbf{x}_0) = 0 \tag{3.7}$$

for any $\|\mathbf{x}_0\| < \infty$ and $\delta < \infty$. □

If $\{\tau_k\}$ is not equi-distant, we have the following theorem.

Theorem 3.3 *If* $0 < \alpha = \inf_k(\tau_{k+1}-\tau_k) \le \sup_k(\tau_{k+1}-\tau_k) = \beta < \infty$ *then the origin of the impulsive controlled system in Eq. (3.1) is asymptotically stable if* $\|e^{A(\tau_k-\tau_{k-1})}(I+PC)\| < 1$, $k\in\mathbb{N}$.

Proof. From Eq. (3.2) we can express the solution of system (3.1) as

$$\begin{aligned} \mathbf{x}(t_0,\mathbf{x}_0) &= \mathbf{x}_0 e^{A(t-\tau_k)}\prod_{t_0<\tau_j<t}(I+PC)e^{A(\tau_j-\tau_{j-1})}, \\ & \quad \tau_0 = t_0, \quad t\in(\tau_k,\tau_{k+1}]. \end{aligned} \tag{3.8}$$

from which we have

$$\begin{aligned} \|\mathbf{x}(t_0,\mathbf{x}_0)\| &\le \mathbf{x}_0\|e^{A(t-\tau_k)}\|\prod_{t_0<\tau_j<t}\|(I+PC)e^{A(\tau_j-\tau_{j-1})}\| \\ & \quad \tau_0 = t_0, \quad t\in(\tau_k,\tau_{k+1}]. \end{aligned} \tag{3.9}$$

Since $0 < \alpha = \inf_k(\tau_{k+1}-\tau_k) \le \sup_k(\tau_{k+1}-\tau_k) = \beta < \infty$ we know that $\|e^{A(t-\tau_k)}\| < \infty$, therefore we know if $\|e^{A(\tau_k-\tau_{k-1})}(I+PC)\| < 1$, $k\in\mathbb{N}$ then $\lim_{t\to\infty}\|\mathbf{x}(t_0,\mathbf{x}_0)\| = \lim_{k\to\infty}\|\mathbf{x}(t_0,\mathbf{x}_0)\| = 0$ for any $\mathbf{x}_0\in\mathbb{R}^n$. □

Theorem 3.4 *If A and PC are commute and $0 < \theta_1 = \inf_k(\tau_{k+1} - \tau_k) \leq \sup_k(\tau_{k+1} - \tau_k) = \theta_2 < \infty$, $k \in \mathbb{N}$. Let $\alpha = \max_j \mathrm{Re}(\lambda_j(A))$ and*

$$\theta_0 = \begin{cases} \theta_1, & \text{if } \alpha \geq 0 \\ \theta_2, & \text{if } \alpha < 0 \end{cases}$$

and if

$$\alpha + \frac{1}{\theta_0} \ln \max_j |1 + \lambda_j(B)| < 0$$

then the origin of the impulsive controlled system in Eq. (3.1) is asymptotically stable.

Proof. Since A and PC are commute, the state transition matrix of system (3.1) is given by

$$\Psi(t, t_0) = e^{A(t-t_0)}(I + B)^{n(t,t_0)} \tag{3.10}$$

where $n(t, t_0)$ is the number of control impulses in $[t_0, t)$. Let

$$\beta = \max_j |1 + \mathrm{Re}(\lambda_j(B))|$$

then for $t > t_0$ there exist constant $K_1 \geq 1$ and an arbitrary small $\epsilon > 0$ such that

$$\|e^{A(t-t_0)}\| \leq K_1 e^{(\alpha+\epsilon)(t-t_0)},$$

and there exists $K_2 = K_2(\epsilon) \geq 1$ such that

$$\|(I + B)^{n(t,t_0)}\| \leq K_2(\beta + \epsilon)^{n(t,t_0)}.$$

We then have

$$\begin{aligned} \|\Psi(t, t_0)\| &\leq K_1 K_2 e^{(\alpha+\epsilon)(t-t_0)}(\beta + \epsilon)^{n(t,t_0)} \\ &\leq K_1 K_2 e^{(\alpha+\epsilon)\theta_0}[e^{(\alpha+\epsilon)\theta_0}(\beta + \epsilon)]^{n(t,t_0)} \end{aligned} \tag{3.11}$$

from which we know that if

$$e^{\alpha\theta_0}\beta < 1$$

then

$$\|\Psi(t, t_0)\| \to 0 \text{ as } t \to \infty.$$

It is because we can choose $\epsilon > 0$ small enough such that

$$e^{(\alpha+\epsilon)\theta_0}(\beta+\epsilon) < 1.$$

□

Theorem 3.5 *Assume that*

(1) A is in real canonical form and

$$\gamma = \max_j \mathrm{Re}(\lambda_j(A));$$

(2) let $n(t+T,t)$ be the number of control impulses in interval $[t, t+T)$ and the limit

$$\lim_{T\to\infty} \frac{n(t+T,t)}{T} = p,$$

exists and is uniform with respect to $t > t_0$;

(3) let $\alpha^2 = \max_j \lambda_j((I+B)^\top(I+B))$ with $\alpha > 0$ and $\gamma + p\ln\alpha < 0$.

Then the impulsive controlled system in Eq. (3.1) is asymptotically stable.

Proof. For any two solutions of system (3.1) $\mathbf{x}(t,\mathbf{x}_0)$ and $\mathbf{x}(t,\mathbf{y}_0)$, we have

$$\mathbf{x}(t,\mathbf{x}_0) - \mathbf{x}(t,\mathbf{y}_0) = e^{A(t-\tau_i)} \prod_{t_0<\tau_i<t} (I+B)e^{A(\tau_j-\tau_{j-1})}(\mathbf{x}_0 - \mathbf{y}_0). \quad (3.12)$$

Since A is in real canonical form, we have

$$\|e^{At}\mathbf{x}\|_2 \le e^{(\gamma+\epsilon)t}\|\mathbf{x}\|_2$$

with $\epsilon > 0$. Then from Eq. (3.12) we have

$$\|\mathbf{x}(t,\mathbf{x}_0) - \mathbf{x}(t,\mathbf{y}_0)\|_2 \le e^{(\gamma+\epsilon)(t-t_0)}\alpha^{n(t,t_0)}\|\mathbf{x}_0 - \mathbf{y}_0\|_2, \quad t > t_0. \quad (3.13)$$

From assumption (2) we know that for any $\xi > 0$ there exists $K = K(\xi)$ such that

$$\alpha^{n(t,t_0)} \le K e^{(p\ln\alpha+\xi)(t-t_0)}$$

and therefore we have

$$\|\mathbf{x}(t,\mathbf{x}_0) - \mathbf{x}(t,\mathbf{y}_0)\|_2 \le e^{(\gamma+p\ln\alpha+\xi+\epsilon)(t-t_0)}\|\mathbf{x}_0 - \mathbf{y}_0\|_2, \quad t > t_0.$$

Since ϵ and ξ can be chosen arbitrarily small, then from assumption (3) we finish the proof. □

Theorem 3.6 *Assume that*

(1) A and PC are commute and $I + PC$ is nonsingular;
(2) let $n(t+T, t)$ be the number of control impulses in interval $[t, t+T)$ and the limit

$$\lim_{T \to \infty} \frac{n(t+T,t)}{T} = p,$$

exists and is uniform with respect to $t > t_0$;
(3) all eigenvalues of $A + p\ln(I + PC)$ are in the left-hand side of s-plane.

Then the impulsive controlled system in Eq. (3.1) is asymptotically stable.

Proof. Since A and PC are commute, for any two solutions of system (3.1) $\mathbf{x}(t, \mathbf{x}_0)$ and $\mathbf{x}(t, \mathbf{y}_0)$, we have

$$\mathbf{x}(t, \mathbf{x}_0) - \mathbf{x}(t, \mathbf{y}_0) = e^{A(t-t_0)}(I + PC)^{n(t,t_0)}(\mathbf{x}_0 - \mathbf{y}_0), \tag{3.14}$$

or

$$\begin{aligned} \mathbf{x}(t, \mathbf{x}_0) - \mathbf{x}(t, \mathbf{y}_0) &= e^{[A+p\ln(I+PC)](t-t_0)} \\ &\quad \times (I + PC)^{[n(t,t_0)-(t-t_0)p]}(\mathbf{x}_0 - \mathbf{y}_0). \end{aligned} \tag{3.15}$$

From assumption (2) we know that there are a constant $K_1 > 0$ and any a small constant $\epsilon > 0$ such that

$$\|(I + PC)^{[n(t,t_0)-(t-t_0)p]}\| \le K_1 e^{\epsilon(t-t_0)}, \quad t \ge t_0. \tag{3.16}$$

From assumption (3) we know there are constants $K_2 > 0$ and $\gamma > 0$ such that

$$\|e^{[A+p\ln(I+PC)]t}\| \le K_2 e^{-\gamma t}, \quad t \ge t_0. \tag{3.17}$$

Therefore, from Eq. (3.15) we have

$$\|\mathbf{x}(t, \mathbf{x}_0) - \mathbf{x}(t, \mathbf{y}_0)\| \le K_1 K_2 e^{-(\gamma-\epsilon)(t-t_0)}\|\mathbf{x}_0 - \mathbf{y}_0\| \quad t \ge t_0. \tag{3.18}$$

□

3.2 Impulsive control of time-varying linear systems

Let us consider the following time-varying linear impulsive control system:

$$\begin{cases} \dot{\mathbf{x}}(t) = A(t)\mathbf{x}, & t \neq \tau_k \\ \Delta\mathbf{x} = B_k\mathbf{x}, & t = \tau_k, \quad k = 1, 2, \cdots, \end{cases} \tag{3.19}$$

where $A(t)$ is continuous matrix bounded for $t \geq t_0$, B_k are matrices uniformly bounded with respect to $k \in \mathbb{N}$ and

$$t_0 < \tau_1 < \tau_2 < \cdots, \qquad \lim_{k\to\infty} \tau_k = \infty.$$

Theorem 3.7 *Assume that*

(1)

$$\inf_i |\det(I + B_i)| \geq \delta > 0; \tag{3.20}$$

(2) let $n(t, t_0)$ be the number of control impulses in the interval $[t_0, t)$ and

$$\lim_{t\to\infty} \frac{n(t, t_0)}{t} = p \tag{3.21}$$

exists and is finite;

(3) the biggest eigenvalue of $\frac{1}{2}(A(t) + A^\top(t))$, λ_n, satisfies

$$\lambda_n \leq \gamma \tag{3.22}$$

for all $t \geq t_0$;

(4) let Λ_i be the biggest eigenvalues of matrices $(I + B_i^\top)(I + B_i)$ such that

$$\Lambda_i^2 \leq \alpha, \qquad i \in \mathbb{N}; \tag{3.23}$$

(5)

$$\gamma + p \ln \sqrt{\alpha} < 0. \tag{3.24}$$

Then the control system (3.19) is asymptotically stable.

Proof. For any $\mathbf{x}_0 \in \mathbb{R}^n$ we have

$$\begin{aligned} \overline{\lim_{t\to\infty}} \frac{\ln \|\mathbf{x}(t, \mathbf{x}_0)\|}{t} &\leq \overline{\lim_{t\to\infty}} \frac{1}{t} \left[\int_{t_0}^{t} \lambda_n(s) ds + \sum_{t_0 < \tau_i < t} \ln \Lambda_i \right] \\ &\leq \gamma + p \ln \sqrt{\alpha} < 0. \end{aligned} \tag{3.25}$$

The proof is completed. □

Let us then consider the following impulsive control system:

$$\begin{cases} \dot{\mathbf{x}} = A\mathbf{x} + P(t)\mathbf{x}, & t \neq \tau_k, \\ \Delta\mathbf{x} = B\mathbf{x} + J_k\mathbf{x}, & t = \tau_k, \quad k = 1, 2, \cdots, \end{cases} \tag{3.26}$$

where $A \in \mathbb{R}^{n\times n}$, $B \in \mathbb{R}^{n\times n}$, $P(t) \in \mathbb{R}^{n\times n}$ is a continuous or piecewise continuous matrix for $t \geq t_0$, $J_k \in \mathbb{R}^{n\times n}$ are constant matrices, $k \in \mathbb{N}$. Let us also consider the following reference system

$$\begin{cases} \dot{\mathbf{x}} = A\mathbf{x}, & t \neq \tau_k, \\ \Delta\mathbf{x} = B\mathbf{x}, & t = \tau_k, \quad k = 1, 2, \cdots. \end{cases} \tag{3.27}$$

We then have the following theorem.

Theorem 3.8 *Assume that*

(1) the solutions of system (3.27) are exponentially stable;
(2) for sufficiently large t and $k \in \mathbb{N}$, we have

$$||P(t)|| < \xi, \qquad ||J_k|| < \xi \tag{3.28}$$

where $\xi > 0$ is sufficiently small;

(3)

$$0 < \theta_1 \leq \tau_{k+1} - \tau_k \leq \theta_2, \qquad k \in \mathbb{N}.$$

Then the solutions of system (3.26) are exponentially stable.

Proof. Let $\Psi(t, s)$ be the state transition matrix of system (3.27), then from assumption (1) we know that there are $K > 0$ and $\gamma > 0$ such that

$$||\Psi(t, s)|| \leq Ke^{-\gamma(t-s)}, \qquad t \geq s. \tag{3.29}$$

For any two solutions of system (3.26) $\mathbf{x}(t, \mathbf{x}_0)$ and $\mathbf{x}(t, \mathbf{y}_0)$, we have

$$\begin{aligned} \mathbf{x}(t, \mathbf{x}_0) - \mathbf{x}(t, \mathbf{y}_0) &= \Psi(t, t_0)(\mathbf{x}_0 - \mathbf{y}_0) \\ &\quad + \int_{t_0}^{t} \Psi(t, s)P(s)[\mathbf{x}(s, \mathbf{x}_0) - \mathbf{x}(s, \mathbf{y}_0)]ds \\ &\quad + \sum_{t_0 < \tau_k < t} \Psi(t, \tau_k)J_k[\mathbf{x}(\tau_k, \mathbf{x}_0) - \mathbf{x}(\tau_k, \mathbf{y}_0)] \end{aligned}$$

we then have

$$
\begin{aligned}
&e^{\gamma(t-t_0)}\|\mathbf{x}(t,\mathbf{x}_0)-\mathbf{x}(t,\mathbf{y}_0)\| \le \\
&K\|\mathbf{x}_0-\mathbf{y}_0\| + \int_{t_0}^{t} Ke^{\gamma(s-t_0)}\|P(s)\|\|\mathbf{x}(s,\mathbf{x}_0)-\mathbf{x}(s,\mathbf{y}_0)\|ds \\
&+\sum_{t_0<\tau_k<t} Ke^{\gamma(\tau_k-t_0)}\|J_k\|\|\mathbf{x}(\tau_k,\mathbf{x}_0)-\mathbf{x}(\tau_k,\mathbf{y}_0)\|.
\end{aligned}
$$

Let us choose $T>t_0$ such that for any $t>T$ and $\tau_k>T$, the assumptions in Eq. (3.28) hold, then we have

$$
\begin{aligned}
&e^{\gamma(t-t_0)}\|\mathbf{x}(t,\mathbf{x}_0)-\mathbf{x}(t,\mathbf{y}_0)\| \le \\
&\phi + \int_{T}^{t} K\xi e^{\gamma(s-t_0)}\|\mathbf{x}(s,\mathbf{x}_0)-\mathbf{x}(s,\mathbf{y}_0)\|ds \\
&+\sum_{T\le\tau_k<t} K\xi e^{\gamma(\tau_k-t_0)}\|\mathbf{x}(\tau_k,\mathbf{x}_0)-\mathbf{x}(\tau_k,\mathbf{y}_0)\|,
\end{aligned}
$$

where

$$
\begin{aligned}
\phi &= K\|\mathbf{x}_0-\mathbf{y}_0\| + \int_{t_0}^{T} Ke^{\gamma(s-t_0)}\|P(s)\|\|\mathbf{x}(s,\mathbf{x}_0)-\mathbf{x}(s,\mathbf{y}_0)\|ds \\
&+\sum_{t_0<\tau_k<T} Ke^{\gamma(\tau_k-t_0)}\|J_k\|\|\mathbf{x}(\tau_k,\mathbf{x}_0)-\mathbf{x}(\tau_k,\mathbf{y}_0)\|. \qquad (3.30)
\end{aligned}
$$

therefore we have

$$
\begin{aligned}
&e^{\gamma(t-t_0)}\|\mathbf{x}(t,\mathbf{x}_0)-\mathbf{x}(t,\mathbf{y}_0)\| \le \\
&\phi + \int_{t_0}^{t} K\xi e^{\gamma(s-t_0)}\|\mathbf{x}(s,\mathbf{x}_0)-\mathbf{x}(s,\mathbf{y}_0)\|ds \\
&+\sum_{t_0<\tau_k<t} K\xi e^{\gamma(\tau_k-t_0)}\|\mathbf{x}(\tau_k,\mathbf{x}_0)-\mathbf{x}(\tau_k,\mathbf{y}_0)\|. \qquad (3.31)
\end{aligned}
$$

Then from Lemma 2.1 we have

$$
e^{\gamma(t-t_0)}\|\mathbf{x}(t,\mathbf{x}_0)-\mathbf{x}(t,\mathbf{y}_0)\| \le \phi(1+K\xi)^{n(t,t_0)}e^{K\xi(t-t_0)}, \qquad (3.32)
$$

and from assumption (3) we have

$$
\begin{aligned}
&\|\mathbf{x}(t,\mathbf{x}_0)-\mathbf{x}(t,\mathbf{y}_0)\| \le \\
&\phi\exp\left\{-\left(\gamma-K\xi-\frac{1}{\theta_1}\ln(1+K\xi)\right)(t-t_0)\right\}. \qquad (3.33)
\end{aligned}
$$

If ξ is so small that

$$\gamma - K\xi - \frac{1}{\theta_1}\ln(1 + K\xi) > 0, \tag{3.34}$$

then we have

$$\|\mathbf{x}(t, \mathbf{x}_0) - \mathbf{x}(t, \mathbf{y}_0)\| \to 0 \text{ as } t \to \infty.$$

□

Similar to Theorem 3.5 we have the following theorem.

Theorem 3.9 *Assume that*

(1) *A is in real canonical form and*

$$\gamma = \max_j \mathrm{Re}(\lambda_j(A));$$

(2) *let $n(t+T, t)$ be the number of control impulses in interval $[t, t+T)$ and the limit*

$$\lim_{T\to\infty} \frac{n(t+T,t)}{T} = p$$

exists and is uniform with respect to $t > t_0$;

(3) *let $\alpha^2 = \max_j \lambda_j((I + B)^\top (I + B))$ with $\alpha > 0$ and $\gamma + p \ln \alpha < 0$;*

(4) *for sufficiently large t and $k \in \mathbb{N}$, we have*

$$\|P(t)\| < \xi, \quad \|J_k\| < \xi \tag{3.35}$$

where $\xi > 0$ is sufficiently small.

Then the solutions of system (3.26) are asymptotically stable.

Similarly, we have the following theorem.

Theorem 3.10 *Assume that*

(1) *A is in real canonical form and*

$$\gamma = \max_j \mathrm{Re}(\lambda_j(A));$$

(2) *$\tau_k, k \in \mathbb{N}$ satisfy*

$$0 < \theta_1 \le \tau_{k+1} - \tau_k \le \theta_2;$$

(3) let $\alpha^2 = \max_j \lambda_j((I+B)^\top(I+B))$ *with* $\alpha > 0$ *and*

$$\gamma + \frac{1}{\theta}\ln\alpha < 0$$

where

$$\theta = \begin{cases} \theta_1, & \text{if } \alpha \geq 1, \\ \theta_2, & \text{if } 0 < \alpha < 1; \end{cases}$$

(4) for sufficiently large t *and* $k \in \mathbb{N}$, *we have*

$$\|P(t)\| < \xi, \qquad \|J_k\| < \xi \tag{3.36}$$

where $\xi > 0$ *is sufficiently small.*

Then the solutions of system (3.26) are asymptotically stable.

3.3 Contrallability of linear impulsive control systems

3.3.1 *Time-varying linear impulsive control systems*

In this section we consider the impulsive control of the following time-varying linear system:

$$\dot{\mathbf{x}}(t) = A(t)\mathbf{x} \tag{3.37}$$

where $\mathbf{x} \in \mathbb{R}^n$ and $A \in C[\mathbb{R}, \mathbb{R}^n \times \mathbb{R}^n]$. The corresponding impulsive controlled system is given by

$$\begin{cases} \dot{\mathbf{x}}(t) = A(t)\mathbf{x}, & t \neq \tau_k \\ \Delta\mathbf{x} = B(\tau_k)\mathbf{u}_k, & t = \tau_k, \quad k = 1, 2, \cdots, \end{cases} \tag{3.38}$$

where $B \in C[\mathbb{R}, \mathbb{R}^n \times \mathbb{R}^m]$ and $\mathbf{u}_k \in \mathbb{R}^m$.

Definition 3.1 The time-varying linear system (3.37) is called *impulsively controllable* if for all vector pair $\mathbf{x}_1 \in \mathbb{R}^n$, $\mathbf{x}_2 \in \mathbb{R}^n$, and any open time interval (t_1, t_2), there exist real numbers $\tau_k \in [t_1, t_2]$, $k = 1, 2, \cdots, p < \infty$, $\tau_1 < \tau_2 < \cdots < \tau_p$, and control vectors $\mathbf{u}_k \in \mathbb{R}^m$, such that the controlled system (3.38) has a solution $\mathbf{x}(t)$ existing on $[t_1, t_2]$ satisfying $\mathbf{x}(t_1) = \mathbf{x}_1$ and $\mathbf{x}(t_2^+) = \mathbf{x}_2$.

Theorem 3.11 *Given two terminal conditions* $(t_1, \mathbf{x}_1)$ *and* $(t_2, \mathbf{x}_2)$, *the control system in (3.38) is controllable at time* t_1 *if and only if*

(1) *the* $n \times m$ *matrix function* $\Phi(t)\Phi^{-1}(t_1)B(t)$ *are linearly independent on* $[t_1, t_2]$,

or,

(2) $\text{Rank}([\Phi^{-1}(\tau_1)B(\tau_1) \quad \Phi^{-1}(\tau_2)B(\tau_2) \quad \cdots \quad \Phi^{-1}(\tau_p)B(\tau_p)]) = n.$

Proof. The proof consists of two steps. In the first step we set $\mathbf{u}_k = 0$ for all $k = 0, 1, 2, \cdots$, then the control system in (3.38) becomes a standard linear control system. Then the control system in (3.38) is controllable at time t_1 if and only if the $n \times \tilde{m}$ matrix function $\Phi(t)\Phi^{-1}(t_1)B(t)$ are linearly independent on $[t_1, t_2]$. We first prove the sufficiency. If the rows of $\Phi(t)\Phi^{-1}(t_1)B(t)$ are linearly independent on $[t_1, t_2]$, then the $n \times n$ Grammian matrix

$$W(t_1, t_2) \triangleq \int_{t_1}^{t_2} \Phi(t_1, t)B(t)B^\top(t)\Phi^\top(t_1, t)dt \tag{3.39}$$

is nonsingular and thus $W^{-1}(t_1, t_2)$ exists. $\Phi(s, p) = \Phi(s)\Phi^{-1}(p)$. Since

$$\begin{aligned} \mathbf{x}(t_2) &= \Phi(t_2, t_1)\left\{\mathbf{x}_1 - \left[\int_{t_1}^{t_2} \Phi(t_1, t)B(t)B^\top(t)\Phi^\top(t_1, t)dt\right]\right. \\ &\quad \left.[W^{-1}(t_1, t_2)(\mathbf{x}_1 - \Phi(t_1, t_2)\mathbf{x}_2)]\right\} \\ &= \Phi(t_2, t_1)[\mathbf{x}_1 - W(t_1, t_2)W^{-1}(t_1, t_2)(\mathbf{x}_1 - \Phi(t_1, t_2)\mathbf{x}_2)] \\ &= \Phi(t_2, t_1)\Phi(t_1, t_2)\mathbf{x}_2 = \mathbf{x}_2 \end{aligned} \tag{3.40}$$

we know that

$$\mathbf{u}(t) = -B^\top(t)\Phi^\top(t_1, t)W^{-1}(t_1, t_2)(\mathbf{x}_1 - \Phi(t_1, t_2)\mathbf{x}_2) \tag{3.41}$$

is a control input that satisfying the terminal conditions $(t_1, \mathbf{x}_1)$ and $(t_2, \mathbf{x}_2)$. Thus the control system in (3.38) is controllable.

We then prove the necessity which implies that if the control system in (3.38) is controllable then the n rows of $\Phi(t)\Phi^{-1}(t_1)B(t)$ are linearly independent. Let us assume that (3.38) is controllable but the rows if $\Phi(t)\Phi^{-1}(t_1)B(t)$ are linearly dependent. Thus, there exists a nonzero and constant $1 \times n$ real vector $\mathbf{w}$ such that

$$\mathbf{w}\Phi(t_1, t)B(t) = 0 \quad \text{for all } t \text{ in } [t_1, t_2] \tag{3.42}$$

We can choose $\mathbf{x}(t_1) = \mathbf{x}_1 = \mathbf{w}^\top$ as an initial state and $\mathbf{x}(t_2) = \mathbf{x}_2 = 0$ as a terminal state because the system is controllable. Then we have

$$0 = \Phi(t_2, t_1)\left[\mathbf{w}^\top + \int_{t_1}^{t_2} \Phi(t_1, t)B(t)\mathbf{u}(t)dt\right] \tag{3.43}$$

from which we have

$$0 = \mathbf{w}\mathbf{w}^\top + \int_{t_1}^{t_2} \mathbf{w}\Phi(t_1, t)B(t)\mathbf{u}(t)dt. \tag{3.44}$$

Followed Eq. (3.42), Eq. (3.44) reduces into

$$\mathbf{w}\mathbf{w}^\top = 0 \tag{3.45}$$

which in turn implies that $\mathbf{w} = 0$. This is a contradiction.

In the second step, we assume that $u(t) = 0$ for $t \in [t_1, t_2]$. For given τ_k, $\mathbf{u}_k$, $k = 1, 2, \cdots, p$, the unique solution of impulsive control system (3.38) satisfying the initial condition $\mathbf{x}(t_1) = \mathbf{x}_1$ is given by

$$\begin{cases} \mathbf{x}(t) = \Phi(t)\Phi^{-1}(t_1)\mathbf{x}_1, & t \in (t_1, \tau_1], \\ \mathbf{x}(t) = \Phi(t)\Phi^{-1}(\tau_k)\mathbf{x}(\tau_k^+), & t \in (\tau_k, \tau_{k+1}], \quad k = 1, 2, \cdots, p, \end{cases} \tag{3.46}$$

where $\Phi(t)$ is the fundamental matrix of Eq. (3.37). From Eqs. (3.38) and (3.46) we have

$$\mathbf{x}(t_2^+) = \Phi(t_2)\left(\Phi^{-1}(t_1)\mathbf{x}_1 + \sum_{k=1}^{p} \Phi^{-1}(\tau_k)B(\tau_k)\mathbf{u}_k\right). \tag{3.47}$$

From the definition of impulsive controllabilty we know that $\mathbf{x}(t_2) = \mathbf{x}_2$ should be satisfied, we then have

$$\sum_{k=1}^{p} \Phi^{-1}(\tau_k)B(\tau_k)\mathbf{u}_k = \Phi^{-1}(t_2)\mathbf{x}_2 - \Phi^{-1}(t_1)\mathbf{x}_1. \tag{3.48}$$

Let us define

$$\begin{aligned} \mathbf{v} &\triangleq \Phi^{-1}(t_2)\mathbf{x}_2 - \Phi^{-1}(t_1)\mathbf{x}_1, \\ W &\triangleq [\Phi^{-1}(\tau_1)B(\tau_1) \;\; \Phi^{-1}(\tau_2)B(\tau_2) \;\; \cdots \;\; \Phi^{-1}(\tau_p)B(\tau_p)], \\ U^\top &\triangleq (\mathbf{u}_1^\top, \mathbf{u}_2^\top, \cdots, \mathbf{u}_p^\top), \end{aligned} \tag{3.49}$$

then Eq. (3.48) can be rewritten as

$$WU = \mathbf{v}.$$

Because $\mathbf{x}_1$, $\mathbf{x}_2$ and therefore $\mathbf{v}$ are arbitrary, the impulsive controllability of system (3.37) is equivalent to

$$\text{Rank}(W) = n. \tag{3.50}$$

□

If the condition in Eq. (3.50) is not satisfied, then we can only find *partial impulsive controllability* of impulsive control systems. Then our question becomes as follows: Given two pairs of terminal conditions $(t_1, \mathbf{x}_1)$ and $(t_2, \mathbf{x}_2)$ under which conditions we can find a solution of system (3.38) to satisfy both of them. From Eq. (3.48) we know that the necessary and sufficient condition for the existence of such a solution is given by

$$\mathbf{v} \in \text{col}\{W\}, \tag{3.51}$$

where col$\{W\}$ is the column space of W.

3.3.2 *Time-invariant linear impulsive control systems*

If the $A(t)$ matrix is constant, then the linear impulsive control system (3.38) can be recast into

$$\begin{cases} \dot{\mathbf{x}}(t) = A\mathbf{x}, & t \neq \tau_k \\ \Delta\mathbf{x} = B(\tau_k)\mathbf{u}_k, & t = \tau_k, \quad k = 1, 2, \cdots. \end{cases} \tag{3.52}$$

In this case, the fundamental matrix is given by

$$\Phi(t) = e^{At}.$$

Followed Eq. (3.49) we have

$$W = [e^{-A\tau_1}B(\tau_1) \;\; e^{-A\tau_2}B(\tau_2) \;\; \cdots \;\; e^{-A\tau_p}B(\tau_p)]. \tag{3.53}$$

By using Cayley-Hamilton theorem we can use an nth-order polynomial in A to represent e^{At}. Let

$$\Lambda(\lambda) = \lambda^n + \sum_{i=0}^{n-1} a_i \lambda^i$$

be the characteristic polynomial of A and let the scalar functions $f_i(\lambda)$, $i = 0, 1, \cdots, n-1$, be the unique solution to the following system

$$\begin{pmatrix} \frac{df_0(\lambda)}{d\lambda} \\ \frac{df_1(\lambda)}{d\lambda} \\ \frac{df_2(\lambda)}{d\lambda} \\ \vdots \\ \frac{df_{n-1}(\lambda)}{d\lambda} \end{pmatrix} = \begin{pmatrix} 0 & 0 & \cdots & 0 & -a_0 \\ 1 & 0 & \cdots & 0 & -a_1 \\ 0 & 1 & \cdots & 0 & -a_2 \\ \vdots & \vdots & \ddots & \vdots & \vdots \\ 0 & 0 & \cdots & 1 & -a_{n-1} \end{pmatrix} \begin{pmatrix} f_0(\lambda) \\ f_1(\lambda) \\ f_2(\lambda) \\ \vdots \\ f_{n-1}(\lambda) \end{pmatrix}, \tag{3.54}$$

satisfying

$$\begin{aligned} f_0(0) &= 1, \\ f_i(0) &= 0, \quad i = 1, 2, \cdots, n-1. \end{aligned} \tag{3.55}$$

We then have the following lemmas.

Lemma 3.1

$$e^{A\lambda} = \sum_{i=0}^{n-1} f_i(\lambda) A^i.$$

Proof.

$$\begin{aligned} \frac{d}{d\lambda} \sum_{i=0}^{n-1} f_i(\lambda) A^i &= f_0(\lambda) A + f_1(\lambda) A^2 + \cdots + f_{n-2}(\lambda) A^{n-1} \\ &\quad + f_{n-1}(\lambda)(-a_0 I - a_1 A - \cdots - a_{n-1} A^{n-1}). \end{aligned} \tag{3.56}$$

By the Cayley-Hamilton theorem we know that

$$A^n = -a_0 I - a_1 A - \cdots - a_{n-1} A^{n-1}.$$

We then have

$$\frac{d}{d\lambda} \sum_{i=0}^{n-1} f_i(\lambda) A^i = A \sum_{i=0}^{n-1} f_i(\lambda) A^i. \tag{3.57}$$

Also,

$$\left. \sum_{i=0}^{n-1} f_i(\lambda) A^i \right|_{\lambda=0} = \sum_{i=0}^{n-1} f_i(0) A^i = A^0 = I. \tag{3.58}$$

Thus, the solution to Eqs. (3.57) and (3.58) is unique and $e^{A\lambda}$ satisfies Eqs. (3.57) and (3.58). □

Lemma 3.2 *The functions $f_i(\lambda)$, $i = 0, 1, \cdots, n-1$, defined in Eqs. (3.54) and (3.55) are linearly independent in any open interval (t_1, t_2).*

Proof. Suppose

$$\sum_{i=0}^{n-1} c_i f_i(\lambda) = 0$$

for some constants c_i, then by Eq. (3.55) we have $c_0 = 0$ and we also have

$$\frac{d}{d\lambda} \sum_{i=0}^{n-1} c_i f_i(\lambda) = 0$$

Then followed Eq. (3.54) we have

$$\sum_{i=1}^{n-1} c_i [f_{i-1}(\lambda) - a_i f_{n-1}(\lambda)] = 0. \tag{3.59}$$

If we define $f_i(\lambda) = 0$ if $i < 0$, then Eq. (3.59) can be rewritten as

$$\sum_{i=0}^{n-1} c_i [f_{i-1}(\lambda) - a_i f_{n-1}(\lambda)] = 0. \tag{3.60}$$

Evaluating Eq. (3.60) at $\lambda = 0$ and followed Eq. (3.55) we have $c_1 = 0$. We differentiate Eq. (3.60) and get

$$\sum_{i=0}^{n-1} c_i \{f_{i-2}(\lambda) - a_{i-1} f_{n-1}(\lambda) - a_i [f_{n-2}(\lambda) - a_{n-1} f_{n-1}(\lambda)]\} = 0, \tag{3.61}$$

which upon evaluation at $\lambda = 0$ gives $c_2 = 0$. By using the similar process, it is easy to show that $c_i = 0$, $i = 0, 1, \cdots, n-1$. Thus, the functions $f_i(\lambda)$ are linearly independent on any interval containing zero. Furthermore, since the functions $f_i(\lambda)$ satisfy a linear constant coefficient system of ODEs, it is known that they must be analytic. This implies that these functions are linearly independent in any open interval (t_1, t_2). □

Lemma 3.3 *If $f_1(\lambda)$, $f_2(\lambda)$, $\cdots$, $f_n(\lambda)$ are linearly independent functions in every open interval, then for a given $p \leq n$, and any open interval (t_1, t_2), there exist real numbers τ_i, $i = 1, 2, \cdots, p$, such that $t_1 \leq \tau_1 < \tau_2 < \cdots <$*

$\tau_p \leq t_2$, *and the matrix*

$$\Gamma_p \triangleq \begin{pmatrix} f_1(\tau_1) & f_1(\tau_2) & \cdots & f_1(\tau_p) \\ f_2(\tau_1) & f_2(\tau_2) & \cdots & f_2(\tau_p) \\ \vdots & \vdots & \cdots & \vdots \\ f_n(\tau_1) & f_n(\tau_2) & \cdots & f_n(\tau_p) \end{pmatrix}$$

has rank p.

Proof. We can choose $\tau_1 \in [t_1, t_2)$ such that Γ_1 has rank one. Suppose that we choose time moments τ_i, $i = 1, 2, \cdots, q$, satisfying $t_1 \leq \tau_1 < \tau_2 < \cdots < \tau_q \leq t_2$ such that Γ_q has rank q, $q < p \leq n$. For fixed k we view Γ_k as a linear transformation $T_k : \mathbb{R}^n \to \mathbb{R}^n$ defined by $T_k\mathbf{x} = (\Gamma_k \;\; \mathbf{0}_k)^\top \mathbf{x}$ where $\mathbf{0}_k$ is the $n \times (n-k)$ zero matrix. It follows that $\dim\{\ker\ T_q\} = n - q > 0$. Choose nonzero $\mathbf{w} \in \ker\ T_q$ and suppose $\mathbf{w} \in \ker\ T_{q+1}$ for all $\tau_{q+1} \in (\tau_q, t_2)$. It follows that

$$(\Gamma_{q+1} \;\; \mathbf{0}_{q+1})^\top \mathbf{w} = 0.$$

hence

$$[f_1(\tau_{q+1}), f_2(\tau_{q+1}), \cdots, f_n(\tau_{q+1})]\mathbf{w} = 0, \qquad \forall \tau_{q+1} \in (\tau_q, t_2). \tag{3.62}$$

However, the function $f_i(\tau)$ are linearly independent in (τ_q, t_2) so that Eq. (3.62) can not hold. This is a contradiction. Therefore, there exists a time $\tau_{q+1} \in (\tau_q, t_2)$ such that $\mathbf{w} \neq \ker\ T_{q+1}$, and since $\ker T_{q+1} \subset \ker\ T_q$, we have $\dim\{\ker\ T_{q+1}\} = \dim\{\ker\ T_q\} - 1 = n - (q+1)$, implying Γ_{q+1} has rank $q+1$. Thus by induction we can construct Γ_p to have rank $p \leq n$. □

Theorem 3.12 *If A and B are constant matrices then system (3.37) is impulsively controllable is and only if*

$$\text{Rank}[B, AB, A^2B, \cdots, A^{n-1}B] = n. \tag{3.63}$$

Furthermore, the maximum number of impulses required to achieve any desired final state for a controllable constant system is $p = \lceil n/m \rceil$, *where* $\lceil x \rceil$ *denote the smallest integer greater than or equal to* x.

Proof. The controbility condition is given by Eqs. (3.49) and (3.50) and followed Lemma 3.1 we have

$$\begin{aligned} W &= \left[\sum_{i=0}^{n-1} f_i(-\tau_1)A^iB, \sum_{i=0}^{n-1} f_i(-\tau_1)A^iB, \cdots, \sum_{i=0}^{n-1} f_i(-\tau_p)A^iB\right] \\ &= [B, AB, \cdots, A^{n-1}B] \\ &\quad \times \begin{bmatrix} f_0(-\tau_1)I_m & f_0(-\tau_2)I_m & \cdots & f_0(-\tau_p)I_m \\ f_1(-\tau_1)I_m & f_1(-\tau_2)I_m & \cdots & f_1(-\tau_p)I_m \\ \vdots & \vdots & \vdots & \vdots \\ f_{n-1}(-\tau_1)I_m & f_{n-1}(-\tau_2)I_m & \cdots & f_{n-1}(-\tau_p)I_m \end{bmatrix}, \end{aligned} \tag{3.64}$$

where I_m is the $m \times m$ identity matrix. We then need to show that the matrix

$$S \triangleq \begin{bmatrix} f_0(-\tau_1)I_m & f_0(-\tau_2)I_m & \cdots & f_0(-\tau_p)I_m \\ f_1(-\tau_1)I_m & f_1(-\tau_2)I_m & \cdots & f_1(-\tau_p)I_m \\ \vdots & \vdots & \vdots & \vdots \\ f_{n-1}(-\tau_1)I_m & f_{n-1}(-\tau_2)I_m & \cdots & f_{n-1}(-\tau_p)I_m \end{bmatrix} \tag{3.65}$$

has full rank that is no less than n for some choice of ordered times $\tau_k \in [t_1, t_2]$, $k = 1, 2, \cdots, p$, where $p = \lceil n/m \rceil$. The matrix S has full rank if and only if the matrix

$$Q \triangleq \begin{bmatrix} f_0(-\tau_1) & f_0(-\tau_2) & \cdots & f_0(-\tau_p) \\ f_1(-\tau_1) & f_1(-\tau_2) & \cdots & f_1(-\tau_p) \\ \vdots & \vdots & \vdots & \vdots \\ f_{n-1}(-\tau_1) & f_{n-1}(-\tau_2) & \cdots & f_{n-1}(-\tau_p) \end{bmatrix} \tag{3.66}$$

has full rank. S has pm columns, hence $\text{Rank}(S) \geq n$ implies $p \geq n/m$; this provides the lower bound for p. Let $p = \lceil n/m \rceil$, then since $m \geq 1$, $p \leq n$ and by Lemma 3.2 we can apply Lemma 3.3 so that there exist $\tau_1, \tau_2, \cdots, \tau_p$ such that Q has full rank. □

Corollary 3.1 *If A and B are constant matrices, then the linear impulsive control system (3.38) is impulsively controllable if and only if*

$$\text{Rank}[sI - A, B] = n, \qquad \forall s \in \mathbb{C}. \tag{3.67}$$

Proof. By using simple linear algebra we know

$$\text{Rank}[B, AB, A^2B, \cdots, A^{n-1}B] = n \tag{3.68}$$

if and only if

$$\text{Rank}[sI - A, B] = n, \qquad \forall s \in \mathbb{C}. \tag{3.69}$$

Then by using Theorem 3.12, we finish the proof. □

Notes

The controllability of linear impulsive system was presented in [Liu95].

Chapter 4

Impulsive Control of Nonlinear Systems

Let a plant be a nonlinear system

$$\dot{\mathbf{x}} = \mathbf{f}(t, \mathbf{x}) \tag{4.1}$$

where $\mathbf{f} : \mathbb{R}_+ \times \mathbb{R}^n \mapsto \mathbb{R}^n$ is continuous. $\mathbf{x} \in \mathbb{R}^n$ is the state variable. Let

$$U(k, \mathbf{x}) = \Delta \mathbf{x}|_{t=\tau_k} \triangleq \mathbf{x}(\tau_k^+(\mathbf{x})) - \mathbf{x}(\tau_k) \tag{4.2}$$

be the *impulsive control law* and $\mathbf{u}(t, \mathbf{x})$ be the *continuous control law* then a general nonlinear impulsive control system is given by

$$\begin{cases} \dot{\mathbf{x}} = \mathbf{f}(t, \mathbf{x}) + \mathbf{u}(t, \mathbf{x}), & t \neq \tau_k(\mathbf{x}), \\ \Delta \mathbf{x} = U(k, \mathbf{x}), & t = \tau_k(\mathbf{x}), \\ \mathbf{x}(t_0^+) = \mathbf{x}_0, & k = 1, 2, \cdots. \end{cases} \tag{4.3}$$

4.1 Methods based on first-order approximation

In this section we consider the stability of zero solution of the following impulsive control system:

$$\begin{cases} \dot{\mathbf{x}} = A(t)\mathbf{x} + \mathbf{g}(t, \mathbf{x}), & t \neq \tau_k, \\ \Delta \mathbf{x} = B_k \mathbf{x} + \mathbf{j}_k(\mathbf{x}), & t = \tau_k, \quad k \in \mathbb{N} \end{cases} \tag{4.4}$$

where $\mathbf{g}(t, 0) = 0$ and $\mathbf{j}_k(0) = 0, k \in \mathbb{N}$. We assume that

$$\tau_{k+1} - \tau_k \geq \theta, \quad k \in \mathbb{N}, \quad \theta > 0. \tag{4.5}$$

We also need the following reference system

$$\begin{cases} \dot{\mathbf{x}} = A(t)\mathbf{x}, & t \neq \tau_k, \\ \Delta\mathbf{x} = B_k\mathbf{x}, & t = \tau_k, \quad k \in \mathbb{N}. \end{cases} \tag{4.6}$$

Theorem 4.1 *Assume that*

(1) the state transition matrix $\Psi(t,s)$ *of reference system (4.6) satisfy*

$$\|\Psi(t,s)\| \leq Ke^{-\gamma(t-s)}, \qquad K \geq 1, \quad \gamma > 0 \tag{4.7}$$

for all t *and* s *with* $t_0 \leq s \leq t$;

(2) for all $t \geq t_0$, $k \in \mathbb{N}$, $\|\mathbf{x}\| \leq h$, $h > 0$ *we have*

$$\|\mathbf{g}(t,\mathbf{x})\| \leq a\|\mathbf{x}\|, \qquad \|\mathbf{j}_k(\mathbf{x})\| \leq a\|\mathbf{x}\|; \tag{4.8}$$

(3)

$$\gamma - Ka - \frac{1}{\theta}\ln(1+Ka) > 0$$

where θ *is given by Eq. (4.5).*

Then the zero solution of system (4.4) is asymptotically stable.

Proof. Let the initial condition be $\mathbf{x}(t_0, \mathbf{x}_0) = \mathbf{x}_0$ then every solution of system (4.4) can be written as

$$\begin{aligned} \mathbf{x}(t,\mathbf{x}_0) &= \Psi(t,t_0)\mathbf{x}_0 + \int_{t_0}^{t} \Psi(t,s)\mathbf{g}(s,\mathbf{x}(s,\mathbf{x}_0))ds \\ &\quad + \sum_{t_0<\tau_i<t} \Psi(t,\tau_i)\mathbf{j}_i(\mathbf{x}(\tau_i,\mathbf{x}_0)). \end{aligned} \tag{4.9}$$

From Eqs. (4.7) and (4.8) we have

$$\begin{aligned} \|\mathbf{x}(t,\mathbf{x}_0)\| &\leq Ke^{-\gamma(t-t_0)}\|\mathbf{x}_0\| + \int_{t_0}^{t} Ke^{-\gamma(t-s)}a\|\mathbf{x}(s,\mathbf{x}_0)\|ds \\ &\quad + \sum_{t_0<\tau_i<t} Ke^{-\gamma(t-\tau_i)}a\|\mathbf{x}(\tau_i,\mathbf{x}_0)\|, \end{aligned} \tag{4.10}$$

or in the following form:

$$\begin{aligned} e^{\gamma(t-t_0)}\|\mathbf{x}(t,\mathbf{x}_0)\| &\leq K\|\mathbf{x}_0\| + \int_{t_0}^{t} Kae^{\gamma(s-t_0)}\|\mathbf{x}(s,\mathbf{x}_0)\|ds \\ &\quad + \sum_{t_0<\tau_i<t} Kae^{\gamma(\tau_i-t_0)}\|\mathbf{x}(\tau_i,\mathbf{x}_0)\|. \end{aligned} \tag{4.11}$$

By using Lemma 2.1, we have

$$e^{\gamma(t-t_0)}||\mathbf{x}(t,\mathbf{x}_0)|| \le K||\mathbf{x}_0||(1+Ka)^{n(t,t_0)}e^{Ka(t-t_0)}. \tag{4.12}$$

Then from the assumption in Eq. (4.5) we have

$$||\mathbf{x}(t,\mathbf{x}_0)|| \le K||\mathbf{x}_0||\exp\left\{-\left(\gamma - Ka - \frac{1}{\theta}\ln(1+Ka)\right)(t-t_0)\right\}. \tag{4.13}$$

Then from assumption (3) we know that for all $||\mathbf{x}_0|| < h/K$,

$$\lim_{t\to\infty} ||\mathbf{x}(t,\mathbf{x}_0)|| = 0.$$

□

Theorem 4.2 *Assume that*

(1) the largest eigenvalue of matrix $\frac{1}{2}(A(t)+A^\top(t))$, $\lambda_n(t)$, *satisfies* $\lambda_n(t) \le \gamma$ *for all* $t \ge t_0$ *and the largest eigenvalue of the matrix* $(I+B_i^\top)(I+B_i)$, Λ_i, *satisfies* $\Lambda_i \le \alpha^2$ *for all* $i \in \mathbb{N}$;

(2) the limit

$$\lim_{T\to\infty} \frac{n(t,t+T)}{T} = p$$

exists and is uniform for all $t \ge t_0$;

(3) $\gamma + p\ln\alpha < 0$;

(4) for all $t \ge t_0$, $k \in \mathbb{N}$, $||\mathbf{x}|| \le h$, $h > 0$ *we have*

$$||\mathbf{g}(t,\mathbf{x})|| \le a||\mathbf{x}||, \qquad ||\mathbf{j}_k(\mathbf{x})|| \le a||\mathbf{x}||. \tag{4.14}$$

Then the zero solution of system (4.4) is asymptotically stable with a sufficiently small constant a.

Proof. Let the initial condition be $\mathbf{x}_r(t_0,\mathbf{x}_0) = \mathbf{x}_0$ then every solution of reference system (4.6) satisfy

$$||\mathbf{x}_r(t_0,\mathbf{x}_0)|| \le \alpha^{n(t,t_0)}e^{\gamma(t-t_0)}||\mathbf{x}_0||. \tag{4.15}$$

Based on assumptions (2) and (3) and Eq. (4.15) we can find $K \ge 1$ and $\mu > 0$ with $0 < \mu < |\gamma + p\ln\alpha|$ such that for all $t_0 \le s \le t$, the state transition matrix of system (4.6), $\Psi(t,s)$, satisfies

$$||\Psi(t,s)\mathbf{z}|| \le Ke^{-\mu(t-s)}||\mathbf{z}||, \qquad \mathbf{z} \in \mathbb{R}^n. \tag{4.16}$$

In a sufficiently small neighborhood of the point $\mathbf{x}_0$ the solution of system (4.4) is given by

$$\begin{aligned}\mathbf{x}(t,\mathbf{x}_0) &= \Psi(t,t_0)\mathbf{x}_0 + \int_{t_0}^{t}\Psi(t,s)\mathbf{g}(s,\mathbf{x}(s,\mathbf{x}_0))ds \\ &\quad + \sum_{t_0<\tau_i<t}\Psi(t,\tau_i)\mathbf{j}_i(\mathbf{x}(\tau_i,\mathbf{x}_0)). \end{aligned} \tag{4.17}$$

Using the similar process in the proof of Theorem 4.1 we have

$$e^{\mu(t-t_0)}||\mathbf{x}(t,\mathbf{x}_0)|| \leq K||\mathbf{x}_0||(1+Ka)^{n(t,t_0)}e^{Ka(t-t_0)}, \tag{4.18}$$

from which we have

$$||\mathbf{x}(t,\mathbf{x}_0)|| \leq K_1||\mathbf{x}_0||\exp\{-(\mu - p\ln(1+Ka) - Ka + \epsilon)(t-t_0)\} \tag{4.19}$$

for any $\epsilon > 0$ with $K_1 = K_1(\epsilon) > 0$. Therefore, if a is sufficiently small such that

$$\mu - p\ln(1+Ka) - Ka > 0$$

then $||\mathbf{x}(t,\mathbf{x}_0)|| \to 0$ as $t \to 0$. □

Similarly, we have the following theorem.

Theorem 4.3 *Assume that*

(1) the largest eigenvalue of matrix $\frac{1}{2}(A(t)+A^{\top}(t))$, $\lambda_n(t)$, *satisfies* $\lambda_n(t) \leq \gamma$ *for all* $t \geq t_0$ *and the largest eigenvalue of the matrix* $(I+B_i^{\top})(I+B_i)$, Λ_i, *satisfies* $\Lambda_i \leq \alpha^2$ *for all* $i \in \mathbb{N}$;

(2) $\tau_k, k \in \mathbb{N}$ *satisfy* $0 < \theta_1 \leq \tau_k - \tau_{k-1} \leq \theta_2$;

(3)

$$\gamma + \frac{1}{\theta}\ln\alpha < 0,$$

where

$$\theta = \begin{cases} \theta_2, & \text{if } 0 < \alpha < 1, \\ \theta_1, & \text{if } \alpha \geq 1; \end{cases}$$

(4) for all $t \geq t_0$, $k \in \mathbb{N}$, $||\mathbf{x}|| \leq h$, $h > 0$ *we have*

$$||\mathbf{g}(t,\mathbf{x})|| \leq a||\mathbf{x}||, \qquad ||\mathbf{j}_k(\mathbf{x})|| \leq a||\mathbf{x}||. \tag{4.20}$$

Then the zero solution of system (4.4) is asymptotically stable with a sufficiently small constant a.

If in system (4.4) $A(t)$ and $B_k, k \in \mathbb{N}$ are time-invariant, then system (4.4) can be rewritten as

$$\begin{cases} \dot{\mathbf{x}} = A\mathbf{x} + \mathbf{g}(t, \mathbf{x}), & t \neq \tau_k, \\ \Delta\mathbf{x} = B\mathbf{x} + \mathbf{j}_k(\mathbf{x}), & t = \tau_k, \quad k \in \mathbb{N} \end{cases} \tag{4.21}$$

where $\mathbf{g}(t, 0) = 0$ and $\mathbf{j}_k(0) = 0, k \in \mathbb{N}$. We then have the following theorems.

Theorem 4.4 *Assume that*

(1) A is in real canonical form and

$$\gamma = \max_j \mathrm{Re}\lambda_j(A), \quad \alpha^2 = \max_j \mathrm{Re}\lambda_j((I + B^\top)(I + B));$$

(2) the limit

$$\lim_{T\to\infty} \frac{n(t, t+T)}{T} = p$$

exists and is uniform for all $t \geq t_0$;

(3) $\gamma + p \ln \alpha < 0$;

(4) for all $t \geq t_0$, $k \in \mathbb{N}$, $||\mathbf{x}|| \leq h$, $h > 0$ we have

$$||\mathbf{g}(t, \mathbf{x})|| \leq a||\mathbf{x}||, \qquad ||\mathbf{j}_k(\mathbf{x})|| \leq a||\mathbf{x}||. \tag{4.22}$$

Then the zero solution of system (4.21) is asymptotically stable with a sufficiently small constant a.

Theorem 4.5 *Assume that*

(1) A is in real canonical form and

$$\gamma = \max_j \mathrm{Re}(\lambda_j(A)), \quad \alpha^2 = \max_j \mathrm{Re}(\lambda_j((I + B^\top)(I + B)));$$

(2) $\tau_k, k \in \mathbb{N}$ satisfy $0 < \theta_1 \leq \tau_k - \tau_{k-1} \leq \theta_2$;

(3)

$$\gamma + \frac{1}{\theta} \ln \alpha < 0,$$

where

$$\theta = \begin{cases} \theta_2, & \text{if } 0 < \alpha < 1, \\ \theta_1, & \text{if } \alpha \geq 1; \end{cases}$$

(4) for all $t \geq t_0$, $k \in \mathbb{N}$, $\|\mathbf{x}\| \leq h$, $h > 0$ we have

$$\|\mathbf{g}(t,\mathbf{x})\| \leq a\|\mathbf{x}\|, \qquad \|\mathbf{j}_k(\mathbf{x})\| \leq a\|\mathbf{x}\|. \tag{4.23}$$

Then the zero solution of system (4.21) is asymptotically stable with a sufficiently small constant a.

4.2 Methods based on linearization

In this section, we consider the stability of the following impulsive control system:

$$\begin{cases} \dot{\mathbf{x}} = \mathbf{f}(\mathbf{x}), & t \neq \tau_k, \\ \Delta\mathbf{x} = U(k,\mathbf{x}), & t = \tau_k, \\ \mathbf{x}(t_0^+) = \mathbf{x}_0, & t_0 \geq 0, \quad k = 1, 2, \cdots, \end{cases} \tag{4.24}$$

where $\mathbf{f}(0) = 0$, $U(k,0) = 0$ and $\mathbf{f} \in C^1(\mathbb{R}^n, \mathbb{R}^n)$. Let matrix $A \in \mathbb{R}^{n\times n}$ be the Jacobian of $\mathbf{f}$ at $\mathbf{x} = 0$; namely,

$$A = \begin{bmatrix} \frac{\partial f_1(0)}{\partial x_1} & \frac{\partial f_1(0)}{\partial x_2} & \cdots & \frac{\partial f_1(0)}{\partial x_n} \\ \frac{\partial f_2(0)}{\partial x_1} & \frac{\partial f_2(0)}{\partial x_2} & \cdots & \frac{\partial f_2(0)}{\partial x_n} \\ \vdots & \vdots & \vdots & \vdots \\ \frac{\partial f_n(0)}{\partial x_1} & \frac{\partial f_n(0)}{\partial x_2} & \cdots & \frac{\partial f_n(0)}{\partial x_n} \end{bmatrix}, \tag{4.25}$$

then the linearized version of system (4.24) is given by

$$\begin{cases} \dot{\mathbf{x}} = A\mathbf{x}, & t \neq \tau_k, \\ \Delta\mathbf{x} = U(k,\mathbf{x}), & t = \tau_k, \\ \mathbf{x}(t_0^+) = \mathbf{x}_0, & t_0 \geq 0, \quad k = 1, 2, \cdots, \end{cases} \tag{4.26}$$

Theorem 4.6 *Assume that*

(1) A is negative-definite and there is a symmetric positive-definite matrix P that is the unique solution of the following equation:

$$A^\top P + PA = -Q. \tag{4.27}$$

Let λ_1 and λ_2 be the smallest and the largest eigenvalues of P and λ_3 be the smallest eigenvalue of $P^{-1}Q$ such that

$$-\lambda_3 + 2\sqrt{\frac{\lambda_2}{\lambda_1}} < 0; \tag{4.28}$$

(2)

$$(\mathbf{x}(\tau_k) + U(k, \mathbf{x}))^\top P(\mathbf{x}(\tau_k) + U(k, \mathbf{x})) \le \mathbf{x}(\tau_k)^\top P\mathbf{x}(\tau_k).$$

Then the origin $\mathbf{x} = 0$ of system (4.24) is asymptotically stable.

Proof. Since $\mathbf{f} \in C^1[\mathbb{R}^n, \mathbb{R}^n]$ and $\mathbf{f}(0) = 0$ we have

$$\mathbf{f}(\mathbf{x}) = A\mathbf{x} + \mathbf{g}(\mathbf{x}) \tag{4.29}$$

where $\mathbf{g} \in C[\mathbb{R}^n, \mathbb{R}^n]$ satisfies

$$\lim_{||\mathbf{x}|| \to 0} \frac{||\mathbf{g}(\mathbf{x})||}{||\mathbf{x}||} = 0 \tag{4.30}$$

from which we know that there is a $\delta > 0$ such that

$$||\mathbf{g}(\mathbf{x})|| \le ||\mathbf{x}||, \quad \text{provided } ||\mathbf{x}|| \le \delta. \tag{4.31}$$

For any $t \in (\tau_k, \tau_{k+1}]$ we have

$$\mathbf{x}(t) = \mathbf{x}(\tau_k^+) + \int_{\tau_k}^{t} [A\mathbf{x}(s) + \mathbf{g}(\mathbf{x}(s))]ds \tag{4.32}$$

from which we have

$$||\mathbf{x}(t)|| \le ||\mathbf{x}(\tau_k^+)|| + \int_{\tau_k}^{t} [||A||||\mathbf{x}(s)|| + ||\mathbf{g}(\mathbf{x}(s))||]ds, \qquad t \in (\tau_k, \tau_{k+1}]. \tag{4.33}$$

Let $\delta_1 = e^{-(||A||+1)(\tau_{k+1}-\tau_k)}$, then we claim that $||\mathbf{x}(t)|| \le \delta$ for all $t \in (\tau_k, \tau_{k+1}]$ whenever $\mathbf{x}(\tau_k^+) \le \delta_1$. For otherwise there should be a $t^* \in (\tau_k, \tau_{k+1}]$ such that $||\mathbf{x}(t^*)|| = \delta$ and $||\mathbf{x}(t)|| \le \delta$ for all $t \in (\tau_k, t^*]$. We will prove that it is not the case.

Since $\|\mathbf{x}(t)\| \le \delta$ for all $t \in (\tau_k, t^*]$ by assumption and in view of Eqs. (4.31) and (4.33) we have

$$\begin{aligned}\|\mathbf{x}(t)\| &\le \|\mathbf{x}(\tau_k^+)\| + \int_{\tau_k}^{t} [\|A\|\|\mathbf{x}(s)\| + \|\mathbf{g}(\mathbf{x}(s))\|]ds \\ &\le \|\mathbf{x}(\tau_k^+)\| + \int_{\tau_k}^{t} [\|A\|\|\mathbf{x}(s)\| + \|\mathbf{x}(s)\|]ds, \\ &\quad t \in (\tau_k, t^*]. \end{aligned} \tag{4.34}$$

By Gronwell-Bellman inequalities, Eq. (4.34) implies that

$$\|\mathbf{x}(t)\| \le \|\mathbf{x}(\tau_k^+)\| e^{(\|A\|+1)(t-\tau_k)}, \qquad t \in (\tau_k, t^*]. \tag{4.35}$$

Therefore

$$\begin{aligned}\|\mathbf{x}(t^*)\| &\le \|\mathbf{x}(\tau_k^+)\| e^{(\|A\|+1)(t^*-\tau_k)} \\ &\le \delta_1 e^{(\|A\|+1)(t^*-\tau_k)} \\ &= \delta e^{-(\|A\|+1)(\tau_{k+1}-\tau_k)} e^{(\|A\|+1)(t^*-\tau_k)} \\ &= \delta e^{(\|A\|+1)(t^*-\tau_{k+1})} \\ &< \delta \end{aligned} \tag{4.36}$$

is a contradiction to the assumption that $\|\mathbf{x}(t^*)\| = \delta$. Thus, we conclude that $\|\mathbf{x}(t)\| \le \delta$ for all $t \in (\tau_k, \tau_{k+1}]$ whenever $\mathbf{x}(\tau_k^+) \le \delta_1$. Then from Eq. (4.31) we have

$$\|\mathbf{g}(\mathbf{x}(t))\| \le \|\mathbf{x}(t)\| \tag{4.37}$$

for all $t \in (\tau_k, \tau_{k+1}]$ and $\|\mathbf{x}(\tau_k^+)\| \le \delta_1$.

Let us choose the Lyapunov function as

$$V(t, \mathbf{x}) = \mathbf{x}(t)^\top P \mathbf{x}(t) \tag{4.38}$$

then for $(t, \mathbf{x}) \in \mathfrak{G}$ we have

$$\begin{aligned}\dot{V}(t, \mathbf{x}) &= \mathbf{x}^\top (A^\top P + PA)\mathbf{x} + (\mathbf{g}(\mathbf{x})^\top P\mathbf{x} + \mathbf{x}^\top P\mathbf{g}(\mathbf{x})) \\ &= -\mathbf{x}^\top Q\mathbf{x} + (\mathbf{g}(\mathbf{x})^\top P\mathbf{x} + \mathbf{x}^\top P\mathbf{g}(\mathbf{x})) \\ &\le \left(-\lambda_3 + 2\sqrt{\frac{\lambda_2}{\lambda_1}}\right) V(t, \mathbf{x}), \qquad (t, \mathbf{x}) \in \mathfrak{G}, \end{aligned} \tag{4.39}$$

whenever $\|\mathbf{x}(\tau_k^+)\| \leq \delta_1$. By choosing

$$\varrho(w) = \left(\lambda_3 - 2\sqrt{\frac{\lambda_2}{\lambda_1}}\right) w$$

the conclusion is straightforward from Corollary 2.1. □

For $t \in (\tau_k, \tau_{k+1}]$, the solution of Eq. (4.24) is given by

$$\mathbf{x}(t) = e^{A(t-\tau_k)}\mathbf{x}(\tau_k^+) + \int_{\tau_k}^{t} e^{A(t-s)}\mathbf{g}(\mathbf{x}(s))ds. \tag{4.40}$$

When $t = \tau_{k+1}$, we have

$$\mathbf{x}(\tau_{k+1}) = e^{A(\tau_{k+1}-\tau_k)}\mathbf{x}(\tau_k^+) + \int_{\tau_k}^{\tau_{k+1}} e^{A(\tau_{k+1}-s)}\mathbf{g}(\mathbf{x}(s))ds. \tag{4.41}$$

Let us use the following notation:

$$\xi_k \triangleq \int_{\tau_k}^{\tau_{k+1}} e^{A(\tau_{k+1}-s)}\mathbf{g}(\mathbf{x}(s))ds. \tag{4.42}$$

For any given $\epsilon > 0$, we choose $\epsilon_1 > 0$ such that

$$\epsilon = \epsilon_1 e^{(2\|A\|+1)(\tau_{k+1}-\tau_k)}. \tag{4.43}$$

From Eq. (4.30) we know that there is a $\delta_2 > 0$ such that

$$\|\mathbf{g}(\mathbf{x})\| \leq \epsilon_1\|\mathbf{x}\|, \quad \text{provided } \|\mathbf{x}\| < \delta_2. \tag{4.44}$$

Let us choose

$$\delta_3 = \min\left(\delta_1, \frac{\delta_2}{e^{(\|A\|+1)(\tau_{k+1}-\tau_k)}}\right)$$

then in case of $\|\mathbf{x}\| \leq \delta_3$, from Eq. (4.35) we have

$$\begin{aligned}\|\mathbf{x}(t)\| &\leq \|\mathbf{x}(\tau_k)^+\|e^{(\|A\|+1)(\tau_{k+1}-\tau_k)}\\ &\leq \delta_3 e^{(\|A\|+1)(\tau_{k+1}-\tau_k)}\\ &\leq \delta_2, \quad t \in (\tau_k, \tau_{k+1}].\end{aligned} \tag{4.45}$$

From Eqs. (4.35), (4.44) and (4.45) we know that

$$\begin{aligned}\|\mathbf{g}(\mathbf{x}(t))\| &\leq \epsilon_1\|\mathbf{x}(t)\|\\ &\leq \epsilon_1\|\mathbf{x}(\tau_k^+)\|e^{(\|A\|+1)(\tau_{k+1}-\tau_k)},\\ &\quad \text{provided } \|\mathbf{x}(\tau_k^+)\| \leq \delta_3, \quad t \in (\tau_k, \tau_{k+1}].\end{aligned} \tag{4.46}$$

We then have

$$\begin{aligned} ||\xi_k|| &\leq \int_{\tau_k}^{\tau_{k+1}} e^{||A||(\tau_{k+1}-\tau_k)}||\mathbf{g}(\mathbf{x}(s))||ds \\ &\leq e^{||A||(\tau_{k+1}-\tau_k)}\epsilon_1||\mathbf{x}(\tau_k^+)||e^{(||A||+1)(\tau_{k+1}-\tau_k)} \\ &= \epsilon||\mathbf{x}(\tau_k^+)||, \quad \text{provided } ||\mathbf{x}(\tau_k^+)|| \leq \delta_3. \end{aligned} \tag{4.47}$$

Thus for any given $\epsilon > 0$, there is a $\delta_3 > 0$, $\delta_3 \leq \delta_1$, such that $||\xi_k|| \leq \epsilon||\mathbf{x}(\tau_k^+)||$ whenever $||\mathbf{x}(\tau_k^+)|| < \delta_3$.

Then from Eq. (4.41) we have for any given $\epsilon > 0$

$$\begin{aligned} ||\mathbf{x}(\tau_{k+1})|| &\leq e^{||A||(\tau_{k+1}-\tau_k)}||\mathbf{x}(\tau_k^+)|| + ||\xi|| \\ &\leq \left(e^{||A||(\tau_{k+1}-\tau_k)} + \epsilon I\right)||\mathbf{x}(\tau_k^+)|| \\ &\quad \text{provided } ||\mathbf{x}(\tau_k^+)|| \leq \delta_3. \end{aligned} \tag{4.48}$$

If the impulsive control law is given by $U(k,\mathbf{x}) = B_k\mathbf{x}$, then we immediately have the following theorem:

Theorem 4.7 *Assume that*

(1)

$$0 < \beth_1 = \inf_{k\in\mathbb{N}}(\tau_{k+1} - \tau_k) \leq \sup_{k\in\mathbb{N}}(\tau_{k+1} - \tau_k) = \beth_2 < \infty;$$

(2)

$$||(I + B_k)e^{A(\tau_{k+1}-\tau_k)}|| \leq \beth_3 < 1, \qquad k \in \mathbb{N}.$$

Then the origin $\mathbf{x} = 0$ of system (4.24) is asymptotically stable.

If

$$\Lambda = \lim_{k\to\infty}(I + B_k)e^{A(\tau_{k+1}-\tau_k)}$$

exists, then we have the following theorem

Theorem 4.8 *Assume that assumption (1) of Theorem 4.7 is satisfied and the spectral radius of Λ, $\rho(\Lambda) < 1$, then the origin $\mathbf{x} = 0$ of system (4.24) is asymptotically stable.*

4.3 Comparison method

Definition 4.1 *Comparison system*
Let $V \in \mathcal{V}_0$ and assume that

$$\begin{cases} D^+V(t,\mathbf{x}) \leq g(t,V(t,\mathbf{x})), & t \neq \tau_k, \\ V(t,\mathbf{x}+U(k,\mathbf{x})) \leq \psi_k(V(t,\mathbf{x})), & t = \tau_k, \end{cases} \tag{4.49}$$

where $g : \mathbb{R}_+ \times \mathbb{R}_+ \to \mathbb{R}$ is continuous and $\psi_k : \mathbb{R}_+ \to \mathbb{R}_+$ is nondecreasing. Then the following system

$$\begin{cases} \dot{w} = g(t,w), & t \neq \tau_k, \\ w(\tau_k^+) = \psi_k(w(\tau_k)), \\ w(t_0^+) = w_0 \geq 0 \end{cases} \tag{4.50}$$

is the comparison system of Eq.(4.3).

Assumptions: $\mathbf{f}(t,0) = 0$, $U(k,0) = 0$ and $g(t,0) = 0$ for all k.

Then the following theorem gives sufficient conditions in a unified way for various stability criteria.

Theorem 4.9 *(Theorem 3.2.1, [LBS89]):*
Assume that

(1) $V : \mathbb{R}_+ \times S_\rho \to \mathbb{R}_+, \rho > 0$, $V \in \mathcal{V}_0$, $D^+V(t,\mathbf{x}) \leq g(t,V(t,\mathbf{x}))$, $t \neq \tau_k$*;*

(2) There exists a $\rho_0 > 0$ *such that* $\mathbf{x} \in S_{\rho_0}$ *implies that* $\mathbf{x}+U(k,\mathbf{x}) \in S_{\rho_0}$ *for all* k *and* $V(t,\mathbf{x}+U(k,\mathbf{x})) \leq \psi_k(V(t,\mathbf{x}))$, $t = \tau_k$, $\mathbf{x} \in S_{\rho_0}$*;*

(3) $\beta(\|\mathbf{x}\|) \leq V(t,\mathbf{x}) \leq \alpha(\|\mathbf{x}\|)$ *on* $\mathbb{R}_+ \times S_\rho$ *where* $\alpha(\cdot), \beta(\cdot) \in \mathcal{K}$.

Then the stability properties of the trivial solution of comparison system (4.50) imply the corresponding stability properties of the trivial solution of (4.3).

Remarks: The system in (4.50) is a scalar impulsive differential equation. If we can find the comparison system of a high order impulsive control system, then the stability analysis may be simplified to that of a scalar impulsive differential equation.

Theorem 4.10 *(Corollary 3.2.1., [LBS89]):*
Let $g(t,w) = \dot{\lambda}(t)w$, $\lambda \in C^1[\mathbb{R}_+, \mathbb{R}_+]$, $\psi_k(w) = d_k w$, $d_k \geq 0$ *for all* k*, then the origin of system (4.3) is asymptotically stable if conditions*

$$\lambda(\tau_{k+1}) + \ln(\gamma d_k) \leq \lambda(\tau_k), \text{ for all } k, \text{ where } \gamma > 1, \tag{4.51}$$

and

$$\dot{\lambda}(t) \geq 0 \tag{4.52}$$

are satisfied.

Let the plant be a nonlinear system as

$$\begin{cases} \dot{\mathbf{x}} = A\mathbf{x} + \phi(\mathbf{x}) \\ \mathbf{y} = C\mathbf{x} \end{cases} \tag{4.53}$$

where $\mathbf{x} \in \mathbb{R}^n$ is state variable, A is an $n \times n$ constant matrix, $\mathbf{y} \in \mathbb{R}^m$ is output and C is an $m \times n$ constant matrix. $\phi : \mathbb{R}^n \to \mathbb{R}^n$ is a nonlinear function. The control instant is given by τ_k, $k = 1, 2, \cdots$. Then the nonlinear impulsive control system is given by

$$\begin{cases} \dot{\mathbf{x}} = A\mathbf{x} + \phi(\mathbf{x}), & \\ \mathbf{y} = C\mathbf{x}, & t \neq \tau_k, \\ \Delta\mathbf{x} = B\mathbf{y}, & t = \tau_k, \quad k = 1, 2, \cdots, \end{cases} \tag{4.54}$$

where B is an $n \times m$ constant matrix. Then the impulsive control system can be rewritten as

$$\begin{cases} \dot{\mathbf{x}} = A\mathbf{x} + \phi(\mathbf{x}), & t \neq \tau_k, \\ \Delta\mathbf{x} = BC\mathbf{x}, & t = \tau_k, \quad k = 1, 2, \cdots, \\ \mathbf{x}(t_0^+) = \mathbf{x}_0 & \end{cases} \tag{4.55}$$

From which we know that $U(k, \mathbf{x}) = BC\mathbf{x}$.

Theorem 4.11 *Let the $n \times n$ matrix Γ be symmetric and positive definite, and $\lambda_1 > 0$, $\lambda_2 > 0$ are respectively, the smallest and the largest eigenvalues of Γ. Let*

$$Q = \Gamma A + A^\top \Gamma \tag{4.56}$$

and λ_3 be the largest eigenvalue of $\Gamma^{-1}Q$. $\phi(\mathbf{x})$ is continuous and $\|\phi(\mathbf{x})\| \leq L\|\mathbf{x}\|$ where $L > 0$ is a constant. λ_4 is the largest eigenvalue of the matrix

$$\Gamma^{-1}(I + (BC)^\top)\Gamma(I + BC). \tag{4.57}$$

We also suppose that $\|I + BC\| \leq 1$. Then the origin of impulsive control system (4.55) is asymptotically stable if

$$\left(\lambda_3 + 2L\sqrt{\frac{\lambda_2}{\lambda_1}}\right)(\tau_{k+1} - \tau_k) \leq -\ln(\gamma\lambda_4), \qquad \gamma > 1 \tag{4.58}$$

and

$$\lambda_3 + 2L\sqrt{\frac{\lambda_2}{\lambda_1}} \geq 0. \tag{4.59}$$

Proof. Construct a Lyapunov function $V(\mathbf{x}) = \mathbf{x}^\top \Gamma \mathbf{x}$, when $t \neq \tau_k$ we have

$$\begin{aligned} D^+V(\mathbf{x}) &= \mathbf{x}^\top (A^\top \Gamma + \Gamma A)\mathbf{x} + (\phi^\top(\mathbf{x})\Gamma\mathbf{x} + \mathbf{x}^\top \Gamma \phi(\mathbf{x})) \\ &= \mathbf{x}^\top Q\mathbf{x} + (\phi^\top(\mathbf{x})\Gamma\mathbf{x} + \mathbf{x}^\top \Gamma \phi(\mathbf{x})) \\ &\leq \left(\lambda_3 + 2L\sqrt{\frac{\lambda_2}{\lambda_1}}\right) V(\mathbf{x}), \quad t \neq \tau_k. \end{aligned} \tag{4.60}$$

Hence, condition 1 of Theorem 4.9 is satisfied with $g(t,w) = \left(\lambda_3 + 2L\sqrt{\frac{\lambda_2}{\lambda_1}}\right) w$.

Given any $\rho_0 > 0$ and $\mathbf{x} \in \mathfrak{S}_{\rho_0}$, we have

$$\begin{aligned} \|\mathbf{x} + U(k,\mathbf{x})\| &= \|\mathbf{x} + BC\mathbf{x}\| \\ &\leq \|I + BC\|\|\mathbf{x}\| \\ &\leq \|\mathbf{x}\|. \end{aligned} \tag{4.61}$$

From which we know that $\mathbf{x} + U(k,\mathbf{x}) \in \mathfrak{S}_{\rho_0}$.

When $t = \tau_k$, we have

$$\begin{aligned} V(\mathbf{x} + BC\mathbf{x})|_{t=\tau_k} &= (\mathbf{x} + BC\mathbf{x})^\top \Gamma (\mathbf{x} + BC\mathbf{x}) \\ &= \mathbf{x}^\top (I + (BC)^\top)\Gamma(I + BC)\mathbf{x} \\ &\leq \lambda_4 V(\mathbf{x}). \end{aligned} \tag{4.62}$$

Hence condition 2 of Theorem 4.9 is satisfied with $\psi_k(w) = \lambda_4 w$. And we have

$$\lambda_1\|\mathbf{x}\|^2 \leq V(\mathbf{x}) \leq \lambda_2\|\mathbf{x}\|^2. \tag{4.63}$$

From which one can see that condition 3 of Theorem 4.9 is also satisfied with $\beta(x) = \lambda_1 x$ and $\alpha(x) = \lambda_2 x$. It follows from Theorem 4.9 that the asymptotic stability of the impulsive control system in Eq.(4.55) is implied by that of the following comparison system:

$$\begin{cases} \dot{w}(t) = (\lambda_3 + 2L\sqrt{\frac{\lambda_2}{\lambda_1}})w(t), & t \neq \tau_k, \\ w(\tau_k^+) = \lambda_4 w(\tau_k), \\ w(t_0^+) = w_0 \geq 0. \end{cases} \tag{4.64}$$

It follows from Theorem 4.10 that if

$$\int_{\tau_k}^{\tau_{k+1}} \left(\lambda_3 + 2L\sqrt{\frac{\lambda_2}{\lambda_1}}\right) dt + \ln(\gamma\lambda_4) \leq 0, \qquad \gamma > 1 \tag{4.65}$$

i.e.,

$$\left(\lambda_3 + 2L\sqrt{\frac{\lambda_2}{\lambda_1}}\right)(\tau_{k+1} - \tau_k) \leq -\ln(\gamma\lambda_4), \qquad \gamma > 1 \tag{4.66}$$

and

$$\lambda_3 + 2L\sqrt{\frac{\lambda_2}{\lambda_1}} \geq 0 \tag{4.67}$$

are satisfied, then the origin of (4.55) is asymptotically stable. □

Remark: If BC is symmetric we know that $I + BC$ is also symmetric. By using Euclidean norm we have

$$\rho(I + BC) = ||I + BC||. \tag{4.68}$$

Then the condition of $||I + BC|| \leq 1$ can be replaced by $\rho(I + BC) \leq 1$.

We then present some design examples to show how to perform impulsive control on chaotic dynamical systems. Impulsive control of chaotic systems has important applications to nonlinear secure communication systems and nonlinear spread spectrum communication systems.

4.4 Example: the Lorenz system

Since the Lorenz system is a benchmark of chaotic systems, in this section we present the impulsive control of the Lorenz system.

4.4.1 *Theory*

The Lorenz (chaotic) system[Lor63] is given by

$$\begin{cases} \dot{x} = -\sigma x + \sigma y \\ \dot{y} = rx - y - xz \\ \dot{z} = xy - bz \end{cases} \tag{4.69}$$

where σ, r, and b are three real positive parameters. When these parameters are chosen as $\sigma = 10$, $r = 28$, and $b = \frac{8}{3}$, the Lorenz system is chaotic.

By decomposing the linear and nonlinear parts of Lorenz system in Eq.(4.69), we can rewrite it as

$$\dot{\mathbf{x}} = A\mathbf{x} + \Phi(\mathbf{x}) \tag{4.70}$$

where $\mathbf{x}^\top = (x, y, z)$ and

$$A = \begin{pmatrix} -\sigma & \sigma & 0 \\ r & -1 & 0 \\ 0 & 0 & -b \end{pmatrix}, \quad \Phi(\mathbf{x}) = \begin{pmatrix} 0 \\ -xz \\ xy \end{pmatrix}. \tag{4.71}$$

The impulsive control of Lorenz system is then given by

$$\begin{cases} \dot{\mathbf{x}} = A\mathbf{x} + \Phi(\mathbf{x}), & t \neq \tau_k, \\ \Delta\mathbf{x} = B\mathbf{x}, & t = \tau_k, \quad k = 1, 2, \cdots, \end{cases} \tag{4.72}$$

where τ_k denote the moments when impulsive control occurs. Without loss of generality, we assume that $\tau_1 > t_0$. We say $\{\tau_i\}$ is *equidistant* if exists a constant $\delta > 0$ such that $\tau_{i+1} - \tau_i = \delta$, $i = 1, 2, \cdots$. Then we use the following theorem to guarantee that the impulsive control system will be asymptotically stabilized at origin.

Theorem 4.12 *Let d_1 be the largest eigenvalue of $(I + B^\top)(I + B)$. B is Hermitian with $\rho(I + B) \leq 1$. Let q be the largest eigenvalue of $(A + A^\top)$ and impulses be equidistant with an interval $\delta > 0$. If*

$$0 \leq q \leq -\frac{1}{\delta}\ln(\xi d_1), \quad \xi > 1 \tag{4.73}$$

then the origin of the impulsively controlled Lorenz system in Eq.(4.72) is asymptotically stable.

Proof. Let $V(t, \mathbf{x}) = \mathbf{x}^\top\mathbf{x}$. For $t \neq \tau_i$, we have

$$\begin{aligned} D^+V(t, \mathbf{x}) &= \mathbf{x}^\top A\mathbf{x} + \mathbf{x}^\top A^\top\mathbf{x} + \mathbf{x}^\top\Phi(\mathbf{x}) + \Phi^\top(\mathbf{x})\mathbf{x} \\ &\leq q\mathbf{x}^\top\mathbf{x} + 2(-xyz + xyz) \\ &= qV(t, \mathbf{x}). \end{aligned} \tag{4.74}$$

Hence condition 1 of Theorem 4.9 is satisfied with $g(t, w) = qw$. Since B is Hermitian we know that $(I + B)$ is also Hermitian. By using Euclidean norm we have

$$\rho(I + B) = \|I + B\|. \tag{4.75}$$

Given any $\rho_0 > 0$ and $\mathbf{x} \in \mathfrak{S}_{\rho_0}$, we have

$$\begin{aligned} \|\mathbf{x} + U(i, \mathbf{x})\| &= \|\mathbf{x} + B\mathbf{x}\| \\ &\leq \|I + B\|\|\mathbf{x}\| \\ &= \rho(I + B)\|\mathbf{x}\| \\ &\leq \|\mathbf{x}\|. \end{aligned} \tag{4.76}$$

The last inequality is in view of $\rho(I + B) \leq 1$. From which we know that $\mathbf{x} + U(i, \mathbf{x}) \in \mathfrak{S}_{\rho_0}$.

For $t = \tau_i$ we have

$$\begin{aligned} V(\tau_i, \mathbf{x} + B\mathbf{x}) &= (\mathbf{x} + B\mathbf{x})^\top (\mathbf{x} + B\mathbf{x}) \\ &= \mathbf{x}^\top (I + B^\top)(I + B)\mathbf{x} \\ &\leq d_1 V(\tau_i, \mathbf{x}). \end{aligned} \tag{4.77}$$

Hence condition 2 of Theorem 4.9 is satisfied with $\psi_i(w) = d_1 w$. We can see that condition 3 of Theorem 4.9 is also satisfied. It follows Theorem 4.9 that the asymptotic stability of system (4.72) is implied by that of the following comparison system:

$$\begin{cases} \dot{\omega} = q\omega, \ t \neq \tau_i, \\ \omega(\tau_i) = d_1 \omega(\tau_i), \\ \omega(t_0^+) = \omega_0 \geq 0. \end{cases} \tag{4.78}$$

Since $0 \leq q \leq -\frac{1}{\delta} \ln(\xi d_1)$, we have

$$\int_{\tau_i}^{\tau_{i+1}} q dt + \ln(\xi d_1) \leq 0, \qquad \xi > 1, \qquad \text{for all } i \tag{4.79}$$

It follows Theorem 4.10 that the trivial solution of Eq. (4.72) is asymptotically stable. □

The above Theorem also gives an estimation of the upper boundary of δ, $\delta_{\max}$:

$$\delta_{\max} = -\frac{\ln(\xi d_1)}{q} > 0, \quad \xi \to 1^+. \tag{4.80}$$

From above we can see that d_1 should satisfy $0 < d_1 < 1$.

4.4.2 *Numerical experimental results*

We the use some simulation results to show how the impulsive control scheme works.

Example 4.1

In this experiment, we chose the matrix B as

$$B = \begin{pmatrix} k & 0 & 0 \\ 0 & -1 & 0 \\ 0 & 0 & -1 \end{pmatrix} \tag{4.81}$$

where B is Hermitian. To satisfy the condition $\rho(I + B) \leq 1$ in Theorem 4.12, we find that k should satisfy $-2 \leq k \leq 0$. It is easy to see that

$$d_1 = (k+1)^2. \tag{4.82}$$

Let the parameters be $\sigma = 10$, $r = 28$ and $b = \frac{8}{3}$, then we have

$$A = \begin{pmatrix} -10 & 10 & 0 \\ 28 & -1 & 0 \\ 0 & 0 & -\frac{8}{3} \end{pmatrix}, \quad A + A^{\top} = \begin{pmatrix} -20 & 38 & 0 \\ 38 & -2 & 0 \\ 0 & 0 & -\frac{16}{3} \end{pmatrix}. \tag{4.83}$$

We find $q = 28.051249$. Then estimations of boundaries of stable regions are given by

$$0 \leq \delta \leq -\frac{(\ln \xi + \ln(k+1)^2)}{q}, \quad -2 \leq k \leq 0. \tag{4.84}$$

Figure 4.1 shows the stable region for different ξ's. The entire region under the curve of $\xi = 1$ is the stable region. When $\xi \to \infty$, the stable region approaches a vertical line $k = -1$.

The numerical experimental results are shown in Fig. 4.2. The solid, dash-dotted, and dotted waveforms show $x(t)$, $y(t)$ and $z(t)$, respectively. Figure 4.2(a) shows the stable results within the stable region with $k = -1.5$ and $\delta = 0.04$. One can see that the system asymptotically approaches the origin with a settling time about 0.2. Figure 4.2(b) shows the unstable results with $k = -1.5$ and $\delta = 0.17$.

Example 4.2

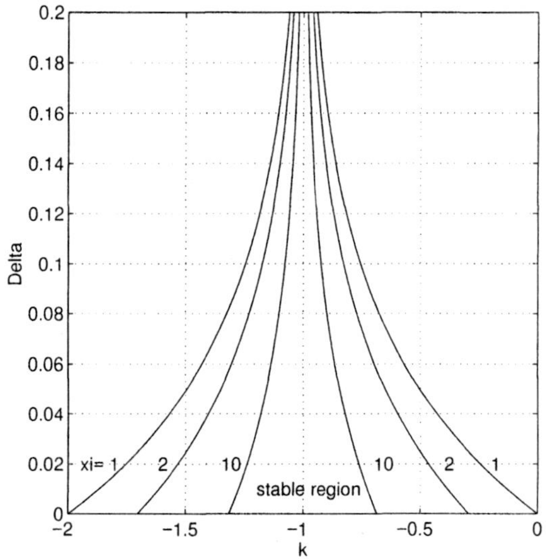

Fig. 4.1 The estimation of boundaries of stable region with different ξ's used in experiment 4.1.

In this experiment, we chose the matrix B as

$$B = \begin{pmatrix} k & 0 & 0 \\ 0 & -0.6 & 0 \\ 0 & 0 & -0.6 \end{pmatrix}. \tag{4.85}$$

Similarly, k should satisfy $-2 \leq k \leq 0$. Then we have

$$d_1 = \begin{cases} (k+1)^2, & (k+1)^2 \geq 0.16, \\ 0.16, & \text{else.} \end{cases} \tag{4.86}$$

Then an estimation of the boundaries of the stable region is given by

$$0 \leq \delta \leq \begin{cases} -\frac{\ln \xi + \ln(k+1)^2}{q}, & (k+1)^2 \geq 0.16 \\ -\frac{\ln \xi + \ln(0.16)}{q}, & \text{else} \end{cases}, \quad -2 \leq k \leq 0. \tag{4.87}$$

Figure 4.3 shows the stable region for different ξ's. The entire region under the curve of $\xi = 1$ is the stable region.

The simulation results are shown in Fig. 4.4. The solid, dash-dotted, and dash-dotted waveforms show $x(t)$, $y(t)$ and $z(t)$, respectively. Figure 4.4(a) shows the stable results within the stable region with $k = -0.6$ and

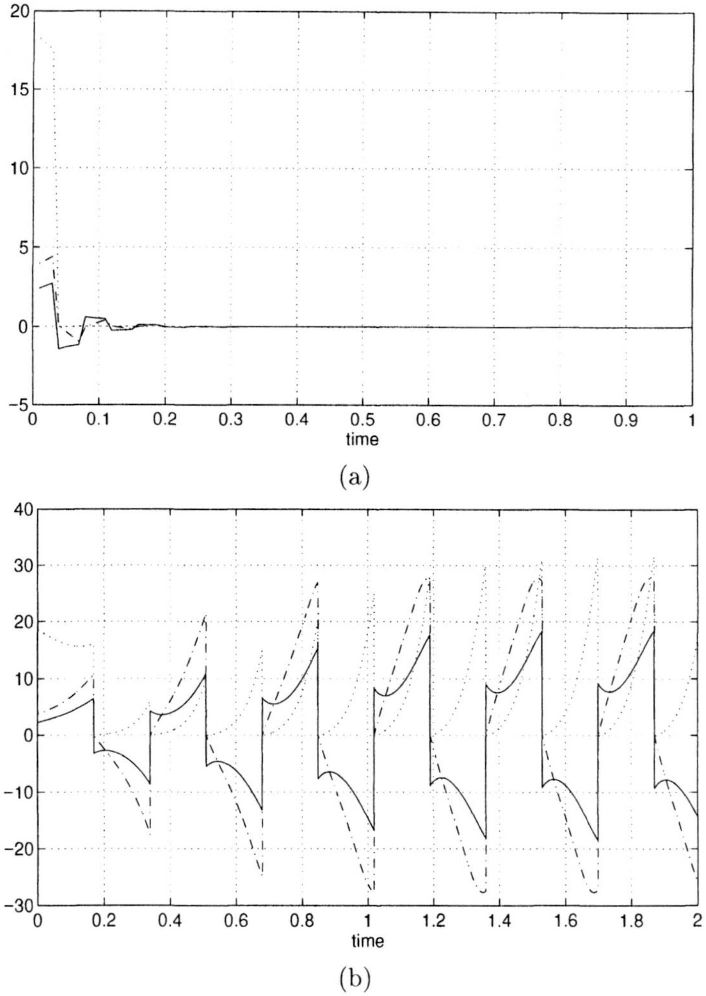

Fig. 4.2 Simulation results. (a) Stable results in stable region. (b) Unstable results out of stable region.

$\delta = 0.06$. One can see that the system asymptotically approaches the origin with a settling time around 1 time unit. Figure 4.4(b) shows the unstable result with $k = -0.6$ and $\delta = 0.09$.

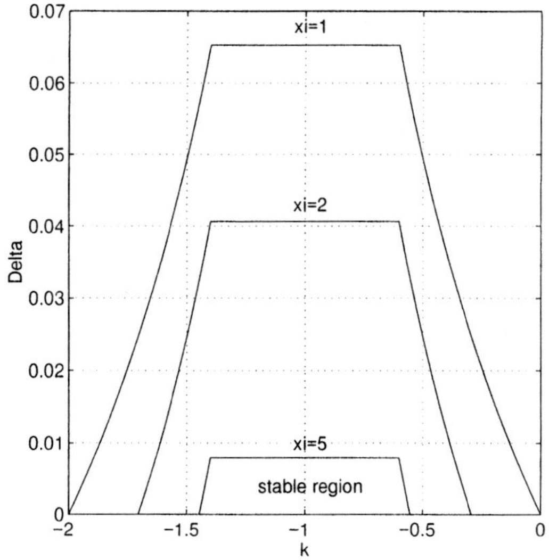

Fig. 4.3 The estimation of boundaries of stable region with different ξ's used in experiment 4.2.

4.5 Linear approximation method

4.5.1 *Theory*

Consider the following impulsive differential system

$$\begin{cases} \dot{\mathbf{x}} = A(t)\mathbf{x} + \mathbf{g}(t, \mathbf{x}), & t \neq \tau_k, \\ \Delta \mathbf{x} = B_k \mathbf{x}, & t = \tau_k, \quad k = 1, 2, \cdots \end{cases} \tag{4.88}$$

where $\mathbf{x} \in \mathbb{R}^n$, $A(t) \in \mathbb{R}^{n \times n}$ and $\mathbf{g} : \mathbb{R}_+ \times \mathbb{R}^n \to \mathbb{R}^n$.

Let $\Phi(t, s)$ be the fundamental matrix of the following ODE:

$$\dot{\mathbf{x}} = A(t)\mathbf{x}. \tag{4.89}$$

Theorem 4.13 *Let the following conditions hold:*

(1) $A(t) \in \mathcal{PC}(\mathbb{R}_+, \mathbb{R}^{n \times n})$;
(2) $\|\Phi(t, s)\| \leq \phi(t)\psi(s)$ *for* $\tau_{k-1} < s \leq t \leq \tau_k, k \in \mathbb{N}$, *where* $\phi(\cdot), \psi(\cdot) \in PC(\mathbb{R}_+, \mathbb{R})$ *and* $\phi(t) > 0$, $\psi(t) > 0$, $\phi(\tau_k^+) > 0$, $\psi(\tau_k^+) > 0$ *for* $t \in \mathbb{R}_+$, $k \in \mathbb{N}$;
(3) The function $\mathbf{g}(t, \mathbf{x})$ *is continuous and locally Lipschitzian continuous with respect to* $\mathbf{x}$ *in the sets* $(\tau_{k-1}, \tau_k] \times \mathbb{R}^n$, $k \in \mathbb{N}$, *and for*

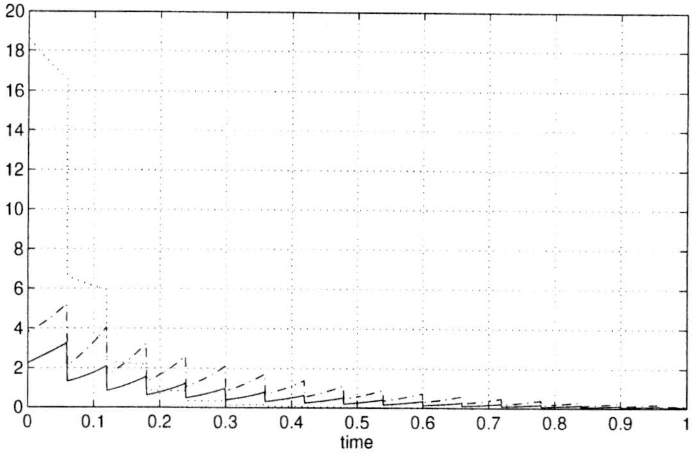

(a)

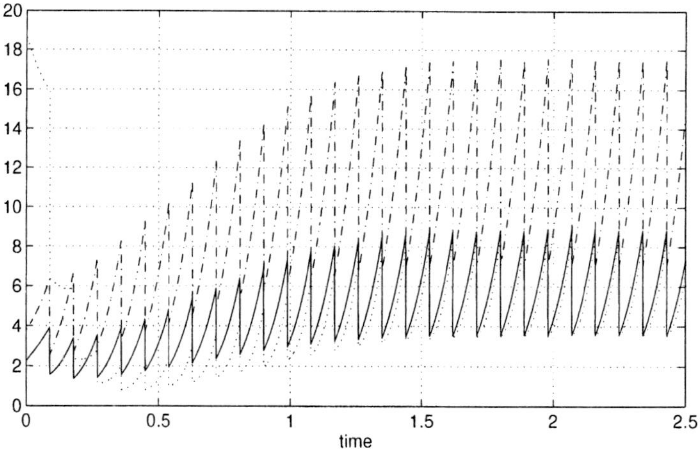

(b)

Fig. 4.4 Simulation results. (a) Stable results in stable region. (b) Unstable results out of stable region.

each k and $\mathbf{y} \in \mathbb{R}^n$ the limit of $\mathbf{g}(t, \mathbf{x})$ exists as $(t, \mathbf{x}) \to (\tau_k, \mathbf{y})$, $t > \tau_k$;

(4) *In the domain $\mathbb{R}_+ \times \mathbb{R}^n$, the following inequality holds: $\|\mathbf{g}(t, \mathbf{x})\| \leq \alpha(t)\|\mathbf{x}\|^m$, where $m > 1$ is a constant and the function $\alpha(\cdot) \in \mathcal{PC}(\mathbb{R}_+, \mathbb{R})$ is nonnegative;*

(5) $\|I + B_k\| \le \gamma_k$, *where* $\gamma_k \ge 0$ *is a constant;*

(6)

$$D(0,\infty) \triangleq \int_0^{\infty} \left(\prod_{0<\tau_k<s} r_k \right)^{m-1} \phi^m(s)\psi(s)\alpha(s)ds < \infty \quad (4.90)$$

where $r_k = \gamma_k \phi(\tau_k)\psi(\tau_k^+)$;

(7) *For any* $\epsilon > 0$ *and* $t_0 \in \mathbb{R}_+$ *there exists* $\sigma = \sigma(\epsilon, t_0)$ *such that*

$$\phi(t) \prod_{0<\tau_k<s} r_k < \epsilon, \text{ for } t \ge t_0 + \sigma. \quad (4.91)$$

Then the solution $\mathbf{x} = 0$ *of system in Eq.(4.88) is asymptotically stable.*

Proof. It is straightforward from Theorem 2.15. □

4.5.2 *Application to advertising strategies*

Our experiences tell us that if we want to introduce a new product into a market, we had to advertise it. But how often should we advertise the new product in order to attract the biggest number of consumers? The answer of this question makes a significant sense when advertisements become very expensive. In this section, the theory of impulsive control is used to show how an unstable equilibrium point of an advertising strategy can be stabilized by advertising at discrete moments.

Consider a financial model[Hul83] which deals with introducing a new product into a market of size N. N consists of the following three components:

(1) $x(t)$—the number of people who are unaware of the existence of the product;
(2) $y(t)$—the number of potential customers who are aware of the product but have not yet purchased it, and
(3) $z(t)$—the number of current customers who have purchased the product.

We can easily see that

$$x(t) + y(t) + z(t) = N. \quad (4.92)$$

Advertising consists of two components:

(1) *Awareness*, u, which informs consumers about the product and thus transfers them from the unaware group x into the potential group y, and

(2) *trial advertising*, v, which persuades consumers to purchase the product and transfers them from the potential consumers y into the current consumers z.

Since the advertisements are broadcast through a network, e.g., a national wide TV network, at discrete moments, we can model the advertising effect by impulsive effect. In this sense, the flows of consumers out and into different groups are given in the following impulsive model:

$$\begin{aligned} \begin{pmatrix} \dot{x} \\ \dot{y} \\ \dot{z} \end{pmatrix} &= \begin{pmatrix} -k & 0 & 0 \\ k & -a & \delta \\ 0 & a & -\delta \end{pmatrix} \begin{pmatrix} x \\ y \\ z \end{pmatrix} + \begin{pmatrix} \frac{k}{N} \\ -\frac{k}{N} \\ 0 \end{pmatrix} x^2, \; t \neq \tau_k, \\ \Delta \begin{pmatrix} x \\ y \\ z \end{pmatrix} &= \begin{pmatrix} -k_u & 0 & 0 \\ k_u & -k_v & 0 \\ 0 & k_v & 0 \end{pmatrix} \begin{pmatrix} x - x_e \\ y - y_e \\ z - z_e \end{pmatrix}, \; t = \tau_k \end{aligned} \tag{4.93}$$

where $k < 0$ is the contact rate, $a > 0$ is the trial(first purchase) rate, $\delta > 0$ is the switching rate. (x_e, y_e, z_e) is the objective equilibrium point. k_u and k_v are the awareness rate and trial rate, respectively.

Let $e_x = x - x_e$, $e_y = y - y_e$, $e_z = z - z_e$, we can rewrite Eq.(4.93) as

$$\begin{aligned} \begin{pmatrix} \dot{e}_x \\ \dot{e}_y \\ \dot{e}_z \end{pmatrix} &= \begin{pmatrix} -k + 2x_e\frac{k}{N} & 0 & 0 \\ k - 2x_e\frac{k}{N} & -a & \delta \\ 0 & a & -\delta \end{pmatrix} \begin{pmatrix} e_x \\ e_y \\ e_z \end{pmatrix} + \begin{pmatrix} \frac{k}{N} \\ -\frac{k}{N} \\ 0 \end{pmatrix} e_x^2 \\ &+ \begin{pmatrix} -k & 0 & 0 \\ k & -a & \delta \\ 0 & a & -\delta \end{pmatrix} \begin{pmatrix} x_e \\ y_e \\ z_e \end{pmatrix} + \begin{pmatrix} \frac{k}{N} \\ -\frac{k}{N} \\ 0 \end{pmatrix} x_e^2, \; t \neq \tau_k, \\ \Delta \begin{pmatrix} e_x \\ e_y \\ e_z \end{pmatrix} &= \begin{pmatrix} -k_u & 0 & 0 \\ k_u & -k_v & 0 \\ 0 & k_v & 0 \end{pmatrix} \begin{pmatrix} e_x \\ e_y \\ e_z \end{pmatrix}, \; t = \tau_k. \end{aligned} \tag{4.94}$$

The system in Eq.(4.93) has two equilibrium points:

$$(N, 0, 0) \text{ and } (0, \frac{\delta}{a+\delta}N, \frac{a}{a+\delta}N).$$

In view of $k < 0$, $a > 0$ and $\delta > 0$, we find that $(0, \frac{\delta}{a+\delta}N, \frac{a}{a+\delta}N)$ is an unstable equilibrium point. Our objective is to control the system to the equilibrium point $(0, \frac{\delta}{a+\delta}N, \frac{a}{a+\delta}N)$ by impulsive advertising. In this case, $(x_e, y_e, z_e) = (0, \frac{\delta}{a+\delta}N, \frac{a}{a+\delta}N)$, then Eq.(4.94) can be rewritten as

$$\begin{aligned}
\begin{pmatrix} \dot{e}_x \\ \dot{e}_y \\ \dot{e}_z \end{pmatrix} &= \begin{pmatrix} -k + 2x_e\frac{k}{N} & 0 & 0 \\ k - 2x_e\frac{k}{N} & -a & \delta \\ 0 & a & -\delta \end{pmatrix} \begin{pmatrix} e_x \\ e_y \\ e_z \end{pmatrix} + \begin{pmatrix} \frac{k}{N} \\ -\frac{k}{N} \\ 0 \end{pmatrix} e_x^2, \ t \neq \tau_k, \\
\Delta \begin{pmatrix} e_x \\ e_y \\ e_z \end{pmatrix} &= \begin{pmatrix} -k_u & 0 & 0 \\ k_u & -k_v & 0 \\ 0 & k_v & 0 \end{pmatrix} \begin{pmatrix} e_x \\ e_y \\ e_z \end{pmatrix}, \ t = \tau_k, \\
(x_e, y_e, z_e) &= \left(0, \frac{\delta}{a+\delta}N, \frac{a}{a+\delta}N\right).
\end{aligned} \tag{4.95}$$

We then give a theorem to guarantee that the impulsively controlled system will be stabilized at the desired equilibrium point.

Theorem 4.14 *Assume that $\{\tau_k\}$ is equi-distant with a fixed interval $\varpi > 0$ and let λ_1 be the largest eigenvalue of the matrix*

$$A = \begin{pmatrix} -k + 2x_e\frac{k}{N} & 0 & 0 \\ k - 2x_e\frac{k}{N} & -a & \delta \\ 0 & a & -\delta \end{pmatrix}.$$

Let

$$r = \|I + B\|_2 \|P\|_2 \|P^{-1}\|_2$$

where $A = PJP^{-1}$, J is the Jordan form of A. If the following condition:

$$\lambda_1 t + \left\lfloor \frac{t}{\varpi} \right\rfloor \ln r < 0, \qquad \text{for } t > \tau_1$$

is satisfied, then the equilibrium point $(x_e, y_e, z_e) = (0, \frac{\delta}{a+\delta}N, \frac{a}{a+\delta}N)$ is asymptotically stable.

Proof. From Eq. (4.94) we have

$$A(t) = \begin{pmatrix} -k + 2x_e\frac{k}{N} & 0 & 0 \\ k - 2x_e\frac{k}{N} & -a & \delta \\ 0 & a & -\delta \end{pmatrix}$$

from which we can see that condition 1 of Theorem 4.13 holds.

And then we have

$$\mathbf{g}(t,\mathbf{x}) = \begin{pmatrix} \frac{k}{N} \\ -\frac{k}{N} \\ 0 \end{pmatrix} e_x^2.$$

Obviously, $\mathbf{g}(t,\mathbf{x})$ is continuous and locally Lipschitzian. Condition 3 of Theorem 4.13 holds.

We can see that $m = 2$ and we have

$$||\mathbf{g}(t,\mathbf{x})||_2 = \frac{\sqrt{2}k}{N} e_x^2 \le \frac{\sqrt{2}k}{N} ||\mathbf{x}||_2.$$

Condition 4 of Theorem 4.13 holds with $\alpha(t) = \frac{\sqrt{2}k}{N}$.

We can see that

$$B = \begin{pmatrix} -k_u & 0 & 0 \\ k_u & -k_v & 0 \\ 0 & k_v & 0 \end{pmatrix}.$$

Then we have

$$||I + B||_2 = \left\| \begin{pmatrix} 1-k_u & 0 & 0 \\ k_u & 1-k_v & 0 \\ 0 & k_v & 1 \end{pmatrix} \right\|_2 = \gamma.$$

We have $e^{At} = Pe^{Jt}P^{-1}$ and

$$\begin{aligned} ||e^{A(t-s)}||_2 &= ||Pe^{J(t-s)}P^{-1}||_2 \\ &\le ||P||_2||P^{-1}||_2||e^{J(t-s)}||_2 \\ &\le ||P||_2||P^{-1}||_2 e^{\lambda_1(t-s)} \end{aligned}$$

where λ_1 is the biggest eigenvalue of A. The last inequality is in view of $||\Gamma||_2 = \rho(\Gamma)$ given Γ is symmetric. $\rho[\cdot]$ denotes spectral radius. Then we have

$$||\Phi(t,s)||_2 \le ||P||_2||P^{-1}||_2 e^{\lambda_1 t} e^{-\lambda_1 s}.$$

We can choose $\phi(t) = ||P||_2 e^{\lambda_1 t}$ and $\psi(s) = ||P^{-1}||_2 e^{-\lambda_1 s}$.

It is easy to see that $r_k = r = \gamma ||P||_2 ||P^{-1}||_2$. Then we have

$$\begin{aligned}
D(0,\infty) &= \int_0^\infty \left(\prod_{0<\tau_k<s} \right)^{m-1} \phi^m(s)\psi(s)\alpha(s)ds \\
&= \int_0^\infty r^{n(s,0)} ||P||_2^2 e^{2\lambda_1 s} ||P^{-1}||_2 e^{-\lambda_1 s} \frac{\sqrt{2}k}{N} ds \\
&= \int_0^\infty r^{n(s,0)} ||P||_2^2 ||P^{-1}||_2 \frac{\sqrt{2}k}{N} e^{\lambda_1 s} ds \\
&= \int_0^\infty ||P||_2^2 ||P^{-1}||_2 \frac{\sqrt{2}k}{N} \exp(\lambda_1 s + n(s,0)\ln r) ds
\end{aligned}$$

We can see that if $\lambda_1 s + n(s,0)\ln r < 0$ holds, then condition 6 of Theorem 4.13 holds.

We then have

$$\begin{aligned}
\phi(t) \prod_{0<\tau_k<t} r_k &= ||P||_2 e^{\lambda_1 t} \prod_{0<\tau_k<t} r \\
&= ||P||_2 e^{\lambda_1 t} \exp(n(t,0)\ln r) \\
&= ||P||_2 \exp(\lambda_1 t + n(t,0)\ln r).
\end{aligned}$$

If $\lambda_1 t + n(t,0)\ln r < 0$ holds, then condition 7 of Theorem 4.13 holds. □

4.6 Control impulses at variable time

Consider the following impulsive control system

$$\begin{aligned}
\dot{\mathbf{x}} &= A(t)\mathbf{x} + \mathbf{g}(t,\mathbf{x}), \quad t \neq \tau_k(\mathbf{x}), \\
\Delta\mathbf{x} &= B_k\mathbf{x} + J_k(\mathbf{x}), \quad t = \tau_k(\mathbf{x}),
\end{aligned} \tag{4.96}$$

where $A(t) \in \mathbb{R}^{n\times n}$ is bounded and continuous for $t \geq t_0$. $\mathbf{g}(t,\mathbf{x})$ is continuous with respect to $\mathbf{x}$ satisfying $||\mathbf{x}|| \leq h, h > 0$. $\mathbf{g}(t,\mathbf{x})$ is continuous or piecewise continuous with respect to t satisfying $t \geq t_0$. $J_k(\mathbf{x}), k \in \mathbb{N}$ is continuous for $||\mathbf{x}|| \leq h$. Observe that the moments of impulsive control inputs are different for different solutions. The corresponding reference system is given by

$$\begin{aligned}
\dot{\mathbf{x}} &= A(t)\mathbf{x}, \quad t \neq \tau_k^*, \\
\Delta\mathbf{x} &= B_k\mathbf{x}, \quad t = \tau_k^*,
\end{aligned} \tag{4.97}$$

where τ_k^* are such that for $\|\mathbf{x}\| \leq h$, $h > 0$ we have

$$|\tau_k^* - \tau_k(0)| \leq \eta \tag{4.98}$$

where $\eta = \eta(h)$ is a positive number satisfying

$$\lim_{h\to 0} \eta(h) = 0. \tag{4.99}$$

Theorem 4.15 *Assume that*

(1) for all $\|\mathbf{x}\| \leq h$, $i \in \mathbb{N}$ and some $\theta > 0$

$$\sup_i \left(\min_{\|\mathbf{x}\|\leq h} \tau_{i+1}(\mathbf{x}) - \max_{\|\mathbf{x}\|\leq h} \tau_i(\mathbf{x}) \right) \geq \theta; \tag{4.100}$$

(2) for all $\|\mathbf{x}\| \leq h$ and $i \in \mathbb{N}$

$$\tau_i(\mathbf{x}) \geq \tau_i((I + B_i)\mathbf{x} + J_i(\mathbf{x})); \tag{4.101}$$

(3) for all $t \geq t_0$, $i \in \mathbb{N}$ and $\|\mathbf{x}\| \leq h$ we have

$$\|\mathbf{g}(t, \mathbf{x})\| \leq a\|\mathbf{x}\|, \quad \|J_i(\mathbf{x})\| \leq a\|\mathbf{x}\|, \quad a > 0; \tag{4.102}$$

(4) the state transition matrix $\Psi(t, s)$ of reference system (4.97) satisfies

$$\|\Psi(t, s)\| \leq Ke^{-\gamma(t-s)}, \quad t \geq s, \quad K \geq 1, \quad \gamma > 0; \tag{4.103}$$

(5)

$$\gamma - Ka - \frac{1}{\theta}\ln(1 + Ka) > 0. \tag{4.104}$$

Then the zero solution of system (4.96) is asymptotically stable.

Proof. Let $\mathbf{x}(t)$ be an arbitrary solution of system (4.96) that at $t = t_0$ passes through a point $\mathbf{x}_0$ which is in a small neighborhood of $\mathbf{x} = 0$. There exist $\tilde{h} < h$ and $T \leq \infty$ such that $\|\mathbf{x}(t)\| < h$ for $t \in (t_0, t_0 + T]$ and $\|\mathbf{x}_0\| \leq \tilde{h}$. Let τ_i^* be a solution of equation $t = \tau_i(\mathbf{x}(t))$ which has a unique solution for all i if assumption (2) holds. Together with assumption (1) we know that $\mathbf{x}(t)$ intersects each surface $t = \tau_i(\mathbf{x})$ only once for $t \in (t_0, t_0+T]$. Then for $t \in (t_0, t_0 + T]$, $\mathbf{x}(t)$ is also a solution of the following system:

$$\begin{aligned} \dot{\mathbf{x}} &= A(t)\mathbf{x} + \mathbf{g}(t, \mathbf{x}), \quad t \neq \tau_i^*, \\ \Delta\mathbf{x} &= B_i\mathbf{x} + J_i(\mathbf{x}), \quad t = \tau_i^* \end{aligned} \tag{4.105}$$

Then $\mathbf{x}(t), t \in (t_0, t_0+T]$ can be represented as

$$\begin{aligned}\mathbf{x}(t) &= \Psi(t,t_0)\mathbf{x}_0 + \int_{t_0}^{t} \Psi(t,s)\mathbf{g}(s,\mathbf{x}(s))ds \\ &+ \sum_{t_0<\tau_i^*<t} \Psi(t,\tau_i^*)J_i(\mathbf{x}(\tau_i^*)),\end{aligned} \tag{4.106}$$

Based on assumptions (3) and (4) we have

$$\begin{aligned}\|\mathbf{x}(t)\| &\leq Ke^{-\gamma(t-t_0)}\|\mathbf{x}_0\| + Ka\int_{t_0}^{t} e^{-\gamma(t-s)}\|\mathbf{x}(s)\|ds \\ &+ \sum_{t_0<\tau_i^*<t} Kae^{-\gamma(t-\tau_i^*)}\|\mathbf{x}(\tau_i^*)\|,\end{aligned} \tag{4.107}$$

From which and Lemma 1 of [SP95] we have

$$\|\mathbf{x}(t)\| \leq K\|\mathbf{x}_0\|e^{-(\gamma-Ka)(t-t_0)}(1+Ka)^{n(t,t_0)}. \tag{4.108}$$

From assumption (1) we know that the surfaces $t=\tau_i(\mathbf{x}), i \in \mathbb{N}$, are mutually separated and from Eq. (4.108) we have

$$\|\mathbf{x}(t)\| \leq K\|\mathbf{x}_0\|\exp\left\{-\left(\gamma - Ka - \frac{1}{\theta}\ln(1+Ka)\right)(t-t_0)\right\}. \tag{4.109}$$

If $\mathbf{x}_0$ satisfies $K\|\mathbf{x}_0\| < h$ and assumption (5) holds, then $\mathbf{x}(t)$ will not leave the h-neighborhood of $\mathbf{x}=0$ for all $t \geq t_0$. Therefore $\mathbf{x}(t)$ intersects each surface $t=\tau_i(\mathbf{x})$, $i \in \mathbb{N}$, only once. Then from (4.109) we know that the zero solution of system (4.96) is asymptotically stable. □

Similarly we have the following theorem.

Theorem 4.16 *Assume that*

(1) the largest eigenvalue of $\frac{1}{2}(A(t)+A^\top(t))$, $\lambda_n(t)$, *satisfies* $\lambda_n(t) \leq \gamma$ *for all* $t \geq t_0$ *and the largest eigenvalues of* $(I+B_i^\top)(I+B_i)$, Λ_i, $i \in \mathbb{N}$, *satisfy* $\Lambda_i^2 \leq \alpha^2$*;*

(2) $n_x(t+T,t)$ *is the number of points of* $\tau_i(\mathbf{x})$ *in time interval* $[t,t+T]$ *and for* $\|\mathbf{x}\| \leq h$ *the limit*

$$\lim_{T\to\infty} \frac{n_x(t+T,t)}{T} = p$$

exists and is uniform with respect to $t \geq t_0$*;*

(3) functions $\tau_i(\mathbf{x})$, $i \in \mathbb{N}$, satisfy the Lipschitz conditions

$$\|\tau_i(\mathbf{x}_1) - \tau_i(\mathbf{x}_2)\| \leq L\|\mathbf{x}_1 - \mathbf{x}_2\|, \quad 0 < L < \infty \tag{4.110}$$

for all $\|\mathbf{x}_1\| \leq h$ and $\|\mathbf{x}_2\| \leq h$;

(4) for all $\|\mathbf{x}\| \leq h$ and $i \in \mathbb{N}$

$$\tau_i(\mathbf{x}) \geq \tau_i((I + B_i)\mathbf{x} + J_i(\mathbf{x})); \tag{4.111}$$

(5) $\gamma + p \ln \alpha < 0$;

(6) for all $t \geq t_0$ and $i \in \mathbb{N}$ and $\|\mathbf{x}\| \leq h$ we have

$$\|\mathbf{g}(t, \mathbf{x})\| \leq a\|\mathbf{x}\|, \quad \|J_i(\mathbf{x})\| \leq a\|\mathbf{x}\|. \tag{4.112}$$

Then the zero solution of system (4.96) is asymptotically stable for sufficiently small values of a.

Proof. Let $\mathbf{x}(t)$ be an arbitrary solution of system (4.96) that at $t = t_0$ passes through a point $\mathbf{x}_0$ which is in a small neighborhood of $\mathbf{x} = 0$. There exist $\tilde{h} < h$ and $T \leq \infty$ such that $\|\mathbf{x}(t)\| < h$ for $t \in (t_0, t_0 + T]$ and $\|\mathbf{x}_0\| \leq \tilde{h}$. Let τ_i^* be a solution of equation $t = \tau_i(\mathbf{x}(t))$ which has a unique solution for all i if assumptions (3) and (4) hold. This means that $\mathbf{x}(t)$ intersects each surface $t = \tau_i(\mathbf{x})$ only once for $t \in (t_0, t_0 + T]$. Then for $t \in (t_0, t_0 + T]$, $\mathbf{x}(t)$ is also a solution of the following system:

$$\begin{aligned} \dot{\mathbf{x}} &= A(t)\mathbf{x} + \mathbf{g}(t, \mathbf{x}), \quad t \neq \tau_i^*, \\ \Delta \mathbf{x} &= B_i \mathbf{x} + J_i(\mathbf{x}), \quad t = \tau_i^* \end{aligned} \tag{4.113}$$

Then $\mathbf{x}(t), t \in (t_0, t_0 + T]$ can be represented as

$$\begin{aligned} \mathbf{x}(t) &= \Psi(t, t_0)\mathbf{x}_0 + \int_{t_0}^{t} \Psi(t, s)\mathbf{g}(s, \mathbf{x}(s))ds \\ &\quad + \sum_{t_0 < \tau_i^* < t} \Psi(t, \tau_i^*) J_i(\mathbf{x}(\tau_i^*)), \end{aligned} \tag{4.114}$$

Using the similar proving process of Theorem 4.2 and based on assumptions (1), (2) and (5) we know that there exist $K \geq 1$, $\mu > 0$ with $0 < \mu < |\gamma + p \ln \alpha|$ such that for all $t_0 \leq s \leq t \leq t + T$

$$\|\Psi(t, s)\mathbf{z}\| \leq Ke^{-\mu(t-s)}\|\mathbf{z}\|, \quad \|\mathbf{z}| \leq h. \tag{4.115}$$

From Eqs. (4.114) and (4.115) and assumption (6) we have

$$\|\mathbf{x}(t)\| \leq e^{-\mu(t-t_0)} K\|\mathbf{x}_0\|(1 + Ka)^{n(t,t_0)} e^{Ka(t-t_0)}, \tag{4.116}$$

from which we have

$$\|\mathbf{x}(t)\| \leq K_1 \|\mathbf{x}_0\| \exp\{-[\mu - p\ln(1 + Ka) - Ka + \epsilon](t - t_0)\} \quad (4.117)$$

for any $\epsilon > 0$ with $K_1 = K_1(\epsilon) > 0$. Therefore, if a is sufficiently small such that

$$\mu - p\ln(1 + Ka) - Ka > 0$$

and if $\mathbf{x}_0$ satisfies $K_1\|\mathbf{x}_0\| < h$, then $\mathbf{x}(t)$ will not leave the h-neighborhood of $\mathbf{x} = 0$ for all $t \geq t_0$. Therefore $\mathbf{x}(t)$ intersects each surface $t = \tau_i(\mathbf{x})$, $i \in \mathbb{N}$, only once. Then from (4.117) we know that the zero solution of system (4.96) is asymptotically stable. □

In parallel we have

Theorem 4.17 *Assume that*

(1) the largest eigenvalue of $\frac{1}{2}(A(t) + A^\top(t))$, $\lambda_n(t)$, *satisfies* $\lambda_n(t) \leq \gamma$ *for all* $t \geq t_0$ *and the largest eigenvalues of* $(I + B_i^\top)(I + B_i)$, Λ_i, $i \in \mathbb{N}$, *satisfy* $\Lambda_i^2 \leq \alpha^2$;

(2)

$$0 < \theta_1 \leq \min_{\|\mathbf{x}\| \leq h} \tau_{i+1}(\mathbf{x}) - \max_{\|\mathbf{x}\| \leq h} \tau_i(\mathbf{x}) \leq \theta_2, \qquad i \in \mathbb{N};$$

(3) $\gamma + \frac{1}{\theta}\ln\alpha < 0$, $\theta = \theta_1$ *if* $\alpha \geq 1$, *and* $\theta = \theta_2$ *if* $0 < \alpha < 1$;

(4) functions $\tau_i(\mathbf{x})$, $i \in \mathbb{N}$, *satisfy the Lipschitz conditions*

$$\|\tau_i(\mathbf{x}_1) - \tau_i(\mathbf{x}_2)\| \leq L\|\mathbf{x}_1 - \mathbf{x}_2\|, \qquad 0 < L < \infty \quad (4.118)$$

for all $\|\mathbf{x}_1\| \leq h$ *and* $\|\mathbf{x}_2\| \leq h$;

(5) for all $\|\mathbf{x}\| \leq h$ *and* $i \in \mathbb{N}$

$$\tau_i(\mathbf{x}) \geq \tau_i((I + B_i)\mathbf{x} + J_i(\mathbf{x})); \quad (4.119)$$

(6) for all $t \geq t_0$ *and* $i \in \mathbb{N}$ *and* $\|\mathbf{x}\| \leq h$ *we have*

$$\|\mathbf{g}(t, \mathbf{x})\| \leq a\|\mathbf{x}\|, \qquad \|J_i(\mathbf{x})\| \leq a\|\mathbf{x}\|. \quad (4.120)$$

Then the zero solution of system (4.96) is asymptotically stable for sufficiently small values of a.

If $A(t)$ and B_i are time-invariant, then system (4.96) becomes

$$\begin{aligned} \dot{\mathbf{x}} &= A\mathbf{x} + \mathbf{g}(t,\mathbf{x}), \quad t \neq \tau_k(\mathbf{x}), \\ \Delta\mathbf{x} &= B\mathbf{x} + J_k(\mathbf{x}), \quad t = \tau_k(\mathbf{x}) \end{aligned} \tag{4.121}$$

whose stability can be guaranteed by the following theorem.

Theorem 4.18 *Assume that*

(1) $n_x(t+T,t)$ *is the number of points of* $\tau_i(\mathbf{x})$ *in time interval* $[t,t+T]$ *and for* $||\mathbf{x}|| \leq h$ *the limit*

$$\lim_{T\to\infty} \frac{n_x(t+T,t)}{T} = p$$

exists and is uniform with respect to $t \geq t_0$*;*

(2)

$$\gamma = \max_j \mathrm{Re}(\lambda_j(A)), \qquad \alpha^2 = \max_j \lambda_j((I+B^\top)(I+B));$$

(3) $\gamma + p\ln\alpha < 0$*;*

(4) for all $||\mathbf{x}|| \leq h$ *and* $i \in \mathbb{N}$

$$\tau_i(\mathbf{x}) \geq \tau_i((I+B)\mathbf{x} + J_i(\mathbf{x})); \tag{4.122}$$

(5) functions $\tau_i(\mathbf{x})$, $i \in \mathbb{N}$, *satisfy the Lipschitz conditions*

$$||\tau_i(\mathbf{x}_1) - \tau_i(\mathbf{x}_2)|| \leq L||\mathbf{x}_1 - \mathbf{x}_2||, \qquad 0 < L < \infty \tag{4.123}$$

for all $||\mathbf{x}_1|| \leq h$ *and* $||\mathbf{x}_2|| \leq h$*;*

(6) for all $t \geq t_0$ *and* $i \in \mathbb{N}$ *and* $||\mathbf{x}|| \leq h$ *we have*

$$||\mathbf{g}(t,\mathbf{x})|| \leq a||\mathbf{x}||, \qquad ||J_i(\mathbf{x})|| \leq a||\mathbf{x}||. \tag{4.124}$$

Then the zero solution of system (4.121) is asymptotically stable for sufficiently small values of a.

In parallel we have

Theorem 4.19 *Assume that*

(1)

$$\gamma = \max_j \mathrm{Re}(\lambda_j(A)), \qquad \alpha^2 = \max_j \lambda_j((I+B^\top)(I+B));$$

(2)

$$0 < \theta_1 \leq \min_{\|\mathbf{x}\| \leq h} \tau_{i+1}(\mathbf{x}) - \max_{\|\mathbf{x}\| \leq h} \tau_i(\mathbf{x}) \leq \theta_2, \qquad i \in \mathbb{N};$$

(3) $\gamma + \frac{1}{\theta} \ln \alpha < 0$, $\theta = \theta_1$ *if* $\alpha \geq 1$, *and* $\theta = \theta_2$ *if* $0 < \alpha < 1$;

(4) for all $\|\mathbf{x}\| \leq h$ *and* $i \in \mathbb{N}$

$$\tau_i(\mathbf{x}) \geq \tau_i((I + B)\mathbf{x} + J_i(\mathbf{x})); \tag{4.125}$$

(5) functions $\tau_i(\mathbf{x})$, $i \in \mathbb{N}$, *satisfy the Lipschitz conditions*

$$\|\tau_i(\mathbf{x}_1) - \tau_i(\mathbf{x}_2)\| \leq L\|\mathbf{x}_1 - \mathbf{x}_2\|, \qquad 0 < L < \infty \tag{4.126}$$

for all $\|\mathbf{x}_1\| \leq h$ *and* $\|\mathbf{x}_2\| \leq h$;

(6) for all $t \geq t_0$ *and* $i \in \mathbb{N}$ *and* $\|\mathbf{x}\| \leq h$ *we have*

$$\|\mathbf{g}(t, \mathbf{x})\| \leq a\|\mathbf{x}\|, \qquad \|J_i(\mathbf{x})\| \leq a\|\mathbf{x}\|. \tag{4.127}$$

Then the zero solution of system (4.121) is asymptotically stable for sufficiently small values of a.

Furthermore, if the system (4.121) is simplified as

$$\begin{aligned} \dot{\mathbf{x}} &= A\mathbf{x} + \mathbf{g}(t, \mathbf{x}), \qquad t \neq \tau_k(\mathbf{x}), \\ \Delta\mathbf{x} &= J_k(\mathbf{x}), \qquad t = \tau_k(\mathbf{x}), \end{aligned} \tag{4.128}$$

we have the following theorem.

Theorem 4.20 *Assume that*

(1) $n_x(t+T, t)$ *is the number of points of* $\tau_i(\mathbf{x})$ *in time interval* $[t, t+T]$ *and for* $\|\mathbf{x}\| \leq h$ *the limit*

$$\lim_{T \to \infty} \frac{n_x(t + T, t)}{T} = p$$

exists and is uniform with respect to $t \geq t_0$;

(2) all eigenvalues of A *are in the left-hand side of* s*-plane;*

(3) for all $\|\mathbf{x}\| \leq h$ *and* $i \in \mathbb{N}$

$$\tau_i(\mathbf{x}) \geq \tau_i(\mathbf{x} + J_i(\mathbf{x})); \tag{4.129}$$

(4) functions $\tau_i(\mathbf{x})$, $i \in \mathbb{N}$, *satisfy the Lipschitz conditions*

$$\|\tau_i(\mathbf{x}_1) - \tau_i(\mathbf{x}_2)\| \leq L\|\mathbf{x}_1 - \mathbf{x}_2\|, \qquad 0 < L < \infty \tag{4.130}$$

for all $\|\mathbf{x}_1\| \leq h$ *and* $\|\mathbf{x}_2\| \leq h$;

(5) for all $t \geq t_0$ and $i \in \mathbb{N}$ and $\|\mathbf{x}\| \leq h$ we have

$$\|\mathbf{g}(t,\mathbf{x})\| \leq a\|\mathbf{x}\|, \qquad \|J_i(\mathbf{x})\| \leq a\|\mathbf{x}\|. \tag{4.131}$$

Then the zero solution of system (4.128) is asymptotically stable for sufficiently small values of a.

In parallel we have

Theorem 4.21 *Assume that*

(1) for $i \in \mathbb{N}$

$$0 < \theta_1 \leq \min_{\|\mathbf{x}\| \leq h} \tau_{i+1}(\mathbf{x}) - \max_{\|\mathbf{x}\| \leq h} \tau_i(\mathbf{x}) \leq \theta_2; \tag{4.132}$$

(2) all eigenvalues of A are in the left-hand side of s-plane;
(3) for all $\|\mathbf{x}\| \leq h$ and $i \in \mathbb{N}$

$$\tau_i(\mathbf{x}) \geq \tau_i(\mathbf{x} + J_i(\mathbf{x})); \tag{4.133}$$

(4) functions $\tau_i(\mathbf{x})$, $i \in \mathbb{N}$, satisfy the Lipschitz conditions

$$\|\tau_i(\mathbf{x}_1) - \tau_i(\mathbf{x}_2)\| \leq L\|\mathbf{x}_1 - \mathbf{x}_2\|, \qquad 0 < L < \infty \tag{4.134}$$

for all $\|\mathbf{x}_1\| \leq h$ and $\|\mathbf{x}_2\| \leq h$;
(5) for all $t \geq t_0$ and $i \in \mathbb{N}$ and $\|\mathbf{x}\| \leq h$ we have

$$\|\mathbf{g}(t,\mathbf{x})\| \leq a\|\mathbf{x}\|, \qquad \|J_i(\mathbf{x})\| \leq a\|\mathbf{x}\|. \tag{4.135}$$

Then the zero solution of system (4.128) is asymptotically stable for sufficiently small values of a.

Note

We can also design nonlinear or linear impulsive control systems based on Lyapunov's second method as presented in Sec. 2.6 and the theory of stability in terms of two measures as presented in Sec. 2.7. An earlier result based on Lyapunov second method can be found in [Pav67]. Some recent results can be found in [YMH98]. The comparison method can be found in [Yan99b]. Impulsive control of chaotic systems can be found in [YYY97a; YL97b; YL97c].

Chapter 5

Practical Stability of Impulsive Control

5.1 Introduction

In many cases, some well designed, asymptotically stable control schemes can not work as expected. There are many reasons contributing to this kind of design failure, among which are: the domain of attraction is too small, some physical parameters are not included into the mathematical models and there are implementation errors of the controller. A way to overcome this problem is to use practical stability instead of asymptotic stability. In a practical control problem, we control a system into an objective region instead of an exact point. The reason is that measuring instruments for generating feedback signals have only definite accuracy. If we want to control a system to an exact point, the expense maybe prohibitively high in some cases.

The practical stability only need to stabilize a system into a region of phase space. Form the practical point of view, the controlled system is considered stable if the fluctuation of states remains within some prescribed bounds. From the mathematical point of view, the equilibrium point may be unstable but the performance is practically acceptable.

In this chapter, we study the impulsive control of nonautonomous systems using practical stability.

5.2 Practical stability in terms of two measurements

5.2.1 *Definitions and notations*

We use the following notations:

$$\begin{aligned} \mathcal{G} &\triangleq [h \in C[\mathbb{R}_+ \times \mathbb{R}^n, \mathbb{R}_+] \| \inf h(t, \mathbf{x}) = 0], \\ \mathcal{PC} &\triangleq [\xi : \mathbb{R}_+ \to \mathbb{R} \text{ is continuous in } (\tau_{k-1}, \tau_k] \text{ and } \xi(\tau_{k-1}^+) \text{ exists} \\ &\quad \text{for } k = 1, 2, \cdots]. \end{aligned} \tag{5.1}$$

Let a general impulsive control system be

$$\begin{aligned} \dot{\mathbf{x}} &= \mathbf{f}(t, \mathbf{x}, \tilde{\mathbf{u}}), \quad t \neq \tau_k, \\ \Delta \mathbf{x} &= \mathbf{u}_k(\mathbf{x}), \quad t = \tau_k, \\ \mathbf{x}(t_0^+) &= \mathbf{x}_0, \quad k = 1, 2, \cdots . \end{aligned} \tag{5.2}$$

and Ω be the admissible control input for $\tilde{\mathbf{u}}$. For a $\tilde{\mathbf{u}}_1 \in \Omega$, we denote by $\mathbf{x}(t) = \mathbf{x}(t, t_0, \mathbf{x}_0, \tilde{\mathbf{u}}_1)$ a solution of system (5.2) with $\mathbf{x}(t_0^+) = \mathbf{x}_0$

Definition 5.1 Let $h_0, h \in \mathcal{G}$. Then the system (5.2) is said to be

(1) (h_0, h)-*practically stable* if, given (λ, α) with $0 < \lambda < \alpha$, we have $h_0(t_0, \mathbf{x}_0) < \lambda$ implies $h(t, \mathbf{x}(t)) < \alpha$, $t \geq t_0$ for some $t_0 \in \mathbb{R}_+$;
(2) (h_0, h)-*uniformly practically stable* if (1) holds for every $t_0 \in \mathbb{R}_+$;
(3) (h_0, h)-*practically quasistable* if, given (λ, α, T) with $\lambda > 0$, $\alpha > 0$, $T > 0$ and some $t_0 \in \mathbb{R}_+$, we have $h_0(t_0, \mathbf{x}_0) < \lambda$ implies $h(t, \mathbf{x}(t)) < \alpha$, $t \geq t_0 + T$;
(4) (h_0, h)-*uniformly practically quasistable* if (3) holds for every $t_0 \in \mathbb{R}_+$;
(5) (h_0, h)-*strongly practically stable* if both (1) and (3) hold;
(6) (h_0, h)-*strongly uniformly practically stable* if both (2) and (4) hold;
(7) (h_0, h)-*practically unstable* if (1) does not hold.

Remark: Definition 5.1 is in terms of two measures which is a more general form than that in terms of one measure. Definition 5.1 reduces to

(1) the standard practical stability if $h(t, \mathbf{x}(t)) = h_0(t, \mathbf{x}(t)) = \|\mathbf{x}\|$;
(2) the practical stability of the prescribed $\mathbf{x}_0$ of system (5.2) if $h(t, \mathbf{x}(t)) = h_0(t, \mathbf{x}(t)) = \|\mathbf{x} - \mathbf{x}_0\|$;

(3) the partial practical stability if $h(t, \mathbf{x}(t)) = \|\mathbf{x}\|_s$, $1 \le s < n$, and $h_0(t, \mathbf{x}(t)) = \|\mathbf{x}\|$;
(4) the orbital practical stability of an orbit $\wp$ if $h(t, \mathbf{x}(t)) = h_0(t, \mathbf{x}(t)) = d(\mathbf{x}, \wp)$, $d(\cdot, \cdot)$ being a distance function;
(5) the practical stability of an invariant set $\wp$ is $h(t, \mathbf{x}(t)) = h_0(t, \mathbf{x}(t)) = d(\mathbf{x}, \wp)$, $d(\cdot, \cdot)$ being a distance function;
(6) the practical stability of conditionally invariant set $\wp_1$ with respect to $\wp_2$, where $\wp_2 \subset \wp_1$, if $h(t, \mathbf{x}(t)) = d(\mathbf{x}, \wp_1)$ and $h_0(t, \mathbf{x}(t)) = d(\mathbf{x}, \wp_2)$, $d(\cdot, \cdot)$ being a distance function.

5.2.2 *Comparison system*

Let a *comparison system* be

$$\begin{cases} \dot{w} = g(t, w, w), & t \neq \tau_k, \\ w(\tau_k^+) = \psi_k(w(\tau_k)), & k = 1, 2, \cdots, \\ w(t_0^+) = w_0 \ge 0, \end{cases} \tag{5.3}$$

where $g : \mathbb{R}_+ \times \mathbb{R} \times \mathbb{R} \to \mathbb{R}$ is continuous in $(\tau_{k-1}, \tau_k] \times \mathbb{R} \times \mathbb{R}$ and for each $(w, y) \in \mathbb{R} \times \mathbb{R}$, $k = 1, 2, \cdots$,

$$\lim_{(t, \tilde{w}, \tilde{y}) \to (\tau_{k-1}^+, w, y)} g(t, \tilde{w}, \tilde{y}) = g(\tau_{k-1}^+, w, y)$$

exists, and $\psi_k : \mathbb{R} \to \mathbb{R}$ is nondecreasing for $k = 1, 2, \cdots$.

Let us begin by stating the following lemma (Theorem 3.1 of [Liu92]) for later use.

Lemma 5.1 *Assume that*

(1) $m, v \in \mathcal{PC}$ *and*

$$\begin{aligned} D^+ m(t) &\le g(t, m(t), v(t)), & t \neq \tau_k, \\ m(\tau_k^+) &\le \psi_k(m(\tau_k)), & k = 1, 2, \cdots; \end{aligned} \tag{5.4}$$

(2) $g(t, w, v)$ *is nondecreasing in* v *for each* (t, w) *and* $\gamma(t)$ *is the maximal solution of Eq. (5.3) existing in* $[t_0, \infty)$ *such that* $v(t) \le \gamma(t)$, $t \ge t_0$.

Then $m(t) \le \gamma(t)$, $t \ge t_0$, *provided* $m(t_0) \le w_0$.

We then have the following comparison theorem.

Theorem 5.1 *Assume that*

(1) $0 < \mu < \nu$ *are given and* $\rho > \nu$*;*

(2) $h_0, h \in \mathcal{G}$ *and* $h(t,\mathbf{x}) \leq \phi(h_0(t,\mathbf{x}))$ *with* $\phi \in \mathcal{K}$ *whenever* $h_0(t,\mathbf{x}) < \mu$*;*

(3) $V \in \mathcal{V}_0$ *and there exist* $\alpha, \beta \in \mathcal{K}$ *such that*

$$\begin{aligned} \beta(h(t,\mathbf{x})) &\leq V(t,\mathbf{x}), \quad \text{if} \quad h(t,\mathbf{x}) < \rho, \\ V(t,\mathbf{x}) &\leq \alpha(h_0(t,\mathbf{x})), \quad \text{if} \quad h_0(t,\mathbf{x}) < \mu; \end{aligned} \tag{5.5}$$

(4) let

$$\Omega = \{\tilde{\mathbf{u}} \in \mathbb{R}^m | \ U(t,\tilde{\mathbf{u}}) \leq \gamma(t), t \geq t_0\} \tag{5.6}$$

where $U : \mathbb{R}_+ \times \mathbb{R}^m \to \mathbb{R}_+$ *is continuous on* $(\tau_{k-1}, \tau_k] \times \mathbb{R}^m$ *and for every* $\tilde{\mathbf{u}} \in \mathbb{R}^m, k = 1, 2, \cdots,$

$$\lim_{(t,\mathbf{y}) \to (t_{k-1}^+, \tilde{\mathbf{u}})} U(t,\mathbf{y}) = U(t_{k-1}^+, \tilde{\mathbf{u}})$$

exists, and $\gamma(t)$ *is the maximal solution of Eq. (5.3). For* $(t,\mathbf{x}) \in (\tau_{k-1}, \tau_k) \times \mathbb{R}^n$ *and* $\tilde{\mathbf{u}}(t) \in \Omega$*,*

$$D^+V(t,\mathbf{x}) \leq g(t, V(t,\mathbf{x}), U(t,\tilde{\mathbf{u}}(t))), \quad \text{if} \quad h(t,\mathbf{x}) < \rho, \tag{5.7}$$

where $g(t,w,v)$ *is nondecreasing in* v *for each* $(t,w) \in \mathbb{R}_+ \times \mathbb{R}$ *and*

$$V(\tau_k^+, \mathbf{x}(\tau_k^+)) \leq \psi_k(V(\tau_k, \mathbf{x}(\tau_k))), \quad \text{if} \quad h(\tau_k, \mathbf{x}(\tau_k)) < \rho; \tag{5.8}$$

(5) $\phi(\mu) < \nu$ *and* $\alpha(\mu) < \beta(\nu)$*;*

(6) $h(t,\mathbf{x}) < \nu$ *implies* $h(t, \mathbf{x} + \mathbf{u}_k(\mathbf{x})) < \rho$ *for all* $k = 1, 2, \cdots$.

Then the practical stability properties of (5.3) with respect to $(\alpha(\mu), \beta(\nu))$ *imply the corresponding* (h_0, h)*-practical stability properties of comparison system (5.2) with respect to* (μ, ν)*.*

Proof. Let $t_0 \geq 0$ and $t_0 \in (\tau_i, \tau_{i+1}]$ for some $i \geq 1$. We have the following notation for $k = 1, 2, \cdots$:

$$\tau_k \triangleq \begin{cases} \tau_{i+k}, & \text{if } t_0 \in (\tau_i, \tau_{i+1}), \\ \tau_{i+1+k}, & \text{if } t_0 = \tau_{i+1}. \end{cases} \tag{5.9}$$

(1) (h_0, h)-practical stability

Suppose that the comparison system system (5.3) is practically stable with respect to $(\alpha(\mu), \beta(\nu))$, then $w_0 < \alpha(\mu)$ implies that

$$w(t, t_0, w_0) < \beta(\nu), \quad t \geq t_0, \tag{5.10}$$

where $w(t, t_0, w_0)$ is any solution of Eq. (5.3) existing on $[t_0, \infty)$. Let us choose $(t_0, \mathbf{x}_0) \in \mathbb{R}_+ \times \mathbb{R}^n$ such that $h_0(t_0, \mathbf{x}_0) < \mu$. Then from assumption (2) and (5), we have

$$h(t_0, \mathbf{x}_0) \leq \phi(h_0(t_0, \mathbf{x}_0)) \leq \phi(\mu) < \nu. \tag{5.11}$$

We then have

$$h(t, \mathbf{x}(t)) < \nu, \forall t \geq t_0, \tag{5.12}$$

where $\mathbf{x}(t) = \mathbf{x}(t, t_0, \mathbf{x}_0, \tilde{\mathbf{u}}^*)$ is any solution of Eq. (5.2) with $h_0(t_0, \mathbf{x}_0) < \mu$. If this is not the case, then there exists a $\tilde{\mathbf{u}}_1$ and a corresponding solution $\mathbf{x}(t) = \mathbf{x}(t, t_0, \mathbf{x}_0, \tilde{\mathbf{u}}_1)$ of Eq. (5.2) with $h_0(t_0, \mathbf{x}_0) < \mu$ and a $t^* > t_0$ such that $\tau_k < t^* \leq \tau_{k+1}$ for some k, satisfying

$$h(t^*, \mathbf{x}(t^*)) \geq \nu, \quad \text{and} \quad h(t, \mathbf{x}(t)) < \nu, t \in [t_0, \tau_k]. \tag{5.13}$$

From assumption (6) we can find a t_1 such that $\tau_k < t_1 \leq t^*$ and

$$\nu \leq h(t_1, \mathbf{x}(t_1)) < \rho. \tag{5.14}$$

Let $m(t) = V(t, \mathbf{x}(t))$, $t \in [t_0, t_1]$, and $w_0 = V(t_0, \mathbf{x}_0)$, then from assumption (3) we have

$$\begin{aligned} D^+ m(t) &\leq g(t, m(t), U(t, \tilde{\mathbf{u}}_1(t))), \quad t \in [t_0, t_1], \quad t \neq \tau_i, \\ m(\tau_i^+) &\leq \psi_i(m(\tau_i)), \quad i = 1, 2, \cdots, k, \end{aligned} \tag{5.15}$$

from which and in view of $\tilde{\mathbf{u}}_1 \in \Omega$ and $g(\cdot, \cdot, v)$ is nondecreasing in v, we have

$$\begin{aligned} D^+ m(t) &\leq g(t, m(t), \gamma(t), \quad t \in [t_0, t_1], \quad t \neq \tau_i, \\ m(\tau_i^+) &\leq \psi_i(m(\tau_i)), \quad i = 1, 2, \cdots, k. \end{aligned} \tag{5.16}$$

where $\gamma(t) = \gamma(t, t_0, w_0)$ is the maximal solution of Eq. (5.3). It then follows from Lemma 5.1 that

$$m(t) \leq \gamma(t), \quad t \in [t_0, t_1], \tag{5.17}$$

Then from Eqs. (5.10), (5.14), and (5.17) we have the following contradiction:

$$\beta(\nu) \leq V(t_1, \mathbf{x}(t_1)) \leq \gamma(t_1) < \beta(\nu). \tag{5.18}$$

We then conclude that system (5.2) is (h_0, h)-practically stable.

(2) (h_0, h)-strongly practical stability
Suppose that the comparison system (5.3) is strongly practically stable with respect to $(\alpha(\mu), \beta(\nu))$. This implies that system (5.2) is (h_0, h)-practically stable. We then know that $h_0(t_0, \mathbf{x}_0) < \mu$ implies

$$h(t, \mathbf{x}(t)) < \nu, \quad t \geq t_0, \tag{5.19}$$

where $\mathbf{x}(t) = \mathbf{x}(t, t_0, \mathbf{x}_0, \mathbf{x}^*)$ is any solution of system (5.2) with $h_0(t_0, \mathbf{x}_0) < \mu$. Given $0 < b < \nu$ and a $T > 0$ and since system (5.3) is practically quasistable, if $\beta(b) > 0$ we then have

$$w_0 < \alpha(\mu) \text{ implies } w(t, t_0, w_0) < \beta(b), \quad t \geq t_0 + T. \tag{5.20}$$

Let us choose $(t_0, \mathbf{x}_0)$ such that $h_0(t_0, \mathbf{x}_0) < \mu$, then in view of Eq. (5.19), arguments leading to Eq. (5.18) yield

$$V(t, \mathbf{x}(t)) \leq \gamma(t, t_0, \alpha(h_0(t_0, \mathbf{x}_0))), \quad t \geq t_0, \tag{5.21}$$

from which and Eq. (5.20) we have

$$\beta(h(t, \mathbf{x}(t))) \leq V(t, \mathbf{x}(t)) < \beta(b), \quad t \geq t_0 + T, \tag{5.22}$$

from which we have

$$h(t, \mathbf{x}(t)) < b, \quad t \geq t_0 + T. \tag{5.23}$$

Thus, the system (5.2) is (h_0, h)-strongly practically stable.

(3) Other cases can be proven similarly.

□

By choosing different types of functions we have the following corollary of Theorem 5.1.

Corollary 5.1

(1) Assume that

$$g(t, w, w) = 0, \quad \psi_k(w) = d_k w, \quad d_k \geq 0, \quad k = 1, 2, \cdots,$$

and if

$$\prod_{k=1}^{\infty} d_k < \infty$$

then the system (5.2) is (h_0, h)*-uniformly practically stable.*

(2) Assume that

$$g(t,w,w) = \frac{\lambda(t)}{dt}w, \quad \lambda \in C^1[\mathbb{R}_+, \mathbb{R}_+],$$
$$\psi_k(w) = d_k w, \quad d_k \geq 0, \quad k = 1, 2, \cdots,$$
$$\lambda(\tau_k) + \ln d_k \leq \lambda(\tau_{k-1}), \qquad k = 1, 2, \cdots, \tag{5.24}$$

then the system (5.2) is (h_0, h)*-practically stable.*

5.2.3 *Controllability*

The following lemma is Theorem 3.2 of [Liu92].

Lemma 5.2 *Assume that*

(1) $m, v \in \mathcal{PC}$ *and*

$$D^+m(t) \leq g(t, m(t), v(t)), \qquad t \neq \tau_k,$$
$$m(\tau_k^+) \leq \psi_k(m(\tau_k)), \qquad k = 1, 2, \cdots; \tag{5.25}$$

(2) $\gamma(t)$ *is the maximal solution of*

$$\begin{cases} \dot{w} = g(t, w, v(t)), & t \neq \tau_k, \\ w(\tau_k^+) = \psi_k(w(\tau_k)), & k = 1, 2, \cdots, \\ w(t_0^+) = w_0 \geq 0, & \end{cases} \tag{5.26}$$

existing in $[t_0, \infty)$.

Then $m(t) \leq \gamma(t), t \geq t_0$, *provided* $m(t_0) \leq w_0$.

Theorem 5.2 *Assume that*

(1) $0 < \mu < \nu$ *are given and* $\rho > \nu$. *For a given* $v \in \mathcal{PC}$ *we define the following control set*

$$\Omega = \{\tilde{\mathbf{u}} \in \mathbb{R}^m | \ U(t, \tilde{\mathbf{u}}) \leq v(t), t \geq t_0\};$$

(2) $h_0, h \in \mathcal{G}$ *and* $h(t, \mathbf{x}) \leq \phi(h_0(t, \mathbf{x}))$ *with* $\phi \in \mathcal{K}$ *whenever* $h_0(t, \mathbf{x}) < \mu$;

(3) $V \in \mathcal{V}_0$ *and there exist* $\alpha, \beta \in \mathcal{K}$ *such that*

$$\begin{aligned} \beta(h(t, \mathbf{x})) &\leq V(t, \mathbf{x}), \qquad \text{if} \quad h(t, \mathbf{x}) < \rho, \\ V(t, \mathbf{x}) &\leq \alpha(h_0(t, \mathbf{x})), \qquad \text{if} \quad h_0(t, \mathbf{x}) < \mu; \end{aligned} \tag{5.27}$$

(4) *for* $(t, \mathbf{x}) \in (\tau_{k-1}, \tau_k) \times \mathbb{R}^n$ *and* $\tilde{\mathbf{u}}(t) \in \Omega$,

$$D^+V(t, \mathbf{x}) \le g(t, V(t, \mathbf{x}), U(t, \tilde{\mathbf{u}}(t))), \quad \text{if} \quad h(t, \mathbf{x}) < \rho, \tag{5.28}$$

where $g(t, w, v)$ *is nondecreasing in* v *for each* $(t, w) \in \mathbb{R}_+ \times \mathbb{R}$ *and*

$$V(t, \mathbf{x}) \le \alpha(h_0(t, \mathbf{x})), \quad \text{if} \quad h_0(t, \mathbf{x}) < \mu; \tag{5.29}$$

(5) $\phi(\mu) < \nu$, $\alpha(\mu) < \beta(\nu)$ *and* $h(t, \mathbf{x}) < \nu$ *implies* $h(t, \mathbf{x} + \mathbf{u}_k(\mathbf{x})) < \rho$ *for all* $k = 1, 2, \cdots$;

(6) *there exists a control function* $v \in \mathcal{PC}$ *such that any solution* $w(t) = w(t, t_0, w_0, w)$ *of the system (5.26) satisfies*

$$w_0 \le \alpha(\mu) \ \text{ implies } \ w(t) < \beta(\nu), \quad t \ge t_0, \tag{5.30}$$

and

$$w(t_0 + T) \le \beta(b), \ 0 < b < \nu, \ \text{for some } T = T(t_0, w_0) > 0. \tag{5.31}$$

Then there exist admissible controls $\tilde{\mathbf{u}} = \tilde{\mathbf{u}}(t) \in \Omega$ *such that*

(1) *the system (5.2) is* (h_0, h)*-practically stable;*

(2) *all solutions* $\mathbf{x}(t) = \mathbf{x}(t, t_0, \mathbf{x}_0, \tilde{\mathbf{u}})$ *starting in*

$$\Omega_1 = \{\mathbf{x} \in \mathbb{R}^n | \ \ h(t, \mathbf{x}) < \mu, t \ge t_0\}$$

are transferred to the region

$$\Omega_2 = \{\mathbf{x} \in \mathbb{R}^n | \ \ h(t, \mathbf{x}) \le b\}$$

in a finite time $T^* = T^*(t_0, \mathbf{x}_0) = T(t_0, V(t_0, \mathbf{x}_0))$; *namely, the system (5.2) is controllable.*

Proof. Let $h_0(t_0, \mathbf{x}_0) \le \phi(h_0(t_0, \mathbf{x}_0)) < \mu$ and $U(t, \tilde{\mathbf{u}}(t)) \le v(t), t \ge t_0$, then we have $h_0(t_0, \mathbf{x}_0) \le \phi(h_0(t_0, \mathbf{x}_0)) < \nu$. Then for any solution $\mathbf{x}(t) = \mathbf{x}(t, t_0, \mathbf{x}_0, \tilde{\mathbf{u}})$ of system (5.2), we have

$$h_0(t_0, \mathbf{x}_0) < \mu \ \text{ implies } \ h(t, \mathbf{x}(t)) < \nu, \quad t \ge t_0. \tag{5.32}$$

If this is not the case, then there exists a $\tilde{\mathbf{u}}_1 \in \Omega$ and a corresponding solution $\mathbf{x}(t) = \mathbf{x}(t, t_0, \mathbf{x}_0, \tilde{\mathbf{u}}_1)$ of system (5.2) and a $t^* > t_0$ such that $t^* \in (\tau_k, \tau_{k+1}]$ for some k, satisfying

$$h(t^*, \mathbf{x}(t^*)) \ge \nu \ \text{ and } \ h(t, \mathbf{x}(t)) < \nu, \quad t \in [t_0, \tau_k]. \tag{5.33}$$

From assumption (5) we know that there exists a $t_1 \in (\tau_k, t^*]$ such that

$$h(t_1, \mathbf{x}(t_1)) \in [\nu, \rho). \tag{5.34}$$

Let $m(t) = V(t, \mathbf{x}(t))$ for $t \in [t_0, t_1]$ then from assumption (4) we have

$$\begin{aligned} D^+m(t) &\leq g(t, m(t), v(t)), \quad t \neq \tau_i, \quad t \in [t_0, t_1], \\ m(\tau_i^+) &\leq \psi_i(m(\tau_i)), \quad i = 1, 2, \cdots, k, \end{aligned} \tag{5.35}$$

which implies by Lemma 5.2 the estimate

$$m(t) \leq \gamma(t, t_0, w_0, v), \qquad t \in [t_0, t_1], \tag{5.36}$$

with $m(t_0) \leq w_0$, where $\gamma(t, t_0, w_0, v)$ is the maximal solution of Eq. (5.26). Let us choose $w_0 = V(t_0, \mathbf{x}_0)$, then we can get from assumption (3) and Eqs. (5.33)-(5.36) the relation

$$\beta(h(t, \mathbf{x}(t))) \leq V(t, \mathbf{x}(t)) \leq \gamma(t, t_0, w_0, v), \qquad t \in [t_0, t_1]. \tag{5.37}$$

Now we are led to the following contradiction, in view of Eqs. (5.30) and (5.34),

$$\beta(\nu) \leq \beta(h(t_1, \mathbf{x}(t_1))) \leq \gamma(t, t_0, w_0, v) < \beta(\nu), \tag{5.38}$$

which proves the practical stability of system (5.2). Furthermore, Eq. (5.37) holds for all $t \geq t_0$, and from Eq. (5.31) we have

$$h(t_0 + T^*, \mathbf{x}(t_0 + T^*)) \leq b, \tag{5.39}$$

where $T^* = T(t_0, V(t_0, \mathbf{x}_0))$. □

5.2.4 *Examples*

Let us consider the following comparison system:

$$\begin{aligned} \dot{w} &= a(t)w + b(t)v(t), \qquad t \neq \tau_k, \\ \Delta w(\tau_k^+) &= (d_k - 1)w(\tau_k), \quad d_k \geq 0, \qquad k = 1, 2, \cdots, \\ w(t_0) &= w_0. \end{aligned} \tag{5.40}$$

where $a, b \in \mathcal{PC}[\mathbb{R}_+, \mathbb{R}]$. The solution of Eq. (5.40) is given by

$$w(t) = \begin{cases} \exp\left(\int_{t_0}^{t} a(s)ds\right) \int_{t_0}^{t} \exp\left(-\int_{t_0}^{s} a(\kappa)d\kappa\right) b(s)v(s)ds \\ +w_0 \exp\left(\int_{t_0}^{t} a(s)ds\right), \qquad t \in [t_0, \tau_1], \\ \exp\left(\int_{\tau_{k-1}}^{t} a(s)ds\right) \int_{\tau_{k-1}}^{t} \exp\left(-\int_{\tau_{k-1}}^{s} a(\kappa)d\kappa\right) b(s)v(s)ds \\ +d_{k-1}w(\tau_{k-1}) \exp\left(\int_{\tau_{k-1}}^{t} a(s)ds\right), \\ \quad t \in (\tau_{k-1}, \tau_k], \quad k = 2, 3, \cdots. \end{cases} \tag{5.41}$$

Given $0 < \beta(\mu) < \beta(\nu)$ and $0 < \beta(\mu_1) < \beta(\nu)$ and let us choose $v \in \mathcal{PC}[\mathbb{R}_+, \mathbb{R}]$ such that

$$\int_{\tau_{k-1}}^{\tau_k} \exp\left(-\int_{\tau_{k-1}}^{s} a(\kappa)d\kappa\right) b(s)v(s)ds \le \gamma_k.$$

If $a(t) \le 0$ for $t \ge t_0$ and

$$\beta(\mu) \prod_{k=1}^{\infty} d_k + \sum_{k=1}^{\infty} \gamma_k \prod_{i=k}^{\infty} d_i \le \beta(\nu),$$

therefore, if $w_0 < \beta(\mu)$ from Eq. (5.41) we have

$$w(t, t_0, w_0) < \beta(\nu), \qquad t \le t_0.$$

It means that system (5.40) is practically stable.

Furthermore, if there is a $T > 0$ such that $t_0 + T \ne \tau_k$ and

$$\exp\left(\int_{\tau_{i-1}}^{\tau_i} a(s)ds\right) \le \sigma_{i-1}, \qquad i = 1, 2, \cdots, k+1,$$

and

$$\beta(\mu) \prod_{i=1}^{k} d_i\sigma_{i-1} + \sum_{i=1}^{k} \prod_{j=1}^{k} d_j\sigma_{j-1}\gamma_i + \sigma_{k+1}\gamma_{k+1} < \beta(\mu_1)$$

then

$$w(t_0 + T, t_0, w_0) < \beta(\mu_1),$$

which shows that the system (5.40) is controllable.

5.3 Practical stability of linear impulsive control systems

Consider the following linear nonautonomous impulsive system:

$$\begin{aligned} \dot{\mathbf{x}} &= A\mathbf{x} + \tilde{B}\tilde{\mathbf{u}} + \sigma(t), \quad t \neq \tau_k, \\ \Delta\mathbf{x} &= B_k\mathbf{x}, \quad t = \tau_k, \quad k = 1, 2, \cdots, \\ \mathbf{x}(t_0^+) &= \mathbf{x}_0, \end{aligned} \tag{5.42}$$

where A and $\tilde{B}$ are $n \times n$ and $n \times m$ matrices, respectively. B_k is an $n \times n$ matrix for each k, and $\sigma : \mathbb{R}_+ \to \mathbb{R}^n$ is piecewise continuous.

Theorem 5.3 *Assume that*

(1) $0 < \mu < \nu$ *are given;*
(2)

$$\lim_{h \to 0} \frac{1}{h}(||I + hA|| - 1) \leq -a, \qquad a > 0,$$

$||\tilde{B}|| = b$, $||\sigma(t)|| \leq l$ *and* $||B_k|| = b_k$ *for* $k = 1, 2, \cdots$*;*
(3) $a - b = \delta > 0$ *and*

$$\mu \prod_{k=1}^{\infty}(1 + b_k) + \frac{1}{\delta}\sum_{k=1}^{\infty}\prod_{i=k}^{\infty}(1 + b_i) + \frac{1}{\delta} < \nu.$$

Then the system (5.42) is practically stable.

Proof. Let us choose $V(t, \mathbf{x}) = ||\mathbf{x}||$ and $h_0(t, \mathbf{x}) = h(t, \mathbf{x}) = ||\mathbf{x}||$ then we have

$$g(t, w, w) = (-a + b)w + l \quad \text{and} \quad \psi_k(w) = (1 + b_k)w.$$

We then construct the following comparison system:

$$\begin{aligned} \dot{w} &= -\delta w + l, \quad t \neq \tau_k, \\ w(\tau_k^+) &= (1 + b_k)w(\tau_k), \quad k = 1, 2, \cdots, \\ w(t_0^+) &= w_0 \end{aligned} \tag{5.43}$$

.

whose solutions are given by

$$\begin{aligned} w(t,t_0,w_0) &= w_0\prod_{j=1}^{k}(1+b_j)e^{-\delta(t-t_0)}+\frac{1}{\delta}\sum_{j=1}^{k}\prod_{i=j}^{k}(1+b_i)e^{-\delta(t-t_j)} \\ &\quad -\frac{1}{\delta}\sum_{j=1}^{k}\prod_{i=j}^{k}(1+b_i)e^{-\delta(t-t_{j-1})}+\frac{1}{\delta}(1-e^{-\delta(t-t_k)}), \\ &\quad t\in(\tau_k,\tau_{k+1}]. \end{aligned} \tag{5.44}$$

From assumption (3) we know that $w_0<\mu$ implies $w(t,t_0,w_0)<\nu$ for $t\geq t_0$. Thus, followed Theorem 5.1 the proof is complete. □

5.4 Practical stability in terms of multicomparison systems

In this section we consider the practical stability of the following impulsive control system:

$$\begin{aligned} \dot{\mathbf{x}} &= \mathbf{f}(t,\mathbf{x},\tilde{\mathbf{u}}), \quad t\neq\tau_k, \\ \Delta\mathbf{x} &= \mathbf{u}_k(\mathbf{x},\tilde{\mathbf{u}}), \quad t=\tau_k, \\ \mathbf{x}(t_0^+) &= \mathbf{x}_0, \quad k=1,2,\cdots \end{aligned} \tag{5.45}$$

where $\mathbf{f}\in\mathcal{PC}[\mathbb{R}_+\times\mathbb{R}^n\times\mathbb{R}^m,\mathbb{R}^n]$ and $\mathbf{u}_k\in C[\mathbb{R}^n\times\mathbb{R}^m,\mathbb{R}^n], k=1,2,\cdots$.

Let us construct the following D comparison systems:

$$\begin{aligned} \dot{\mathbf{w}} &= \mathbf{g}(t,\mathbf{w},\mathbf{w}), \quad t\neq\tau_k, \\ \mathbf{w}(\tau_k^+) &= \psi_k(\mathbf{w}(\tau_k),\mathbf{w}(\tau_k)), \quad k=1,2,\cdots, \\ \mathbf{w}(t_0) &= \mathbf{w}(0) \end{aligned} \tag{5.46}$$

where $\mathbf{w}\in\mathbb{R}^D$ and

(1) $\mathbf{g}\in\mathcal{PC}[\mathbb{R}_+\times\mathbb{R}_+^D\times\mathbb{R}_+^D,\mathbb{R}^D]$, $\mathbf{g}(t,\mathbf{w},\mathbf{v})$ is quasimonotone nondecreasing in $\mathbf{w}$ for each $(t,\mathbf{v})$ and nondecreasing in $\mathbf{v}$ for each $(t,\mathbf{w})$;

(2) $\psi_k:\mathbb{R}_+^D\times\mathbb{R}_+^D\to\mathbb{R}_+^D$ and $\psi_k(\mathbf{w},\mathbf{v})$ is nondecreasing in $(\mathbf{w},\mathbf{v})$ for each k.

Let Ω be the admissible control input for $\tilde{\mathbf{u}}$ as

$$\Omega=\{\tilde{\mathbf{u}}\in\mathbb{R}^m|\ U(t,\tilde{\mathbf{u}})\leq\gamma(t),t\geq t_0\},$$

where $U \in C[\mathbb{R}_+ \times \mathbb{R}^m, \mathbb{R}_+^D]$ and $\gamma(t)$ is the maximal solution of the comparison system (5.46).

From Theorem 2.1 of [McR94] we have the following lemma:

Lemma 5.3 *Assume that*

(1) $\gamma(t)$ *is the maximal solution of comparison system (5.46) in* $[t_0, \infty)$*;*

(2) $\gamma^*(t)$ *is the maximal solution of the following system*

$$\begin{aligned} \dot{\mathbf{w}} &= \mathbf{g}(t, \mathbf{w}, \gamma(t)), \qquad t \neq \tau_k, \\ \mathbf{w}(\tau_k^+) &= \psi_k(\mathbf{w}(\tau_k), \gamma(\tau_k)), \qquad k = 1, 2, \cdots, \\ \mathbf{w}(t_0) &= \mathbf{w}(0) \geq 0, \end{aligned} \tag{5.47}$$

in $[t_0, \infty)$*;*

(3) $\mathbf{m}, \mathbf{v} \in \mathcal{PC}[\mathbb{R}_+, \mathbb{R}_+^D]$ *and*

$$\begin{aligned} D^+\mathbf{m}(t) &\leq \mathbf{g}(t, \mathbf{m}(t), \mathbf{v}(t)), \qquad t \neq \tau_k, \\ \mathbf{m}(\tau_k^+) &\leq \psi_k(\mathbf{m}(\tau_k), \mathbf{v}(\tau_k)), \qquad k = 1, 2, \cdots; \end{aligned} \tag{5.48}$$

(4) $\mathbf{v}(t) \leq \gamma(t),\ t \geq t_0$*;*

(5) $\mathbf{m}(t_0) < \mathbf{w}_0$.

Then

$$\mathbf{m}(t) \leq \gamma^*(t), \qquad t \geq t_0.$$

We have the following theorem:

Theorem 5.4 *Assume that*

(1) $0 < \mu < \nu$ *are given;*

(2) $\mathbf{V} \in \mathcal{PC}[\mathbb{R}_+ \times \mathbb{R}^n, \mathbb{R}_+^D]$, $\mathbf{V}(t, \mathbf{x}) = (V_1(t, \mathbf{x}), V_2(t, \mathbf{x}), \cdots, V_D(t, \mathbf{x}))^\top$ *is locally Lipschitzian in* $\mathbf{x}$*, and there exist* $\alpha, \beta \in \mathcal{K}[\mathbb{R}_+, \mathbb{R}_+]$ *such that*

$$\beta(\|\mathbf{x}\|) \leq \sum_{i=1}^{D} V_i(t, \mathbf{x}) \leq \alpha(\|\mathbf{x}\|), \quad (t, \mathbf{x}) \in \mathbb{R}_+ \times \mathfrak{S}_\rho, \quad \rho > \nu;$$

(3) for $(t, \mathbf{x}) \in (\tau_k, \tau_{k+1}] \times \mathfrak{S}_\rho$ *and* $\tilde{\mathbf{u}} \in \Omega$ *we have*

$$D^+\mathbf{V}(t, \mathbf{x}) \leq g(t, \mathbf{V}(t, \mathbf{x}), U(t, \tilde{\mathbf{u}}));$$

(4) $\mathbf{x} \in \mathfrak{S}_\nu$ *implies* $\mathbf{x} + \mathbf{u}_k(\mathbf{x}, \tilde{\mathbf{u}}) \in \mathfrak{S}_\rho$ *and*

$$\mathbf{V}(\tau_k^+, \mathbf{x} + \mathbf{u}_k(\mathbf{x}, \tilde{\mathbf{u}})) \leq \psi_k(\mathbf{V}(\tau_k, \mathbf{x}), U(\tau_k, \tilde{\mathbf{u}})), \quad \mathbf{x} \in \mathfrak{S}_\rho;$$

(5) $\alpha(\mu) < \beta(\nu)$.

Then the practical stability properties of comparison system (5.46) with respect to $(\alpha(\mu), \beta(\nu))$ *imply those of system (5.45) with respect to* (μ, ν) *for every* $\tilde{\mathbf{u}} \in \Omega$.

Proof.

(1) Practical stability
Let us suppose that system (5.46) is practical stable with respect to $(\alpha(\mu), \beta(\nu))$, then we have

$$\sum_{i=1}^{D} w_i(0) < \alpha(\mu) \text{ implies } \sum_{i=1}^{D} w_i(t, t_0, \mathbf{w}(0)) < \beta(\nu), \ t \geq t_0, \tag{5.49}$$

where $\mathbf{w}(t, t_0, \mathbf{w}(0))$ is any solution of system (5.46) in $[t_0, \infty)$. Suppose that $\|\mathbf{x}_0\| < \mu$ and we claim that $\|\mathbf{x}(t, t_0, \mathbf{x}_0, \mathbf{u}^*)\| < \nu$ with $\mathbf{u}^* \in \Omega$. If this is not the case, then there exists a $\mathbf{u}_1 \in \Omega$ such that

$$\|\mathbf{x}(t_2, t_0, \mathbf{x}_0, \mathbf{u}_1)\| \geq \nu, \ \ \|\mathbf{x}(t, t_0, \mathbf{x}_0, \mathbf{u}_1)\| < \nu, \tag{5.50}$$

where $t_2 \in (\tau_k, \tau_{k+1}]$ for some k and $t \in [t_0, \tau_k]$. From assumption (4) we know that there exists a $t_1 \in (\tau_k, t_2]$ such that

$$\nu \leq \|\mathbf{x}(t_1, t_0, \mathbf{x}_0, \mathbf{u}_1)\| < \rho. \tag{5.51}$$

We have $\|\mathbf{x}(t)\| < \rho, t \in [t_0, t_1]$. Let $\mathbf{m}(t) = \mathbf{V}(t, \mathbf{x}(t)), t \in [t_0, t_1]$, $\mathbf{w}(0) = \mathbf{V}(t_0, \mathbf{x}_0)$ and in view of assumption (3) we have

$$\begin{aligned} D^+\mathbf{m}(t) &\leq \mathbf{g}(t, \mathbf{m}(t), U(t, \mathbf{u}_1(t))), \ \ t \in [t_0, t_1], \quad t \neq \tau_k, \\ \mathbf{m}(\tau_i^+) &\leq \psi_i(\mathbf{m}(\tau_i), U(\tau_i, \mathbf{u}_1(\tau_i))), \quad i = 1, 2, \cdots, k, \\ \mathbf{m}(t_0) &= \mathbf{w}(0). \end{aligned} \tag{5.52}$$

Since $\mathbf{u}_1 \in \Omega$ and $\mathbf{g}(t, \mathbf{w}, \mathbf{v})$ is nondecreasing in $\mathbf{v}$, we have

$$\begin{aligned} D^+\mathbf{m}(t) &\leq \mathbf{g}(t, \mathbf{m}(t), \gamma(t)), \ \ t \in [t_0, t_1], \quad t \neq \tau_i, \\ \mathbf{m}(\tau_i^+) &\leq \psi_i(\mathbf{m}(\tau_i), \gamma(\tau_i)), \quad i = 1, 2, \cdots, k. \end{aligned} \tag{5.53}$$

Thus, by Lemma 5.3 we have

$$\mathbf{m}(t) \leq \gamma(t), \quad t \in [t_0, t_1]. \tag{5.54}$$

Assumption (2) and Eqs (5.49), (5.51), (5.54) lead to the contradiction

$$\beta(\nu) \leq \sum_{i=1}^{D} V_i(t_1, \mathbf{x}(t_1)) \leq \sum_{i=1}^{D} \gamma_i(t_1) < \beta(\nu), \tag{5.55}$$

which proves that the control system (5.45) is practically stable.

(2) Strong practical stability
Let us suppose that system (5.46) is strongly practical stable with respect to $(\alpha(\mu), \beta(\nu))$. Thus system (5.45) is practically stable and we have

$$\|\mathbf{x}(0)\| < \mu \text{ implies } \|\mathbf{x}(t, t_0, \mathbf{x}(0), \mathbf{u}^*)\| < \nu, \ t \geq t_0. \tag{5.56}$$

Given $0 < \nu_1 < \nu$ and $T > 0$ and since system (5.46) is quasipractically stable, we have

$$\begin{aligned}&\sum_{i=1}^{D} w_i(0) < \alpha(\mu) \text{ implies}\\ &\sum_{i=1}^{D} w_i(t, t_0, \mathbf{w}(0)) < \beta(\nu), \ t \geq t_0 + T.\end{aligned} \tag{5.57}$$

Let us suppose that $\|\mathbf{x}_0\| \leq \mu$ and because of Eq. (5.56), arguments leading to Eq. (5.55) yield

$$\mathbf{V}(t, \mathbf{x}(t)) \leq \gamma(t, t_0, \mathbf{w}(0)), \ t \geq t_0. \tag{5.58}$$

Then in views of Eqs. (5.57), (5.58), and assumption (2), it follows that

$$\begin{aligned}\beta(\|\mathbf{x}(t, t_0, \mathbf{x}_0, \mathbf{u}^*)\|) &\leq \sum_{i=1}^{D} V_i(t, \mathbf{x}(t, t_0, \mathbf{x}_0, \mathbf{u}^*))\\ &\leq \sum_{i=1}^{D} \gamma_i(t, t_0, \mathbf{w}(0)) < \beta(\nu_1),\\ \|\mathbf{x}(t, t_0, \mathbf{x}_0, \mathbf{u}^*)\| &< \nu_1, \quad t \geq t_0 + T.\end{aligned} \tag{5.59}$$

Thus, system (5.45) is practically stable.

(3) Other practical stability
Proofs are similar.

□

Let $\mathbf{V}^{[1]}, \mathbf{V}^{[2]} \in \mathcal{PC}[\mathbb{R}_+ \times \mathbb{R}^n, \mathbb{R}_+^D]$, $\mathbf{V}^{[1]}(t,\mathbf{x})$ and $\mathbf{V}^{[2]}(t,\mathbf{x})$ are locally Lipschitzian in $\mathbf{x}$. And we have the following two sets of comparison systems:

$$\begin{aligned} \dot{\mathbf{w}} &= \mathbf{g}_1(t,\mathbf{w},\mathbf{w}), \quad t \neq \tau_k \\ \mathbf{w}(\tau_k^+) &= \psi_k(\mathbf{w}(\tau_k),\mathbf{w}(\tau_k)), \quad k = 1,2,\cdots, \\ \mathbf{w}(t_0) &= \mathbf{w}(0) \geq 0, \end{aligned} \tag{5.60}$$

where $\psi_k : \mathbb{R}_+^D \times \mathbb{R}_+^D \to \mathbb{R}_+^D$ and $\mathbf{g}_1 \in \mathcal{PC}[\mathbb{R}_+ \times \mathbb{R}_+^D \times \mathbb{R}_+^D, \mathbb{R}^D]$ and $\mathbf{g}_1(t,\mathbf{w},\mathbf{v})$ is quasimonotone nondecreasing in $\mathbf{w}$ for each $(t,\mathbf{v})$ and monodecreasing in $\mathbf{v}$ for each $(t,\mathbf{w})$.

$$\begin{aligned} \dot{\mathbf{z}} &= \mathbf{g}_2(t,\mathbf{z},\mathbf{z}), \quad t \neq \tau_k \\ \mathbf{z}(\tau_k^+) &= \phi_k(\mathbf{z}(\tau_k),\mathbf{z}(\tau_k)), \quad k = 1,2,\cdots, \\ \mathbf{z}(t_0) &= \mathbf{z}(0) \geq 0, \end{aligned} \tag{5.61}$$

where $\phi_k : \mathbb{R}_+^D \times \mathbb{R}_+^D \to \mathbb{R}_+^D$ and $\mathbf{g}_2 \in \mathcal{PC}[\mathbb{R}_+ \times \mathbb{R}_+^D \times \mathbb{R}_+^D, \mathbb{R}^D]$ and $\mathbf{g}_2(t,\mathbf{w},\mathbf{v})$ is quasimonotone nondecreasing in $\mathbf{w}$ for each $(t,\mathbf{v})$ and monodecreasing in $\mathbf{v}$ for each $(t,\mathbf{w})$.

Theorem 5.5 *Assume that*

(1) $0 < \mu < \nu$ *are given;*
(2) for $(t,\mathbf{x}) \in \mathbb{R}_+ \times \mathfrak{S}_\nu$,

$$\sum_{i=1}^{D} V_i^{[1]}(t,\mathbf{x}) \leq \alpha_1(t,\|\mathbf{x}\|)$$

and for $(t,\mathbf{x}) \in \mathbb{R}_+ \times \overline{\mathfrak{S}_\mu} \cap \mathfrak{S}_\nu$

$$\beta(\|\mathbf{x}\|) \leq \sum_{i=1}^{D} V_i^{[2]}(t,\mathbf{x}) \leq \alpha_2(\|\mathbf{x}\|) + \sum_{i=1}^{D} V_i^{[1]}(t,\mathbf{x}),$$

where $\alpha_1 \in \mathcal{CK}[\mathbb{R}_+,\mathbb{R}_+]$ *and* $\alpha_2 \in \mathcal{K}[\mathbb{R}_+,\mathbb{R}_+]$*;*
(3) for $(t,\mathbf{x}) \in (\tau_{k-1},\tau_k] \times \mathfrak{S}_\nu$ *and* $\tilde{\mathbf{u}} \in \Omega$

$$D^+\mathbf{V}^{[1]}(t,\mathbf{x}) \leq \mathbf{g}_1(t,\mathbf{V}^{[1]}(t,\mathbf{x}),U(t,\tilde{\mathbf{u}}));$$

(4) $\|\mathbf{x} + \mathbf{u}_k(\mathbf{x},\tilde{\mathbf{u}})\| \leq \|\mathbf{x}\|$ *for* $\mathbf{x} \in \mathfrak{S}_\nu$*;*

(5) *for* $\mathbf{x} \in \mathfrak{S}_\nu$,

$$\mathbf{V}_1(\tau_k^+, \mathbf{x} + \mathbf{u}_k(\mathbf{x}, \tilde{\mathbf{u}})) \le \psi_k(\mathbf{V}_1(\tau_k, \mathbf{x}), U(\tau_k, \tilde{\mathbf{u}})), \quad k = 1, 2, \cdots,$$

where $\psi_k(\mathbf{w}, \mathbf{v})$ *is nondecreasing in* $(\mathbf{w}, \mathbf{v})$ *for all* $k = 1, 2, \cdots$;

(6) *for* $(t, \mathbf{x}) \in (\tau_{k-1}, \tau_k] \times \overline{\mathfrak{S}_\mu} \cap \mathfrak{S}_\nu$ *and* $\tilde{\mathbf{u}} \in \Omega$,

$$D^+\mathbf{V}^{[2]}(t, \mathbf{x}) \le \mathbf{g}_2(t, \mathbf{V}^{[2]}(t, \mathbf{x}), U(t, \tilde{\mathbf{u}}));$$

(7) *for* $\mathbf{x} \in \overline{\mathfrak{S}_\mu} \cap \mathfrak{S}_\nu$

$$\mathbf{V}_2(\tau_k^+, \mathbf{x} + \mathbf{u}_k(\mathbf{x}, \tilde{\mathbf{u}}) \le \phi_k(\mathbf{V}_2(\tau_k, \mathbf{x}), U(t, \tilde{\mathbf{u}})), \quad k = 1, 2, \cdots,$$

where $\phi_k(\mathbf{w}, \mathbf{v})$ *is nondecreasing in* $(\mathbf{w}, \mathbf{v})$;

(8) $\alpha_1(t_0, \mu) + \alpha_2(\mu) < \beta(\nu)$, *for all* $t_0 \in \mathbb{R}_+$;

(9)

$$\sum_{i=1}^{D} w_i(0) < \alpha_1(t_0, \mu) \text{ implies } \sum_{i=1}^{D} w_i(t, t_0, \mathbf{w}_0) < \alpha_1(t_0, \mu)$$

for $t \ge t_0$ *with some* $t_0 \in \mathbb{R}_+$, *where* $\mathbf{w}(t, t_0, \mathbf{w}_0)$ *is any solution of Eq. (5.60) and*

$$\sum_{i=1}^{D} z_i(0) < \alpha_2(\mu) + \alpha_1(t_0, \mu) \text{ implies } \sum_{i=1}^{D} z_i(t, t_0, \mathbf{z}_0) < \beta(\nu)$$

for $t \ge t_0$ *with all* $t_0 \in \mathbb{R}_+$, *where* $\mathbf{z}(t, t_0, \mathbf{z}_0)$ *is any solution of Eq. (5.61).*

Then the impulsive control system (5.45) is practically stable.

Proof. If for any solution $\mathbf{x}(t, t_0, \mathbf{x}_0, \tilde{\mathbf{u}})$ of system (5.45)

$$\|\mathbf{x}_0\| < \mu \text{ implies } \|\mathbf{x}(t, t_0, \mathbf{x}_0, \tilde{\mathbf{u}})\| < A, \quad t \ge t_0,$$

is false, then there exists a $\tilde{\mathbf{u}}^* = \tilde{\mathbf{u}}^*(t) \in \Omega$ and a corresponding solution $\mathbf{x}(t, t_0, \mathbf{x}_0, \tilde{\mathbf{u}}^*)$, and owing to assumption (4), there should exist $t_2 > t_1 > t_0$ such that

$$\begin{aligned} &\|\mathbf{x}(t_1, t_0, \mathbf{x}_0, \tilde{\mathbf{u}}^*)\| = \mu, \quad \|\mathbf{x}(t_2, t_0, \mathbf{x}_0, \tilde{\mathbf{u}}^*)\| = \nu, \\ &\mu \le \|\mathbf{x}(t, t_0, \mathbf{x}_0, \tilde{\mathbf{u}}^*)\| \le \nu, \\ &t_1 \in (\tau_i, \tau_{i+1}], \ t_2 \in (\tau_j, \tau_{j+1}], \ t \in [t_1, t_2]. \end{aligned} \tag{5.62}$$

Then in view of assumptions (3) and (5), we get by Lemma 5.3,

$$\mathbf{V}^{[1]}(t, \mathbf{x}(t)) \leq \gamma_1(t, t_0, \mathbf{V}^{[1]}(t_0, \mathbf{x}_0)), \qquad t \in [t_0, t_2], \tag{5.63}$$

where $\gamma_1(t, t_0, \mathbf{w}_0)$ is the maximal solution of Eq. (5.60) passing though $(t_0, \mathbf{w}_0)$. Based on the assumptions (6) and (7) we have

$$\mathbf{V}^{[2]}(t, \mathbf{x}(t)) \leq \gamma_2(t, t_1, \mathbf{V}^{[2]}(t_1, \mathbf{x}(t_1))), \qquad t \in [t_1, t_2], \tag{5.64}$$

where $\gamma_2(t, t_1, \mathbf{z}_1)$ is the maximal solution of Eq. (5.61) passing through $(t_1, \mathbf{z}_1)$ and $\mathbf{z}_1 = \mathbf{V}^{[2]}(t_1, \mathbf{x}(t_1))$. Thus, from the assumptions (2), (9), and Eq. (5.63) we get

$$\sum_{i=1}^{D} V_i^{[1]}(t_1, \mathbf{x}(t_1)) \leq \sum_{i=1}^{D} \gamma_{1i}(t_1, t_0, \mathbf{V}^{[1]}(t_0, \mathbf{x}_0)) < \alpha_1(t_0, \mu). \tag{5.65}$$

From Eqs. (5.62), (5.65), and assumption (2), we get

$$\begin{aligned} \sum_{i=1}^{D} V_i^{[2]}(t_1, \mathbf{x}(t_1)) &\leq \alpha_2(\|\mathbf{x}(t_1, t_0, \mathbf{x}_0, \tilde{\mathbf{u}}^*)\|) + \sum_{i=1}^{D} V_i^{[1]}(t_1, \mathbf{x}(t_1)) \\ &< \alpha_2(\mu) + \alpha_1(t_0, \mu). \end{aligned} \tag{5.66}$$

Thus, from Eqs. (5.62), (5.64), (5.66), and assumptions (2) and (9), we get

$$\begin{aligned} \beta(\nu) &= \beta(\|\mathbf{x}(t_2, t_0, \mathbf{x}_0, \tilde{\mathbf{u}}^*)\|) \leq \sum_{i=1}^{D} V_i^{[2]}(t_2, \mathbf{x}(t_2)) \\ &\leq \sum_{i=1}^{D} \gamma_{2i}(t_2, t_1, \mathbf{V}^{[2]}(t_1, \mathbf{x}(t_1))) < \beta(\nu), \end{aligned} \tag{5.67}$$

which is a contradiction. □

5.4.1 *Example*

As a demonstrating example, let us consider the following system[McR94]:

$$
\begin{aligned}
\begin{pmatrix} \dot{x}_1 \\ \dot{x}_2 \end{pmatrix} &= \begin{bmatrix} \eta(t) & \zeta(t) \\ \zeta(t) & \eta(t) \end{bmatrix} \begin{pmatrix} x_1 \\ x_2 \end{pmatrix} \\
&\quad + \begin{pmatrix} -\sin^2 t(x_1^3 + x_1 x_2^2) + ag(u_1, u_2) \\ -\sin^2 t(x_1^2 x_2 + x_2^3) + bg(u_1, u_2) \end{pmatrix}, \qquad t \neq \tau_k, \\
\Delta \begin{pmatrix} x_1 \\ x_2 \end{pmatrix} &= \begin{bmatrix} e_k & f_k \\ f_k & e_k \end{bmatrix} \begin{pmatrix} x_1(\tau_k) \\ x_2(\tau_k) \end{pmatrix} \\
&\quad + \begin{pmatrix} h_k g(u_1(\tau_k), u_2(\tau_k)) \\ i_k g(u_1(\tau_k), u_2(\tau_k)) \end{pmatrix}, \qquad t = \tau_k, \quad k = 1, 2, \cdots, \\
\begin{pmatrix} x_1(t_0^+) \\ x_2(t_0^+) \end{pmatrix} &= \begin{pmatrix} x_1(0) \\ x_2(0) \end{pmatrix}.
\end{aligned} \tag{5.68}
$$

where a, b, e_k, f_k, h_k, i_k are real constants, $g(u_1, u_2) \geq 0$, $\eta(t) \geq 0$, $\zeta(t) \geq 0$.

Let us choose

$$V_1 = \frac{1}{2}(x_1 + x_2)^2, \qquad V_2 = \frac{1}{2}(x_1 - x_2)^2,$$

then we have

$$V_1 + V_2 = x_1^2 + x_2^2 = \|\mathbf{x}\|^2.$$

Thus we can choose $\alpha(x) = (1 + \epsilon)x^2$ and $\beta(x) = (1 - \epsilon)x^2$ with a constant $\epsilon \in (0, 1]$. We then have

$$
\begin{aligned}
\dot{V}_1 &= (x_1 + x_2)^2(\eta(t) + \zeta(t)) - \sin^2 t(x_1 + x_2)^2(x_1^2 + x_2^2) \\
&\quad + (a + b)(x_1 + x_2)g(u_1, u_2) \\
&\leq (\eta(t) + \zeta(t))V_1 + |a + b| g(u_1, u_2)\sqrt{2V_1}.
\end{aligned} \tag{5.69}
$$

Similar we have

$$\dot{V}_2 \leq (\eta(t) - \zeta(t))V_2 + |a - b| g(u_1, u_2)\sqrt{2V_2}. \tag{5.70}$$

Then we have

$$\begin{aligned} V_1(\tau_k^+, \mathbf{x}_k(\tau_k^+)) &= \frac{1}{2}(x_1(\tau_k^+) + x_2(\tau_k^+))^2 \\ &= \frac{1}{2}[(e_k + f_k)(x_1 + x_2) + (h_k + i_k)g(u_1(\tau_k), u_2(\tau_k))]^2 \\ &\leq \frac{1}{2}[|e_k + f_k|\sqrt{2V_1} + |h_k + i_k|g(u_1(\tau_k), u_2(\tau_k))]^2 \end{aligned}$$

and similarly

$$V_2(\tau_k^+, \mathbf{x}_k(\tau_k^+)) \leq \frac{1}{2}[|e_k - f_k|\sqrt{2V_2} + |h_k - i_k|g(u_1(\tau_k), u_2(\tau_k))]^2.$$

Let us define

$$U_1(\tilde{\mathbf{u}}) \triangleq \frac{1}{2}[(a+b)g(u_1, u_2)]^2, \quad U_2(\tilde{\mathbf{u}}) \triangleq \frac{1}{2}[(a-b)g(u_1, u_2)]^2 \tag{5.71}$$

then we have

$$\begin{aligned} g_1(t, w, U) &= [\eta(t) + \zeta(t)]w_1 + \sqrt{2w_1}\sqrt{2U_1(\tilde{\mathbf{u}})}, \\ g_1(t, w, U) &= [\eta(t) - \zeta(t)]w_2 + \sqrt{2w_1}\sqrt{2U_1(\tilde{\mathbf{u}})}, \\ \psi_k(w(\tau_k), U(\mathbf{u}(\tau_k))) &= \frac{1}{2}\left(|e_k + f_k|\sqrt{2w_1} + \left|\frac{h_k + i_k}{a+b}\right|\sqrt{2U_1(\mathbf{u}(\tau_k))}\right)^2, \\ \phi_k(w(\tau_k), U(\mathbf{u}(\tau_k))) &= \frac{1}{2}\left(|e_k - f_k|\sqrt{2w_2} + \left|\frac{h_k - i_k}{a-b}\right|\sqrt{2U_2(\mathbf{u}(\tau_k))}\right)^2, \end{aligned}$$

from which we have the following comparison system:

$$\begin{aligned} \dot{w}_1 &= g_1(t, w, w) = [\eta(t) + \zeta(t) + 2]w_1 \triangleq \sigma(t)w_1, \quad t \neq \tau_k, \\ \dot{w}_2 &= g_2(t, w, w) = [\eta(t) - \zeta(t) + 2]w_2 \triangleq \gamma(t)w_2, \quad t \neq \tau_k, \\ w_1(\tau_k^+) &= \psi_k(w(\tau_k), w(\tau_k)) \\ &= \left(|e_k + f_k| + \frac{h_k + i_k}{a+b}\right)^2 w_1(\tau_k^+) \triangleq c_k w_1(\tau_k^+), k = 1, 2, \cdots, \\ w_2(\tau_k^+) &= \phi_k(w(\tau_k), w(\tau_k)) \\ &= \left(|e_k - f_k| + \frac{h_k - i_k}{a-b}\right)^2 w_2(\tau_k^+) \triangleq d_k w_2(\tau_k^+), k = 1, 2, \cdots, \\ w_1(t_0) &= \frac{1}{2}(x_1(0) + x_2(0))^2 \triangleq w_1(0), \\ w_2(t_0) &= \frac{1}{2}(x_1(0) - x_2(0))^2 \triangleq w_2(0). \end{aligned} \tag{5.72}$$

It is easy to verify that g, ψ_k, and ϕ_k satisfy the assumptions of Theorem 5.4. The solution to Eq. (5.72) is given as follows:

$$\begin{aligned} w_1(t) &= w_1(0) \prod_{0<\tau_k<t} c_k \exp\left\{\int_{t_0}^{t} \sigma(s)ds\right\}, \\ w_2(t) &= w_2(0) \prod_{0<\tau_k<t} d_k \exp\left\{\int_{t_0}^{t} \gamma(s)ds\right\}, \qquad t \geq t_0 \end{aligned}$$

and thus we have the admissible control set as

$$\Omega = \{\mathbf{u}|\ \ U(\mathbf{u}) \leq w(t, t_0, w_0), t \geq t_0\}.$$

Given $0 < \mu < \nu$, if

(1) $0 \leq c_k \leq \exp\left\{-\int_{\tau_k}^{\tau_k+1} \sigma(s)ds\right\}$ and $0 \leq d_k \leq \exp\left\{-\int_{\tau_k}^{\tau_k+1} \gamma(s)ds\right\}$ for all $k = 1, 2, \cdots$;

(2)

$$\exp\left\{\int_{t_0}^{\tau_1} \sigma(s)ds\right\} < \frac{\beta(\nu)}{\alpha(\mu)} \quad \text{and} \quad \exp\left\{\int_{t_0}^{\tau_1} \gamma(s)ds\right\} < \frac{\beta(\nu)}{\alpha(\mu)}$$

then the comparison system (5.72) is practically stable. Furthermore, given $0 < \nu_1 < \nu$ and $T > 0$ where $t_0 + T \in (\tau_i, \tau_{i+1}]$ for some i, and (1) and (2) hold for all $k \neq i$, and

$$0 \leq c_i \leq \frac{\beta(\nu_1)}{\alpha(\mu)} \exp\left\{-\int_{t_0}^{\tau_1} \sigma(s)ds - \int_{\tau_i}^{\tau_i+1} \sigma(s)ds\right\}$$

and

$$0 \leq d_i \leq \frac{\beta(\nu_1)}{\alpha(\mu)} \exp\left\{-\int_{t_0}^{\tau_1} \gamma(s)ds - \int_{\tau_i}^{\tau_i+1} \gamma(s)ds\right\},$$

then the comparison system (5.72) is strongly practically stable.

5.5 Controllablity in terms of multicomparison systems

From Theorem 2.2 of [McR94] we have the following lemma.

Lemma 5.4 *Assume that*

(1) $D^+\mathbf{m}(t) \le \mathbf{g}(t, \mathbf{m}(t)), t \ne \tau_k$, $\mathbf{m}(\tau_k^+) \le \dot{\psi}_k(\mathbf{m}(\tau_k))$, $k = 1, 2, \cdots$, *where* $\mathbf{g} \in \mathcal{PC}[\mathbb{R}_+ \times \mathbb{R}_+^D, \mathbb{R}^D]$, $\mathbf{g}(t, \mathbf{w})$ *is quasimonotone nondecreasing in* $\mathbf{w}$, *and* $\psi_k : \mathbb{R}_+^D \to \mathbb{R}_+^D$ *is nondecreasing for* $k = 1, 2, \cdots$;

(2) $\gamma(t)$ *is the maximal solution of*

$$\begin{aligned} \dot{\mathbf{w}} &= \mathbf{g}(t, \mathbf{w}), \quad t \ne \tau_k, \\ \mathbf{w}(\tau_k^+) &= \psi_k(\mathbf{w}(\tau_k)), \quad k = 1, 2, \cdots, \\ \mathbf{w}(t_0) &= \mathbf{w}_0 \ge 0, \end{aligned} \tag{5.73}$$

existing in $[t_0, \infty)$.

Then $\mathbf{m}(t) \le \gamma(t)$, $t \ge t_0$, *provided* $\mathbf{m}(t_0) \le \mathbf{w}_0$.

Theorem 5.6 *Assume that*

(1) $0 < \mu < \nu$ *and* $0 < \nu_1 < \nu$ *are given;*

(2) $\mathbf{V} \in \mathcal{PC}[\mathbb{R}_+ \times \mathbb{R}_+^D, \mathbb{R}_+^D]$, $\mathbf{V}(t, \mathbf{x})$ *is locally Lipschitzian in* $\mathbf{x}$, $Q \in \mathcal{K}[\mathbb{R}_+^D, \mathbb{R}_+]$, *and for* $(t, \mathbf{x}) \in \mathbb{R}_+ \times \mathfrak{S}_\rho$, $\rho > \nu$,

$$\beta(\|\mathbf{x}\|) \le Q(\mathbf{V}(t, \mathbf{x})) \le \alpha(\|\mathbf{x}\|), \quad \alpha, \beta \in \mathcal{K}[\mathbb{R}_+, \mathbb{R}_+];$$

(3) *for* $(t, \mathbf{x}) \in (\tau_k, \tau_{k+1}] \times \mathfrak{S}_\rho, k = 1, 2, \cdots$, *and* $\tilde{\mathbf{u}} \in \mathcal{PC}[\mathbb{R}_+, \mathbb{R}^m]$

$$D^+\mathbf{V}(t, \mathbf{x}) \le \mathbf{g}(t, \mathbf{V}(t, \mathbf{x}), U(t, \tilde{\mathbf{u}}));$$

(4) $\mathbf{x} \in \mathfrak{S}_\nu$ *implies* $\mathbf{x} + \mathbf{u}_k(\mathbf{x}, \tilde{\mathbf{u}})) \in \mathfrak{S}_\rho$ *and*

$$\mathbf{V}(\tau_k^+, \mathbf{x} + \mathbf{u}_k(\mathbf{x}, \tilde{\mathbf{u}})) \le \psi_k(\mathbf{V}(\tau_k, \mathbf{x}), U(\tau_k, \tilde{\mathbf{u}})), \quad k = 1, 2, \cdots,$$

where $\mathbf{x} \in \mathfrak{S}_\rho$ *and* $U \in C[\mathbb{R}_+ \times \mathbb{R}^m, \mathbb{R}_+^D]$;

(5) $\alpha(\mu) < \beta(\nu)$;

(6) *given a* $T > 0$ *satisfying* $t_0 + T \ne \tau_k$, *there exists a* $\mathbf{v} \in \mathcal{PC}[\mathbb{R}_+, \mathbb{R}_+^D]$ *such that any solution* $\mathbf{w}(t, t_0, \mathbf{w}_0, \mathbf{v}(t))$ *of*

$$\begin{aligned} \dot{\mathbf{x}} &= \mathbf{g}(t, \mathbf{w}, \mathbf{v}(t)), \quad t \ne \tau_k, \\ \mathbf{w}(\tau_k^+) &= \psi_k(\mathbf{w}(\tau_k), \mathbf{v}(\tau_k)), \quad k = 1, 2, \cdots, \\ \mathbf{w}(t_0) &= \mathbf{w}_0, \end{aligned} \tag{5.74}$$

satisfies

$$Q(\mathbf{w}_0) < \alpha(\mu) \text{ implies } Q(\mathbf{w}(t, t_0, \mathbf{w}_0, \mathbf{v})) < \beta(\nu), \ t \ge t_0, \tag{5.75}$$

and

$$Q(\mathbf{w}(t_0 + T, t_0, \mathbf{w}_0, \mathbf{v})) \leq \beta(\nu_1). \tag{5.76}$$

Then there exists an admissible control $\tilde{\mathbf{u}} = \tilde{\mathbf{u}}(t)$ such that system (5.45) is practically stable and all solutions $\mathbf{x}(t, t_0, \mathbf{x}_0, \tilde{\mathbf{u}})$ starting in $\Omega_1 = \{\mathbf{x} | \ ||\mathbf{x}|| < \mu\}$ are transferred to the region $\Omega_2 = \{\mathbf{x} | \ ||\mathbf{x}|| < \nu_1\}$ in a finite time T; namely, system (5.45) is controllable.

Proof. Assume that $||\mathbf{x}_0|| < \mu$ and there is a $\mathbf{v} = \mathbf{v}(t)$ such that assumption (6) holds. Let $\tilde{\mathbf{u}} = \tilde{\mathbf{u}}(t)$ be any control such that $U(t, \tilde{\mathbf{u}}) = \mathbf{v}(t), t \geq t_0$. If for any solution $\mathbf{x}(t, t_0, \mathbf{x}_0, \tilde{\mathbf{u}})$ of system (5.45)

$$||\mathbf{x}_0|| < \mu \text{ implies } ||\mathbf{x}(t, t_0, \mathbf{x}_0, \tilde{\mathbf{u}})|| < \nu, \qquad t \geq t_0$$

is false, there should exist a $\tilde{\mathbf{u}}^*(t)$ such that $\mathbf{v}(t) = U(t, \tilde{\mathbf{u}}^*)$. Correspondingly, there should be a solution $\mathbf{x}(t, t_0, \mathbf{x}_0, \tilde{\mathbf{u}}^*)$ and a $t^* > t_0, t^* \in (\tau_k, \tau_{k+1}]$ for some k such that

$$||\mathbf{x}(t^*, t_0, \mathbf{x}_0, \tilde{\mathbf{u}}^*)|| \geq \nu, \ ||\mathbf{x}(t, t_0, \mathbf{x}_0, \tilde{\mathbf{u}}^*)|| < \nu, \ t \in [t_0, \tau_k]. \tag{5.77}$$

Then, from assumption (4), there is a $t_1 \in (\tau_k, t^*]$, such that

$$\nu \leq ||\mathbf{x}(t_1, t_0, \mathbf{x}_0, \tilde{\mathbf{u}}^*)|| < \rho. \tag{5.78}$$

Let $m(t) = \mathbf{V}(t, \mathbf{x}(t))$ for $t \in [t_0, t_1]$ and $\mathbf{w}_0 = \mathbf{V}(t_0, \mathbf{x}_0)$, from assumption (3) and Lemma 5.4, we have

$$m(t) \leq \gamma(t, t_0, \mathbf{w}_0, \mathbf{v}), \qquad t \in [t_0, t_1], \tag{5.79}$$

where $\gamma(t, t_0, \mathbf{w}_0, \mathbf{v})$ is the maximal solution of Eq. (5.74). Then, from assumption (2) we have

$$\begin{aligned} Q(||\mathbf{x}(t, t_0, \mathbf{x}_0, \tilde{\mathbf{u}}^*)||) &\leq Q(\mathbf{V}(t, \mathbf{x})) \\ &\leq Q(\gamma(t, t_0, \mathbf{w}_0, \mathbf{v})), \qquad t \in [t_0, t_1]. \end{aligned} \tag{5.80}$$

Thus, Eqs. (5.75), (5.77), and (5.80) lead to the following contradiction:

$$\begin{aligned} \beta(\nu) &\leq \beta(||\mathbf{x}(t_1, t_0, \mathbf{x}_0, \tilde{\mathbf{u}}^*)||) \leq Q(\mathbf{V}(t_1, \mathbf{x}(t_1))) \\ &\leq Q(\gamma(t_1, t_0, \mathbf{w}_0, \mathbf{v})) < \beta(\nu) \end{aligned} \tag{5.81}$$

which proves the practical stability of system (5.45).

Since Eq. (5.80) holds for all $t \geq t_0$ and therefore by using Eq. (5.76), we have

$$\beta(\|\mathbf{x}(t_0+T, t_0, \mathbf{x}_0, \tilde{\mathbf{u}}^*)\|) \leq Q(\gamma(t_0+T, t_0, \mathbf{w}_0, \mathbf{v})) \leq \beta(\nu_1) \tag{5.82}$$

and thus $\|\mathbf{x}(t_0+T, t_0, \mathbf{x}_0, \tilde{\mathbf{u}}^*)\| \leq \nu_1$. This means that system (5.45) is controllable. □

5.5.1 *Examples*

Let a time-varying linear type-II impulsive control system be

$$\begin{aligned}
\dot{\mathbf{w}} &= A(t)\mathbf{w} + B(t)\tilde{\mathbf{u}}(t), \quad t \neq \tau_k, \\
\Delta\mathbf{w}(\tau_k) &= \left(\sum_{i=1}^{m} d_i(\tau_k)\tilde{u}_i(\tau_k)\right)\mathbf{w}(\tau_k), \quad k = 1, 2, \cdots, \\
\mathbf{w}(t_0) &= \mathbf{w}_0 \geq 0,
\end{aligned} \tag{5.83}$$

where $A(t)$ and $B(t)$ are $n \times n$ and $n \times m$ continuous matrices on $\mathbb{R}_+$, respectively. $\tilde{\mathbf{u}} \in \mathcal{PC}[\mathbb{R}_+, \mathbb{R}^m]$ and $d_i(\tau_k) \in \mathbb{R}$ for all k and i. Let $\Phi(t)$ be the fundamental matrix of Eq. (5.83), then use transformation $\mathbf{w} = \Phi(t)\mathbf{z}$ we have

$$\begin{aligned}
\dot{\mathbf{z}} &= \Phi^{-1}(t)B(t)\tilde{\mathbf{u}}(t), \quad t \neq \tau_k, \\
\Delta\mathbf{z}(\tau_k) &= \left(\sum_{i=1}^{m} d_i(\tau_k)\tilde{u}_i(\tau_k)\right)\mathbf{z}(\tau_k), \quad k = 1, 2, \cdots, \\
\mathbf{z}(t_0) &= \mathbf{w}_0.
\end{aligned} \tag{5.84}$$

The solution of Eq. (5.84) is given by

$$\begin{aligned}
\mathbf{z}(t) = &\prod_{t_0<\tau_k<t}\left(I + \sum_{i=1}^{m} d_i(\tau_k)\tilde{u}_i(\tau_k)\right)\mathbf{w}_0 \\
&+ \int_{t_0}^{t} \prod_{s<\tau_k<t}\left(I + \sum_{i=1}^{m} d_i(\tau_k)\tilde{u}_i(\tau_k)\right)\Phi^{-1}(s)B(s)\tilde{\mathbf{u}}(s)ds, \\
&t \geq t_0.
\end{aligned} \tag{5.85}$$

Let us choose $\tilde{\mathbf{u}}(\tau_k) = 0, k = 1, 2, \cdots$, then from Eq. (5.85) we have

$$\mathbf{z}(t) = \mathbf{w}_0 + \int_{t_0}^{t} \Phi^{-1}(s)B(s)\tilde{\mathbf{u}}(s)ds. \tag{5.86}$$

Let us choose $\tilde{\mathbf{u}}(t) = B(t)^\top \Phi^{-1}(t)^\top \mathbf{h} + \mathbf{v}(t)$, $t \neq \tau_k$, with

$$\int_{t_0}^{\infty} B(s)^\top \Phi^{-1}(s)^\top \mathbf{v}(s) ds = 0$$

for some vector $\mathbf{h}$, we have

$$\mathbf{z}(t) = \mathbf{w}_0 + \int_{t_0}^{t} \Phi^{-1}(s) B(s) [B(s)^\top \Phi^{-1}(s)^\top \mathbf{h} + \mathbf{v}(s)] ds. \tag{5.87}$$

Given $0 < \mu < \nu$ and $0 < \nu_1 < \nu$ and choose $\mathbf{h}$ such that

$$\int_{t_0}^{\infty} ||\Phi^{-1}(s) B(s) B(s)^\top \Phi^{-1}(s)^\top|| ds \leq \frac{\nu - \mu}{||\mathbf{h}||}.$$

Then from Eq. (5.87) and if $||\mathbf{w}_0|| < \mu$ we have

$$||\mathbf{z}(t)|| < \mu + \frac{\nu - \mu}{||\mathbf{h}||} = \nu, \qquad t \geq t_0.$$

Since $||\mathbf{w}(t)|| \leq ||\Phi(t)|| ||\mathbf{z}(t)||$ for $t \geq t_0$ by definition, we know that if $||\Phi(t)|| < 1$ for $t \geq t_0$, then system (5.83) is practical stable.

If for a given T satisfying $t_0 + T \neq \tau_k$ and $||\Phi(t_0 + T)|| \leq \nu_1/\nu$, then $\mathbf{w}(t_0 + T) < \nu_1$. Therefore system (5.83) is controllable.

5.6 Impulsive control of nonautonomous chaotic systems

5.6.1 *Theory*

Let a general nonautonomous chaotic system be

$$\dot{\mathbf{x}} = \mathbf{f}(t, \mathbf{x}, \tilde{\mathbf{u}}(t)) \tag{5.88}$$

where $\mathbf{f} : \mathbb{R}_+ \times \mathbb{R}^n \times \mathbb{R}^m \to \mathbb{R}^n$ is continuous. $\mathbf{x} \in \mathbb{R}^n$ is the state variable. $\tilde{\mathbf{u}} : \mathbb{R}_+ \to \mathbb{R}^m$ is an external force which is independent of the system. Suppose that a discrete instant set $\{\tau_i\}$ satisfies

$$0 < \tau_1 < \tau_2 < \cdots < \tau_i < \tau_{i+1} < \cdots, \ \tau_i \to \infty \text{ as } i \to \infty.$$

Let

$$\mathbf{u}_i(\mathbf{x}) = \Delta \mathbf{x}|_{t=\tau_i} \stackrel{\triangle}{=} \mathbf{x}(\tau_i^+) - \mathbf{x}(\tau_i) \tag{5.89}$$

be change of state variable at instant τ_i, then an impulsively controlled system is given by

$$\begin{cases} \dot{\mathbf{x}} = \mathbf{f}(t, \mathbf{x}, \tilde{\mathbf{u}}(t)), & t \neq \tau_i, \\ \Delta \mathbf{x} = \mathbf{u}_i(\mathbf{x}), & t = \tau_i, \\ \mathbf{x}(t_0^+) = \mathbf{x}_0, & t_0 \geq 0, \quad i = 1, 2, \cdots . \end{cases} \tag{5.90}$$

Definition 5.2 *Comparison system*
Let $V \in \mathcal{V}_0$ and assume that

$$\begin{cases} D^+V(t, \mathbf{x}) \leq g(t, V(t, \mathbf{x}), v(t)), & t \neq \tau_i, \\ V(t, \mathbf{x} + \mathbf{u}_i(\mathbf{x})) \leq \psi_i(V(t, \mathbf{x})), & t = \tau_i \end{cases} \tag{5.91}$$

where $g : \mathbb{R}_+ \times \mathbb{R}_+ \times \mathbb{R}_+ \to \mathbb{R}$ is continuous and $\psi_i : \mathbb{R}_+ \to \mathbb{R}_+$ is nondecreasing. Then the system

$$\begin{cases} \dot{w} = g(t, w, v(t)), & t \neq \tau_i, \\ w(\tau_i^+) = \psi_i(w(\tau_i)), & i = 1, 2, \cdots, \\ w(t_0^+) = w_0 \geq 0 \end{cases} \tag{5.92}$$

is the comparison system of Eq.(5.90).

Let the set Ω be

$$\Omega = \{\tilde{\mathbf{u}} \in \mathbb{R}^m \mid \Gamma(t, \tilde{\mathbf{u}}) \leq \gamma(t), t \geq t_0\} \tag{5.93}$$

where $\Gamma \in C[\mathbb{R}_+ \times \mathbb{R}^m, \mathbb{R}_+]$ and $\gamma(t)$ is the maximal solution of the comparison system (5.92).

Theorem 5.7 *Assume that*

(1) $g \in \mathcal{PC}[\mathbb{R}_+ \times \mathbb{R}_+ \times \mathbb{R}_+, \mathbb{R}_+]$, *$g(t, u, v)$ is nondecreasing in u for each (t, v) and nondecreasing in v for each (t, u);*
(2) *ψ_i is nondecreasing for each i;*
(3) *$0 < \mu < \nu$ (resp. $0 < \nu < \mu$) is given;*
(4) *$V \in \mathcal{PC}[\mathbb{R}_+ \times \mathbb{R}^n, \mathbb{R}_+]$, $V(t, \mathbf{x})$ is local Lipschitzian in $\mathbf{x}$, and there exists $\alpha, \beta \in \mathcal{K}$ such that*

$$\begin{aligned} &\beta(\|\mathbf{x}\|) \leq V(t, \mathbf{x}) \leq \alpha(\|\mathbf{x}\|), \\ &(t, \mathbf{x}) \in \mathbb{R}_+ \times \mathfrak{S}_\rho, \qquad \rho > \nu; \end{aligned} \tag{5.94}$$

(5) *for $(t, \mathbf{x}) \in (\tau_i, \tau_{i+1}] \times \mathfrak{S}_\rho$ and $\tilde{\mathbf{u}}(t) \in \Omega$*

$$D^+V(t, \mathbf{x}) \leq g(t, V(t, \mathbf{x}), \Gamma(t, \tilde{\mathbf{u}})); \tag{5.95}$$

(6) $\mathbf{x} \in \mathfrak{S}_\nu$ *implies* $\mathbf{x}+\mathbf{u}_i(\mathbf{x}) \in \mathfrak{S}_\rho$ *and* $V(\tau_i^+, \mathbf{x}+\mathbf{u}_i(\mathbf{x})) \le \psi_i(V(t,\mathbf{x}))$, $\mathbf{x} \in \mathfrak{S}_\rho$*;*

(7) $\alpha(\mu) < \beta(\nu)$ *(resp.,* $\alpha(\mu) > \beta(\nu)$*).*

Then, the practical stability properties of the comparison system (5.92), with respect to $(\alpha(\mu), \beta(\nu))$*, imply the corresponding practical stability properties of system (5.90) with respect to* (μ, ν) *for every* $\tilde{\mathbf{u}}(t) \in \Omega$*.*

Proof. Set $h(t,\mathbf{x}) = h_0(t,\mathbf{x}) = \|x\|$, from Theorem 5.1 we immediately get the conclusion. □

Remark: This theorem can also be viewed as a special case of Theorem 3.1 of [McR94]. Theorem 5.7 is very powerful because it translates the stable problem of an nth-order impulsive differential equation into that of a first-order impulsive differential equation. In many cases, this translation provides us an easy way to study the stability of a high-order impulsive differential equation. In next theorem, we present the stability criterion for a first-order impulsive differential equation, which is the general form of the comparison system of impulsively controlled chaotic systems we will study.

Theorem 5.8 *Let* $\delta_{\max}$ *be*

$$\delta_{\max} \triangleq \sup_{i \in \mathbb{N}} \{\tau_i - \tau_{i-1}\}. \tag{5.96}$$

For a given (μ, ν), $(0 < \mu < \nu)$ *or* $(0 < \nu < \mu)$,

(1) Let

$$\begin{aligned} g(t,w,v) &= \phi w + \theta, \quad \phi > 0, \quad \theta > 0, \quad t \ne \tau_i, \\ \psi_i(w) &= d_i w, \quad d_i > 0, \quad t = \tau_i, i = 1, 2, \cdots, \\ w(t_0^+) &= w_0 \ge 0. \end{aligned} \tag{5.97}$$

(2)

$$\mu \prod_{i=1}^{\infty} d_i e^{\phi(t-t_0)} + \frac{\theta}{\phi} \sum_{j=1}^{\infty} \prod_{i=j}^{\infty} d_i \left| e^{\phi(t-\tau_j)} - e^{\phi(t-\tau_{j-1})} \right| + \left| \frac{\theta}{\phi}(1 - e^{\phi\delta_{\max}}) \right| < \nu \tag{5.98}$$

Then the system (5.97) is practical stable with respect to (μ, ν) *for every* $v(t)$*.*

Proof. The solution of the comparison system (5.97) is given by[SP95]

$$\begin{aligned} w(t,t_0,w_0) &= w_0 \prod_{i=1}^{k} d_i e^{\phi(t-t_0)} + \frac{\theta}{\phi}\sum_{j=1}^{k}\prod_{i=j}^{k} d_i \left(e^{\phi(t-\tau_j)} - e^{\phi(t-\tau_{j-1})}\right) \\ &\quad + \frac{\theta}{\phi}(1-e^{\phi(t-\tau_k)}) \\ &< w_0 \prod_{i=1}^{k} d_i e^{\phi(t-t_0)} + \frac{\theta}{\phi}\sum_{j=1}^{k}\prod_{i=j}^{k} d_i \left|e^{\phi(t-\tau_j)} - e^{\phi(t-\tau_{j-1})}\right| \\ &\quad + \left|\frac{\theta}{\phi}(1-e^{\phi(t-\tau_k)})\right|, \\ &\quad t \in (\tau_k, \tau_{k+1}]. \end{aligned} \tag{5.99}$$

We have

$$\left|\frac{\theta}{\phi}(1-e^{\phi(t-\tau_k)})\right| < \left|\frac{\theta}{\phi}(1-e^{\phi\Delta_{max}})\right|. \tag{5.100}$$

Since $w_0 < \mu$ we have $w(t,t_0,w_0) < \nu$. □

5.7 Examples

5.7.1 *Example 1*

First, we study the following chaotic system:

$$\begin{cases} \dot{x} = y - f(x) \\ \dot{y} = -\epsilon x - \zeta y + \eta\sin(\omega t) \end{cases} \tag{5.101}$$

where $f(\cdot)$ is a piecewise linear function given by

$$f(x) = bx + \frac{1}{2}(a-b)(|x+1| - |x-1|) \tag{5.102}$$.

where $a < b < 0$. We choose the parameters as: $\epsilon = 1$, $\zeta = 1.015$ and $\omega = 0.75$, $\eta = 0.15$, $a = -1.02$, and $b = -0.55$. The uncontrolled system is chaotic with a chaotic attractor shown in Fig. 5.1.

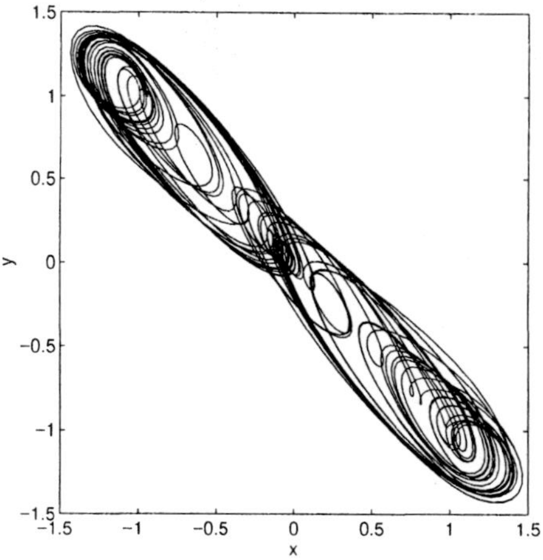

Fig. 5.1 The chaotic attractor of the uncontrolled chaotic system in Eq.(5.101).

The impulsively controlled chaotic system is given by

$$\left.\begin{aligned} \dot{x} &= y - f(x) \\ \dot{y} &= -\epsilon x - \zeta y + \eta \sin(\omega t) \end{aligned}\right\}, \quad t \neq \tau_i,$$
$$\left.\begin{aligned} x(\tau_i^+) &= d_i x(\tau_i) \\ y(\tau_i^+) &= d_i y(\tau_i) \end{aligned}\right\}, \quad d_i > 0, \quad t = \tau_i. \tag{5.103}$$

To give conditions for the practical stability of the impulsively controlled system in Eq.(5.103), we first construct a comparison system of (5.103). We will use Proposition 5.1 to show that the practical stability of the comparison system implies that of (5.103). Finally, we will use Theorem 5.9 to show that the comparison system is practical stable.

Proposition 5.1 *The practical stability of impulsively controlled chaotic system in Eq.(5.103) with respect to (μ, ν) is implied by that of the following comparison system*

$$\begin{aligned} g(t, w, v) &= \phi w + \theta, \phi = \max\{|\epsilon + |a||, |1 - \zeta|\}, \quad \theta = |\eta|, \quad t \neq \tau_i, \\ \psi_i(w) &= d_i w, \quad d_i > 0, \quad t = \tau_i, i = 1, 2, \cdots, \\ w(t_0^+) &= w_0 \geq 0 \end{aligned} \tag{5.104}$$

with respect to $(\sqrt{2}\mu, \nu)$.

Proof. Choose a Lyapunov function as[YY96]

$$V(t, \mathbf{x}) = |x| + |y| \tag{5.105}$$

then we have

$$\begin{aligned} D^+V(t, \mathbf{x}) &= \dot{x}\text{sgn}(x) + \dot{y}\text{sgn}(y) \\ &= y\text{sgn}(x) - f(x)\text{sgn}(x) - \epsilon x\text{sgn}(y) - \zeta|y| + \eta\text{sgn}(y)\sin(\omega t) \\ &\leq |a||x| - \zeta|y| + (y\text{sgn}(x) - \epsilon x\text{sgn}(y)) + \eta\text{sgn}(y)\sin(\omega t) \\ &\leq (\epsilon + |a|)|x| + (1 - \zeta)|y| + |\eta| \\ &\leq \max\{|\epsilon + |a||, |1 - \zeta|) + |\eta|. \end{aligned} \tag{5.106}$$

Notice that $-f(x)\text{sgn}(x) \leq |a||x|$ due to $a < b < 0$, hence

$$g(t, w, v) = \phi w + \theta \tag{5.107}$$

where $\phi = \max\{|\epsilon + |a||, |1 - \zeta|)$ and $\theta = |\eta|$.

Since

$$V(\tau_i^+, \mathbf{x}(\tau_i^+)) = d_i V(\tau_i, \mathbf{x}(\tau_i)) \tag{5.108}$$

we have $\psi_i(w) = d_i w$.

It is well-known that $|x| + |y| \geq \sqrt{x^2 + y^2}$. Since

$$2(x^2 + y^2) = 2(|x|^2 + |y|^2) \geq |x|^2 + 2|x||y| + |y|^2 = (|x| + |y|)^2 \tag{5.109}$$

we have $|x| + |y| \leq \sqrt{2}\sqrt{x^2 + y^2}$. We then have $\alpha(\chi) = \sqrt{2}\chi$ and $\beta(\chi) = \chi$. We find all conditions in Theorem 5.7 are satisfied. □

Theorem 5.9 *Let $d_i = d > 0$ be constant for all $i = 1, 2, \cdots$, the impulses are equi-distant with a fixed interval $\delta = \tau_{i+1} - \tau_i$ for all $i = 1, 2, \cdots$, and*

(1)

$$\frac{1}{\delta}\ln(d) + \phi < 0, \tag{5.110}$$

(2)

$$\frac{\theta}{\phi}|1 - e^{\phi\delta}|de^{\phi\delta}\frac{1}{1 - de^{\phi\delta}} + \frac{\theta}{\phi}|1 - e^{\phi\delta}| < \nu,$$

i.e,

$$\frac{\theta}{\phi}\frac{|1 - e^{\phi\delta}|}{1 - de^{\phi\delta}} < \nu, \tag{5.111}$$

then the impulsive controlled chaotic system (5.103) is practical stable with respect to (μ, ν) for any $\mu < \infty$.

Proof. Let

$$\begin{aligned}\xi(t) &= \left\lfloor \frac{t-t_0}{\delta} \right\rfloor \ln(d) + \phi(t-t_0) \\ &= \left\lfloor \frac{t-t_0}{\delta} \right\rfloor \ln(d) + \phi\left\{\delta\left\lfloor \frac{t-t_0}{\delta} \right\rfloor\right\} + \phi\left\{(t-t_0) - \delta\left\lfloor \frac{t-t_0}{\delta} \right\rfloor\right\} \\ &\leq \left\lfloor \frac{t-t_0}{\delta} \right\rfloor (\ln(d) + \delta\phi) + \delta\phi \end{aligned} \tag{5.112}$$

where $\lfloor \chi \rfloor$ denotes the largest integer less than or equal to χ.

Since $\frac{1}{\delta}\ln(d) + \phi < 0$ we have

$$\lim_{t\to\infty} \xi(t) = -\infty \tag{5.113}$$

then

$$\lim_{t\to\infty} \prod_{i=1}^{k} de^{\phi(t-t_0)} = \lim_{t\to\infty} e^{\lfloor \frac{t-t_0}{\delta} \rfloor \ln(d) + \phi(t-t_0)} = \lim_{t\to\infty} e^{\xi(t)} = 0, \tag{5.114}$$

from which we see that the first term in Eq. (5.98) becomes zero for any $\mu < \infty$.

For $t \in (\tau_k, \tau_{k+1}]$ we have

$$\begin{aligned} &\lim_{k\to\infty} \frac{\theta}{\phi} \sum_{j=1}^{k} \prod_{i=j}^{k} d|e^{\phi(t-\tau_j)} - e^{\phi(t-\tau_{j-1})}| \\ = \; & \frac{\theta}{\phi} \lim_{k\to\infty} |1 - e^{\phi\delta}| \sum_{j=1}^{k} \prod_{i=j}^{k} de^{\phi(t-\tau_j)} \\ = \; & \frac{\theta}{\phi} \lim_{k\to\infty} |1 - e^{\phi\delta}| \sum_{j=1}^{k} \prod_{i=j}^{k} de^{\phi(\tau_{k+1}-\tau_j)} \\ \leq \; & \frac{\theta}{\phi} |1 - e^{\phi\delta}| \lim_{k\to\infty} \sum_{j=1}^{k} d^j e^{j\phi\delta} \\ = \; & \frac{\theta}{\phi} |1 - e^{\phi\delta}| de^{\phi\delta} \frac{1}{1 - de^{\phi\delta}}. \end{aligned} \tag{5.115}$$

The last equation is satisfied if $de^{\phi\delta} < 1$, i.e., $\frac{1}{\delta}\ln(d) + \phi < 0$.

Since the w_0 can be any finite value, it follows from Theorem 5.8 that the comparison system is practical stable with respect to $(\sqrt{2}\mu, \nu)$ for any $\mu < \infty$. It follows from Proposition 5.1 that the impulsively controlled chaotic system (5.103) is practical stable with respect to (μ, ν) for any $\mu < \infty$. □

Remark: Since μ is used to decide the range of initial conditions, Theorem 5.9 guarantees that no matter how far the initial condition is away from the origin, the impulsively controlled chaotic system can be practically stabilized around the origin.

Simulation results

With the given parameters, we have $\phi = \epsilon + |a| = 2.02$. We choose $\delta = 0.1s$. According to Eq. (5.110), for practically stable control we should choose $d < e^{-0.202} \approx 0.817095$. We choose $\theta = 0.15$. The simulation results are show in Fig. 5.2. The fourth-order Runge-Kutta method with fixed time step $0.01s$ is used. The initial condition is $(x(0), y(0)) = (2, -1)$.

In Fig. 5.2(a) the practically stable result with small objective region is shown. We choose $d = 0.2$ and we have $A \approx 0.022010$. It means that under this condition, the maximum amplitudes of x and y of the impulsively controlled system are less than 0.022010. The dashed line and solid line show $x(t)$ and $y(t)$, respectively. We can see that $x(t)$ approach to 0 though $y(t)$ fluctuates around 0 with a small amplitude. The two horizontal dotted straight lines show the range of the states of the controlled system predicted by Theorem 5.9. We can see that Theorem 5.9 gives a very good estimate to the range of the states of the controlled system.

In Fig. 5.2(b) we show an unstable result. We choose $d = 0.95$ which is bigger than 0.817095. The dashed line and solid line show $x(t)$ and $y(t)$, respectively. In this case, we can see that both $x(t)$ and $y(t)$ go far away from the origin.

5.7.2 *Example 2*

We then study the following chaotic system:

$$\begin{cases} \dot{x} = -x + \epsilon f(x) - \zeta f(y) + \eta \sin(\omega t) \\ \dot{y} = -y + \epsilon f(y) - \zeta f(x) \end{cases} \tag{5.116}$$

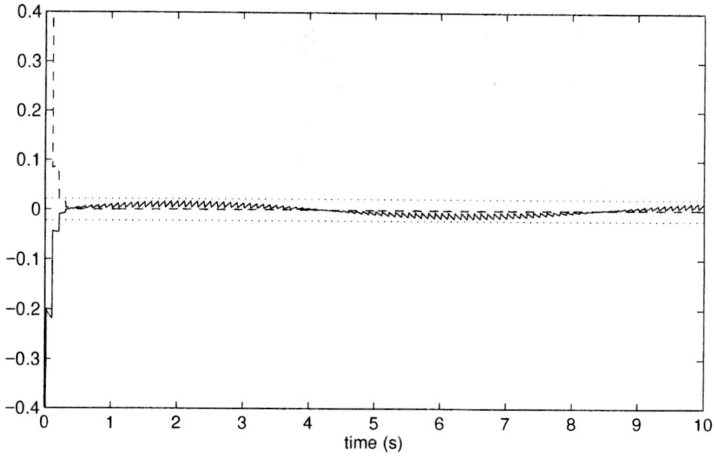

(a)

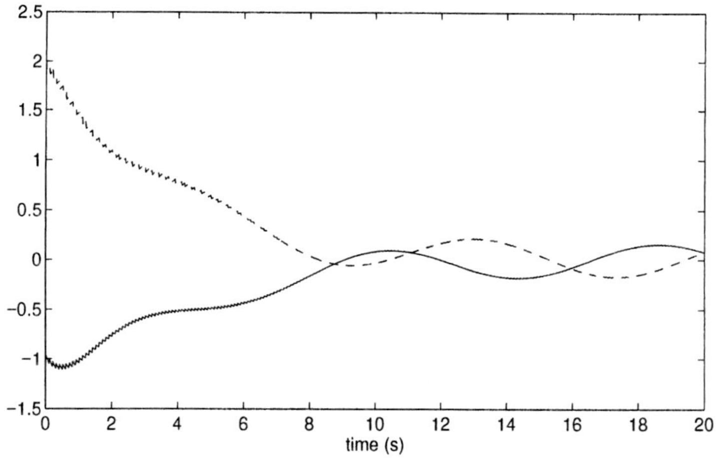

(b)

Fig. 5.2 The behaviors of the impulsively controlled chaotic system in Eq. (5.103). (a) The practically stable result with small objective region. (b) The unstable result.

where $f(\cdot)$ is a piecewise linear function given by

$$f(x) = \frac{1}{2}(|x+1| - |x-1|). \tag{5.117}$$

We choose the parameters as: $\epsilon = 2$, $\zeta = 1.2$, $\omega = \pi/2$, and $\eta = 4.04$. The uncontrolled system is chaotic with a chaotic attractor shown in Fig.5.3.

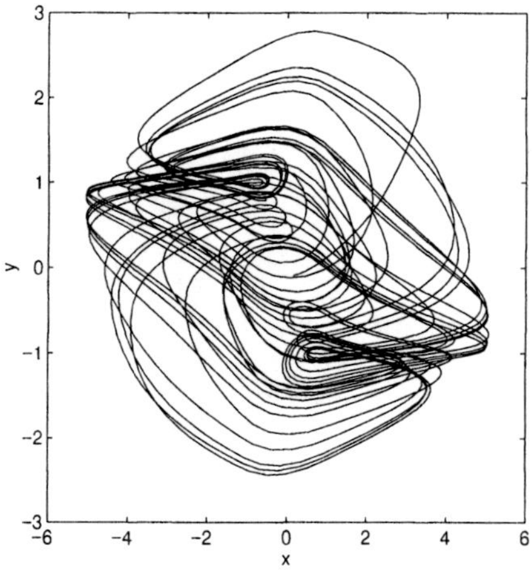

Fig. 5.3 The chaotic attractor of the uncontrolled chaotic system in Eq.(5.116).

The impulsively controlled chaotic system is given by

$$\left.\begin{array}{l}\dot{x} = -x + \epsilon f(x) - \zeta f(y) + \eta \sin(\omega t) \\ \dot{y} = -y + \epsilon f(y) - \zeta f(x)\end{array}\right\}, \quad t \neq \tau_i,$$
$$\left.\begin{array}{l}x(\tau_i^+) = d_i x(\tau_i) \\ y(\tau_i^+) = d_i y(\tau_i)\end{array}\right\}, \quad d_i > 0, \quad t = \tau_i. \tag{5.118}$$

Theorem 5.10 *The practical stability of impulsively controlled chaotic system in Eq.(5.118) with respect to* (μ, ν) *is implied by that of the following comparison system*

$$\begin{array}{rcl} g(t, w, v) & = & \phi w + \theta, \quad \phi = \epsilon - 1 + |\zeta|, \quad \theta = |\eta|, \quad t \neq \tau_i \\ \psi_i(w) & = & d_i w, \quad d_i > 0, \quad t = \tau_i, \quad i = 1, 2, \cdots, \\ w(t_0^+) & = & w_0 \geq 0 \end{array} \tag{5.119}$$

with respect to $(\sqrt{2}\mu, \nu)$.

Proof. Choose a Lyapunov function as[YY96]

$$V(t, \mathbf{x}) = |x| + |y|, \tag{5.120}$$

then we have

$$\begin{aligned} D^+V(t,\mathbf{x}) &= \dot{x}\mathrm{sgn}(x) + \dot{y}\mathrm{sgn}(y) \\ &= -|x| + \epsilon f(x)\mathrm{sgn}(x) - \zeta f(y)\mathrm{sgn}(x) + \eta\mathrm{sgn}(x)\sin(\omega t) \\ &\quad -|y| + \epsilon f(y)\mathrm{sgn}(y) + \zeta f(x)\mathrm{sgn}(y) \\ &\leq (\epsilon - 1 + |\zeta|)(|x| + |y|) + |\eta|, \end{aligned} \tag{5.121}$$

hence

$$g(t, w, v) = \phi w + \theta \tag{5.122}$$

where $\phi = \epsilon - 1 + |\zeta|$ and $\theta = |\eta|$. The rest part of the proof is the same as that in Proposition 5.1. □

Simulation results

With the given parameters, we have $\phi = 2.2$. We choose $\delta = 0.01s$. According to Eq.(5.142), for practically stable control we should choose $d < e^{-0.022} \approx 0.978240$. We choose $\theta = 4.04$. The simulation results are show in Fig. 5.4. The fourth-order Runge-Kutta method with fixed time step $0.002s$ is used. The initial condition is $(x(0), y(0)) = (2, -1)$.

In Fig. 5.4(a) the practically stable result with small objective region is shown. We choose $d = 0.01$ and we have $A \approx 0.041269$. It means that under this condition, the maximum amplitudes of x and y of the impulsively controlled system are less than 0.041269. The solid line and dashed line show $x(t)$ and $y(t)$, respectively. We can see that $y(t)$ approach to 0 though $x(t)$ fluctuates around 0 with a small amplitude. The two horizontal dotted straight lines show the range of the states of the controlled system predicted by Theorem 5.9. We can see that Theorem 5.9 gives a very good estimate to the range of the states of the controlled system.

In Fig. 5.4(b) we show an unstable result. We choose $d = 0.99$ which is bigger than 0.978240. The solid line and dashed line show $x(t)$ and $y(t)$, respectively. In this case, we can see that both $x(t)$ and $y(t)$ oscillate with big amplitudes.

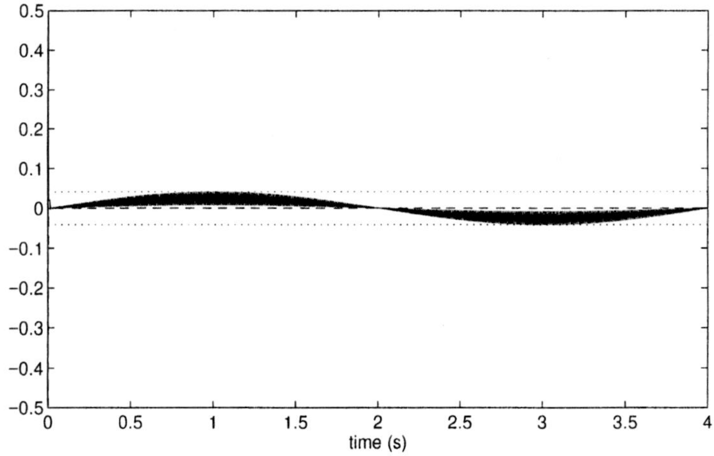

(a)

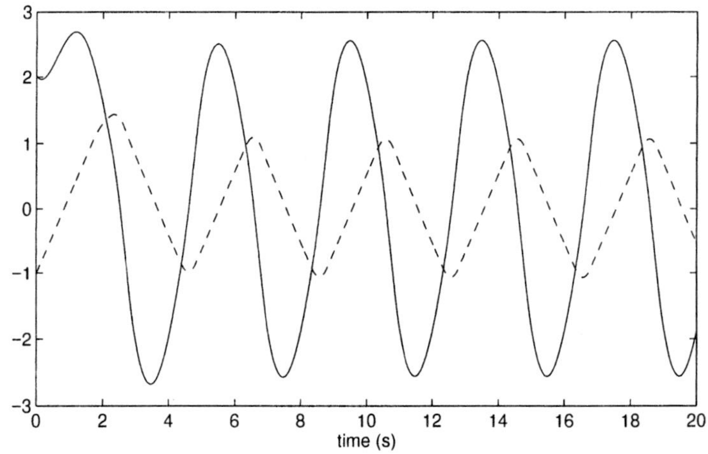

(b)

Fig. 5.4 The behaviors of the impulsively controlled chaotic system in Eq.(5.118). (a) The practically stable result with small objective region. (b) The unstable result.

5.7.3 *Example 3: Duffing's oscillator*

A Duffing's oscillator is given by

$$\begin{aligned} \dot{x} &= y, \\ \dot{y} &= x - x^3 - \epsilon y + \theta \cos(\omega t). \end{aligned} \tag{5.123}$$

We choose the parameters as: $\epsilon = 0.25$, $\theta = 0.3$ and $\omega = 1$. The uncontrolled system is a chaotic system whose chaotic attractor is shown in Fig.5.5.

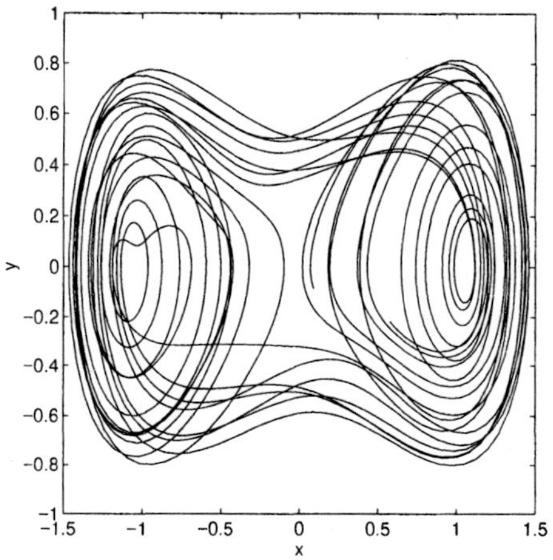

Fig. 5.5 The chaotic attractor of the uncontrolled Duffing's oscillator.

The impulsively controlled Duffing's oscillator is given by

$$\left.\begin{aligned} \dot{x} &= y \\ \dot{y} &= x - x^3 - \epsilon y + \theta\cos(\omega t) \end{aligned}\right\}, \quad t \neq \tau_i,$$
$$\left.\begin{aligned} x(\tau_i^+) &= d_i x(\tau_i) \\ y(\tau_i^+) &= d_i y(\tau_i) \end{aligned}\right\}, \quad d_i > 0, \quad t = \tau_i. \tag{5.124}$$

To give conditions for practical stability of the impulsively controlled system in Eq.(5.124), we first construct a comparison system of (5.124). We then use Proposition 5.2 to show that the practical stability of the comparison system implies that of (5.124). Finally, we use Theorem 5.11 to show that the comparison system is practically stable.

Proposition 5.2 *The practical stability of impulsively controlled Duffing's oscillator in Eq.(5.124) with respect to (μ, ν) is implied by that of the* *following comparison system*

$$\begin{aligned} g(t,w,v) &= \phi w+\theta, \quad \phi=\max\{1-\epsilon,1+x^2\}, \quad t\neq\tau_i, \\ \psi_i(w) &= d_i w, d_i>0, \quad t=\tau_i, \quad i=1,2,\cdots, \\ w(t_0^+) &= w_0\geq 0 \end{aligned} \tag{5.125}$$

with respect to $(\sqrt{2}\mu,\nu)$.

Proof. Choose a Lyapunov function as[YY96]

$$V(t,\mathbf{x})=|x|+|y|. \tag{5.126}$$

Then we have

$$\begin{aligned} D^+V(t,\mathbf{x}) &= \dot{x}\mathrm{sgn}(x)+\dot{y}\mathrm{sgn}(y) \\ &= y\mathrm{sgn}(x)+x\mathrm{sgn}(y)-x^3\mathrm{sgn}(y)-\epsilon|y|+\theta\sin(\omega t)\mathrm{sgn}(y) \\ &\leq (1-\epsilon)|y|+(1+x^2)|x|+|\theta\sin(\omega t)| \\ &\leq \max(1-\epsilon,1+x^2)V(t,\mathbf{x})+|\theta|. \end{aligned} \tag{5.127}$$

Hence

$$g(t,w,v)=\max(1-\epsilon,1+x^2)w+|\theta|=\phi w+\theta \tag{5.128}$$

where $\phi=\max(1-\epsilon,1+x^2)$ and $\theta=|\theta|$ because of $\theta>0$. The rest part of the proof is the same as that in Proposition 5.1. □

Theorem 5.11 *Let $d_i=d>0$ be constant for all $i=1,2,\cdots$, the impulses are equidistant with a fixed interval $\delta=\tau_{i+1}-\tau_i$ for all $i=1,2,\cdots$, and*

(1)

$$\frac{1}{\delta}\ln(d)+\phi<0, \tag{5.129}$$

(2)

$$\theta/\phi|1-e^{\phi\delta}|de^{\phi\delta}\frac{1}{1-de^{\phi\delta}}+\theta/\phi|1-e^{\phi\delta}|<\nu,$$

i.e,

$$\frac{\theta}{\phi}\frac{|1-e^{\phi\delta}|}{1-de^{\phi\delta}}<\nu, \tag{5.130}$$

then the impulsively controlled Duffing's oscillator is practically stable with respect to (μ, ν) *for any* $\mu < \infty$.

Proof. Since the comparison systems in Eqs. (5.104) and (5.125) has the same form, the proof is similar to that of Theorem 5.9. □

· *Numerical experiments*

We first study the relation between time interval δ and d for given values of A. If $\frac{theta}{\phi} > 0$, the conditions in Eqs. (5.129) and (5.130) are reduced to

$$d < e^{-\phi\delta} - \frac{\theta}{A\phi}|1 - e^{-\phi\delta}| \tag{5.131}$$

which specifies how the increase of δ decreases d. From Fig. 5.5 we see that $-1.5 < x < 1.5$ such that $\phi = 3.25$. By choosing $\theta = 0.3$, the boundaries of practical stability for different values of A are shown in Fig. 5.6. In Fig. 5.6 we only show the cases of $A = 0.01, 0.05, 0.1$ and 0.5. The region below the corresponding line is the practically stable region. We see that the lines intersect at the point $(\delta, d) = (0, 1)$. This means that if $d < 1$ is satisfied, we can impulsively control the Duffing's oscillator into any prescribed small region around the origin by choosing δ small enough. When A approaches 0, the curves approach the open interval on the vertical axis, i.e., $d \in (0, 1)$ and $\delta = 0$.

We then present simulation results to verify the conclusions shown in Fig. 5.6. The simulation results are show in Fig. 5.7. The fourth-order Runge-Kutta method with fixed time step $0.01s$ is used. The initial condition is $(x(0), y(0)) = (2, -2)$.

In Fig. 5.7(a) the practically stable result with small objective region is shown. We choose $A = 0.05$. By checking the second curve in Fig. 5.6, we find the point $(\delta, d) = (0.1, 0.2)$ is in the practically stable region. It means that under this condition, the maximal amplitudes of x and y of the impulsively controlled system are less than 0.05. The dashed line and solid line show $x(t)$ and $y(t)$, respectively. We can see that $x(t)$ approaches 0 though $y(t)$ fluctuates around 0 with a small amplitude. The two horizontal dotted straight lines show the range of the states of the controlled system predicted by Fig. 5.6. We see that Theorem 5.11 gives a very good estimate for the range of the states of the controlled system.

Although the Theorem 5.11 guarantees that the controlled trajectory

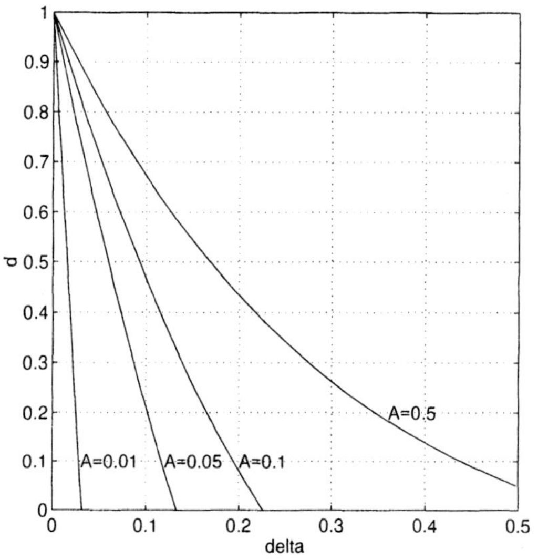

Fig. 5.6 The regions of practical stability under different A's.

will stay in the neighborhood of the origin. In order to give the reader an idea of the controlled attractor, in Fig. 5.7(b) we show the $x-y$ plot of controlled trajectories. We find that after the trajectory enters the controlled attractor, it never escapes.

In Fig. 5.7(c) a practically stable result with large objective region is shown. We choose $A = 0.1$. By checking the third curve in Fig. 5.6, we find that the point $(\delta, d) = (0.1, 0.4663)$ is in the practically stable region. It means that under this condition, the maximal amplitude of x and y of the impulsive controlled system are smaller than 0.1. The dashed line and solid line show $x(t)$ and $y(t)$, respectively. In this case, we see that $x(t)$ fluctuates around 0 with a visible amplitude and $y(t)$ fluctuates around 0 with a small amplitude. The two horizontal dotted straight lines show the range of the states of the controlled system predicted by Theorem 5.11.

5.8 Impulsive synchronization of nonautonomous nonlinear systems

In this section we study the impulsive synchronization of nonautonomous chaotic systems using practical stability. The practical stability of impulsive

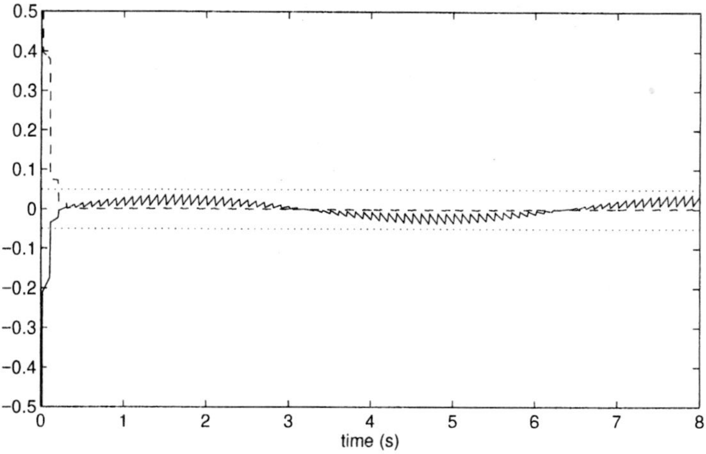

(a)

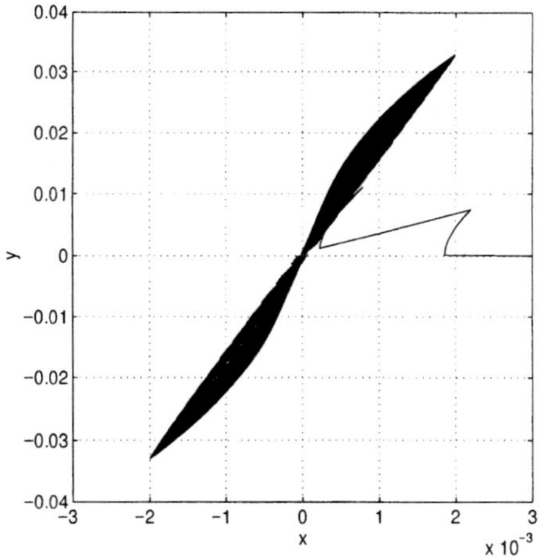

(b)

Fig. 5.7 The behaviors of the impulsively controlled Duffing's oscillator. (a) Practically stable result with small objective region. (b) $x - y$ plot of controlled trajectories. (c) Practically stable result with big objective region.

synchronization between two nonautonomous chaotic systems is equivalent to that of the origin of the synchronization error system, which is modeled by an impulsive differential equation.

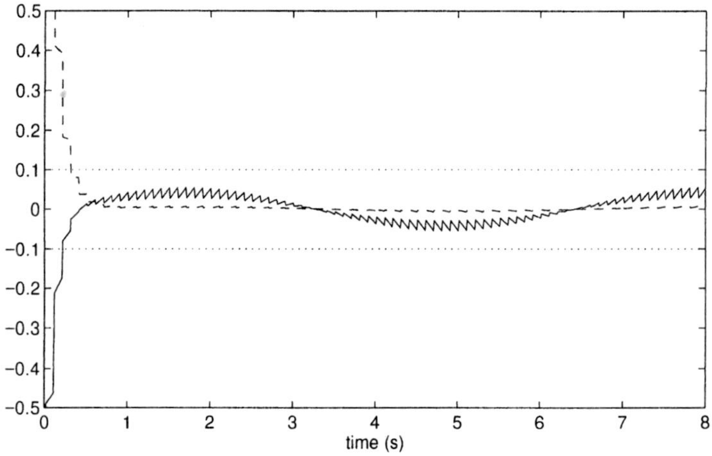

(c)

Fig. 5.7 (*Continued*)

5.8.1 *Practical impulsive synchronization of chaotic systems*

In this section, we study the following chaotic system[MLC94]:

$$\begin{cases} \dot{x} = y - f(x) \\ \dot{y} = -\epsilon x - \zeta y + \eta \sin(\omega t) \end{cases} \tag{5.132}$$

where $f(\cdot)$ is a piecewise linear function given by

$$f(x) = bx + \frac{1}{2}(a - b)(|x+1| - |x-1|) \tag{5.133}$$

where $a < b < 0$. We choose the parameters as: $\epsilon = 1$, $\zeta = 1.015$, $\omega = 0.75$, $\eta = 0.15$, $a = -1.02$, and $b = -0.55$. With these parameters, the system is chaotic with a chaotic attractor shown in Fig.5.1.

We only study the unidirectional synchronization, in which one chaotic system called *driving system* sends its states to the other chaotic system called *driven system* such that the states of the driven system will keep the same as those of the driving system in a practical sense. The driving system is given in Eq. (5.132), and the driven system is given by

$$\left.\begin{array}{l}\dot{\tilde{x}} = \tilde{y} - f(\tilde{x}) \\ \dot{\tilde{y}} = -\epsilon\tilde{x} - \zeta\tilde{y} + \eta\sin(\omega t + \tilde{\vartheta})\end{array}\right\}, \quad t \neq \tau_i,$$
$$\left.\begin{array}{l}\tilde{x}(\tau_i^+) = \tilde{x}(\tau_i) + (1-d_i)e_x(\tau_i) \\ \tilde{y}(\tau_i^+) = \tilde{y}(\tau_i) + (1-d_i)e_y(\tau_i)\end{array}\right\}, \quad d_i > 0, \quad t = \tau_i \tag{5.134}$$

where $\tilde{\vartheta}$ is a random phase difference between the external force of the driving system and that of the driven system. The synchronization error is defined as $\mathbf{e} = (e_x, e_y)^\top \triangleq (x - \tilde{x}, y - \tilde{y})^\top$. In the above equation we use an impulsive feedback control law. To study the practical stability of the synchronization error system around $\mathbf{e} = \mathbf{0}$, we derive the synchronization error system as

$$\left.\begin{array}{l}\dot{e_x} = e_y - [f(x) - f(\tilde{x})] \\ \dot{e_y} = -\epsilon e_x - \zeta e_y + \eta[\sin(\omega t) - \sin(\omega t + \tilde{\vartheta})]\end{array}\right\}, \quad t \neq \tau_i,$$
$$\left.\begin{array}{l}e_x(\tau_i^+) = d_i e_x(\tau_i) \\ e_y(\tau_i^+) = d_i e_y(\tau_i)\end{array}\right\}, \quad d_i > 0, \quad t = \tau_i. \tag{5.135}$$

Notice that

$$\begin{aligned} e_x(\tau_i^+) &= x(\tau_i^+) - \tilde{x}(\tau_i^+) \\ &= x(\tau_i) - [\tilde{x}(\tau_i) + (1-d_i)e_x(\tau_i)] \\ &= d_i e_x(\tau_i). \end{aligned} \tag{5.136}$$

Similarly, we have $e_y(\tau_i^+) = d_i e_y(\tau_i)$.

5.8.2 *Theory*

To find conditions for the practical stability of the synchronization error system in Eq.(5.135), we first construct a comparison system of (5.135). We will use Proposition 5.3 to show that the practical stability of the comparison system implies that of (5.135). Finally, we will use Theorem 5.12 to show that the comparison system is practically stable.

Proposition 5.3 *The practical stability of the synchronization error system in Eq.(5.135) with respect to (μ, ν), $\mu > 0$, and $\nu > 0$, is implied by* *that of the following comparison system:*

$$\begin{aligned} g(t,w,v) &= \phi w+\theta, \quad \phi=\max\{|\epsilon+|a||, |1-\zeta|\}, \\ &\quad \theta = 2|\eta|, \quad t\neq\tau_i, \\ \psi_i(w) &= d_i w, \quad d_i>0, \quad t=\tau_i, \quad i=1,2,\cdots, \\ w(t_0^+) &= w_0\geq 0 \end{aligned} \tag{5.137}$$

with respect to $(\sqrt{2}\mu, \nu)$.

Proof. Choose a Lyapunov function as

$$V(t,\mathbf{e}) = |e_x|+|e_y| \tag{5.138}$$

then we have

$$\begin{aligned} D^+V(t,\mathbf{e}) &= \dot{e}_x\mathrm{sgn}(e_x)+\dot{e}_y\mathrm{sgn}(e_y) \\ &= e_y\mathrm{sgn}(e_x)-[f(x)-f(\tilde{x})]\mathrm{sgn}(e_x)-\epsilon e_x\mathrm{sgn}(e_y) \\ &\quad -\zeta|e_y|+\eta\mathrm{sgn}(e_y)[\sin(\omega t)-\sin(\omega t+\tilde{\vartheta})] \\ &\leq |a||e_x|-\zeta|e_y|+(|e_y|-\epsilon|e_x|)\mathrm{sgn}(x)\mathrm{sgn}(y)+2|\eta| \\ &\leq (\epsilon+|a|)|e_x|+(1-\zeta)|e_y|+2|\eta| \\ &\leq \max\{|\epsilon+|a||, |1-\zeta|)V(t,\mathbf{e})+2|\eta|. \end{aligned} \tag{5.139}$$

Notice that $-[f(x)-f(\tilde{x})]\mathrm{sgn}(e_x)\leq |a||e_x|$ due to $a<b<0$, hence

$$g(t,w,v)=\phi w+\theta \tag{5.140}$$

where $\phi=\max\{|\epsilon+|a||, |1-\zeta|)$ and $\theta=2|\eta|$. The rest part of the proof is the same as that in Proposition 5.1. □

Remark: If we know the value of $\tilde{\vartheta}$ then we can give a more accurate estimate to θ. Observe that

$$\begin{aligned} \sin(\omega t)-\sin(\omega t+\tilde{\vartheta}) &= \sin(\omega t)(1-\cos(\tilde{\vartheta}))-\cos(\omega t)\sin(\tilde{\vartheta}) \\ &\leq |1-\cos(\tilde{\vartheta})|+|\sin(\tilde{\vartheta})|, \end{aligned} \tag{5.141}$$

we then have $\theta=(|1-\cos(\tilde{\vartheta})|+|\sin(\tilde{\vartheta})|)|\eta|$.

Theorem 5.12 *Let* $d_i=d>0$ *be constant for all* $i=1,2,\cdots$, *the impulses are equi-distant with a fixed time interval* $\delta=\tau_{i+1}-\tau_i$, $0<\delta<+\infty$ *for all* $i=1,2,\cdots$, *and*

(1)

$$\frac{1}{\delta}\ln(d) + \phi < 0, \tag{5.142}$$

(2)

$$\frac{\theta}{\phi}|1 - e^{\phi\delta}|de^{\phi\delta}\frac{1}{1 - de^{\phi\delta}} + \frac{\theta}{\phi}|1 - e^{\phi\delta}| < \nu,$$

i.e,

$$\frac{\theta}{\phi}\frac{|1 - e^{\phi\delta}|}{1 - de^{\phi\delta}} < \nu, \tag{5.143}$$

then the synchronization error system (5.135) is practically stable with respect to (μ, ν), $\mu > 0$, *and* $A > 0$, *for any* $\mu < +\infty$.

Proof. Since the comparison systems in Eqs. (5.104) and (5.137) has the same form, the proof is similar to that of Theorem 5.9. □

Remark: Theorem 5.12 guarantees that the practical stability of the impulsive synchronization is global.

5.8.3 *Numerical experimental results*

With the given parameters, we have $\phi = \epsilon + |a| = 2.02$. We choose $\delta = 0.1s$. According to Eq. (5.142), for practically stable synchronization we should choose $d < e^{-0.202} \approx 0.817095$. We choose $\theta = 0.15$. The fourth-order Runge-Kutta method with fixed time step $0.01s$ is used. The initial conditions are $(x(0), y(0)) = (2, -1)$ and $(\tilde{x}(0), \tilde{y}(0)) = (-2, 2)$. The random phase is chosen as $\tilde{\vartheta} = 1.21\pi$.

In Fig. 5.8 we show a stable result. We choose $d = 0.1$ and we have $A \approx 0.037881$. It means that under this condition, the maximum amplitudes of e_x and e_y are less than 0.037881. The simulation result is shown in Fig. 5.8(a). The dashed and solid curves show $e_x(t)$ and $e_y(t)$, respectively. Observe that $e_x(t)$ and $e_y(t)$ fluctuate around 0 with small amplitudes. The two horizontal dotted straight lines show the range of synchronization errors predicted by Theorem 5.12. Observe that Theorem 5.12 gives a very good estimate to the range of synchronization errors. The dynamics of impulsive synchronization is shown in Fig. 5.8(b) and (c) as the plots of x versus $\tilde{x}$ and y versus $\tilde{y}$, respectively.

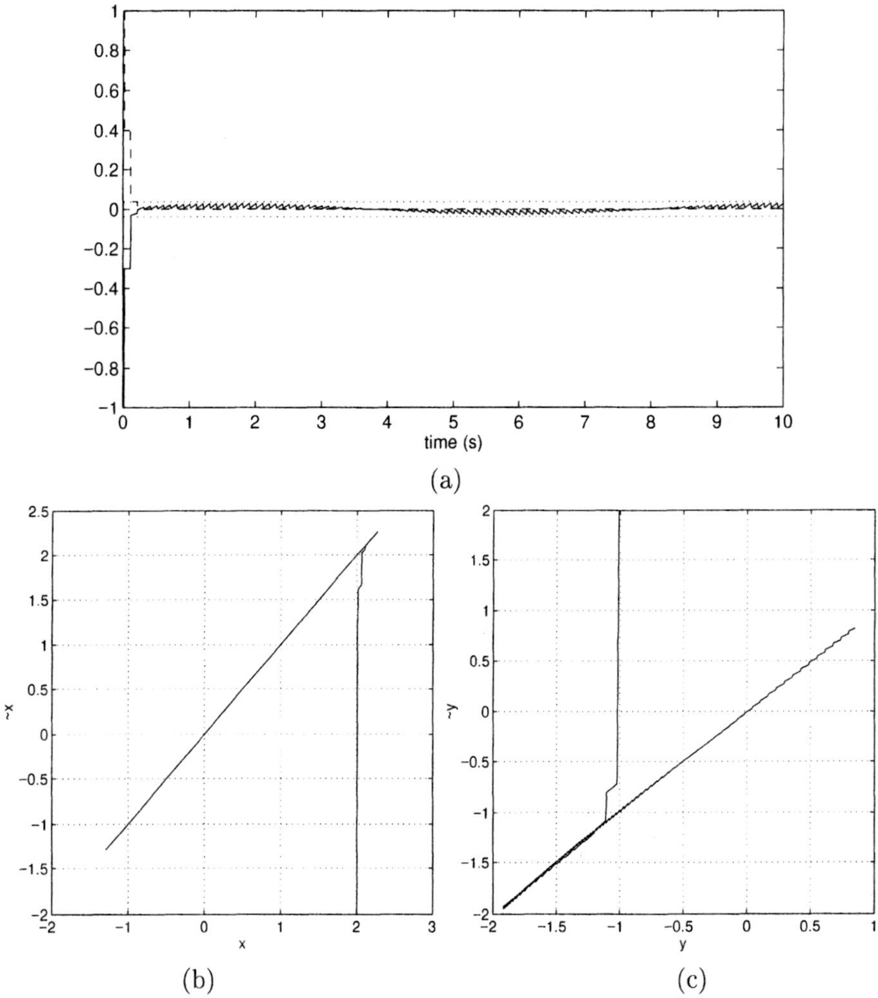

Fig. 5.8 Practically stable impulsive synchronization between two chaotic systems. (a) The dynamics of synchronization errors. (b) The x versus $\tilde{x}$ plot. (c) The y versus $\tilde{y}$ plot.

In Fig. 5.9 we show an unstable result. We choose $d = 0.9$ which is bigger than 0.817095. The simulation result is shown in Fig. 5.9(a). The dashed and solid curves show $e_x(t)$ and $e_y(t)$, respectively. In this case, we can see that both $e_x(t)$ and $e_y(t)$ go far away from the origin. We also show the plots of x versus $\tilde{x}$ and y versus $\tilde{y}$ in Fig. 5.9(b) and (c), respectively.

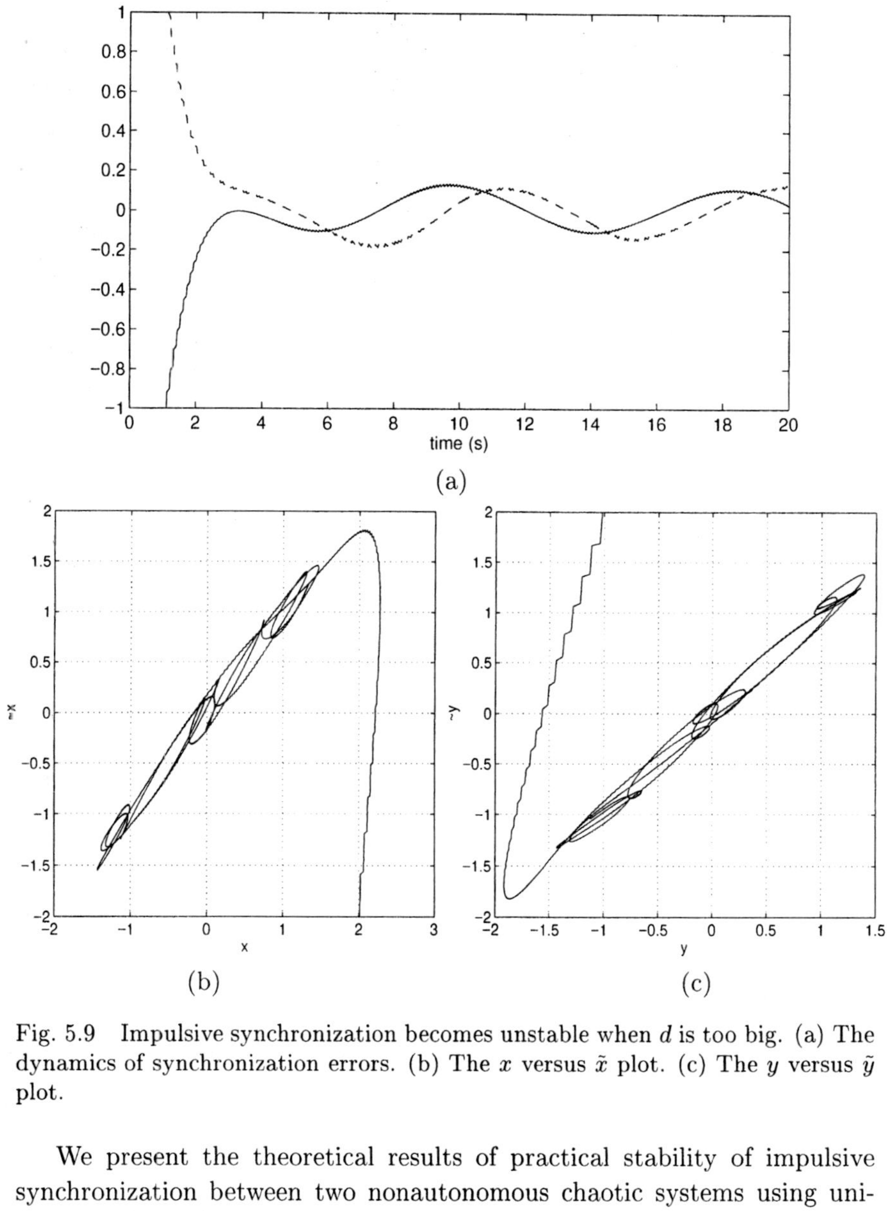

Fig. 5.9 Impulsive synchronization becomes unstable when d is too big. (a) The dynamics of synchronization errors. (b) The x versus $\tilde{x}$ plot. (c) The y versus $\tilde{y}$ plot.

We present the theoretical results of practical stability of impulsive synchronization between two nonautonomous chaotic systems using unidirectional synchronization scheme. The numerical experimental results correspond to what the theory predicts. Our theory is constructive because for any given range of impulsive synchronization error, it can give regions for choosing d and δ. Although some results are given under equidistant condition, varying interval cases can be similarly derived under the theoretical framework provided here.

Note

The definition and theory of practical stability in terms of two measures was presented in [Liu92]. The theory of practical stability in terms of multicomparison systems was presented in [McR94]. The practical stability of control and synchronization of nonautonomous chaotic systems was presented in [YSC98; YC00a; YY00].

Chapter 6

Global Stability of Impulsive Control

In this chapter we consider the global stability of the following impulsive control system:

$$\begin{cases} \dot{\mathbf{x}} = \mathbf{f}(t, \mathbf{x}) + \mathbf{u}(t, \mathbf{x}), & t \neq \tau_k, \\ \Delta \mathbf{x} = U(k, \mathbf{x}), & t = \tau_k, \\ \mathbf{x}(t_0^+) = \mathbf{x}_0, & t_0 \geq 0, \quad k = 1, 2, \cdots, \end{cases} \tag{6.1}$$

where $\mathbf{x} \in \mathbb{R}^n$, $\mathbf{f} \in C[\mathbb{R}^n, \mathbb{R}^n]$.

Definition 6.1 The zero solution of impulsive control system (6.1) is *globally asymptotically stable* (*globally stable* for short) if for any $\epsilon > 0$, there exists $\delta(\epsilon) > 0$ such that if $||\mathbf{x}_0|| < ||\delta(\epsilon)||$

$$||\mathbf{x}(t, t_0, \mathbf{x}_0)|| < \epsilon \quad \text{for all } t \geq t_0 \tag{6.2}$$

and for any $\mathbf{x}_0 \in \mathbb{R}^n$

$$\lim_{t \to +\infty} \mathbf{x}(t, t_0, \mathbf{x}_0) = 0. \tag{6.3}$$

Definition 6.2 The function $V(\mathbf{x})$ is said to be *positive definite*(resp., *negative definite*) if $V(\mathbf{x}) \geq 0$ (resp., $-V(\mathbf{x}) \geq 0$) and $V(\mathbf{x}) = 0$ if and only if $\mathbf{x} = 0$. The function $V(\mathbf{x})$ is said to be *positive semi-definite*(resp., *negative semi-definite*) if $V(\mathbf{x}) \geq 0$ (resp., $-V(\mathbf{x}) \geq 0$).

The positive or negative definite functions are usually called *Lyapunov functions*.

Definition 6.3 The function $V(\mathbf{x})$ is said to be *radially unbounded positive definite* if it is positive definite and $V(\mathbf{x}) \to +\infty$ as $||\mathbf{x}|| \to +\infty$.

Lemma 6.1 *Assume that $V(\mathbf{x})$ is positive definite, then there exist two functions $\alpha, \beta \in \mathcal{K}$ such that*

$$\alpha(\|\mathbf{x}\|) \leq V(\mathbf{x}) \leq \beta(\|\mathbf{x}\|). \tag{6.4}$$

Proof. See proof of Lemma 1.1.5 of [Lia93]. □

6.1 Plants with separable variables

In this section we study the following nonlinear impulsive control system with separable variables:

$$\begin{aligned} \dot{\mathbf{x}} &= \mathrm{col}\left(\sum_{j=1}^{n} f_{1j}(x_j), \sum_{j=1}^{n} f_{2j}(x_j), \cdots, \sum_{j=1}^{n} f_{nj}(x_j)\right), \quad t \neq \tau_k, \\ \Delta\mathbf{x} &= U(k, \mathbf{x}), \quad t = \tau_k, \\ \mathbf{x}(t_0^+) &= \mathbf{x}_0, \quad t_0 \geq 0, \quad k = 1, 2, \cdots, \end{aligned} \tag{6.5}$$

where $f_{ij} \in C[\mathbb{R}, \mathbb{R}]$, $f_{ij}(0) = 0$, $i, j = 1, 2, \cdots, n$, $U(k, \mathbf{x})$ is continuous and $U(k, 0) = 0$, $k \in \mathbb{N}$.

Theorem 6.1 *If there exist functions $\phi_i(x_i)$ on $(-\infty, +\infty)$ ($i = 1, 2, \cdots, n$), which are continuous or have only finite discontinuous points of the first or third kind such that*

(1) $\phi_i(x_i)x_i > 0$ *for* $x_i \neq 0$, $i = 1, 2, \cdots, n$;

(2)

$$\int_0^{\pm\infty} \phi_i(x_i) dx_i = +\infty, \qquad i = 1, 2, \cdots, n;$$

(3)

$$\sum_{i=1}^{n} \phi_i(x_i \pm 0) \sum_{j=1}^{n} f_{ij}(x_j) \leq -\alpha(\|\mathbf{x}\|), \qquad \alpha \in \mathcal{K};$$

(4)

$$\sum_{j=1}^{n} \int_0^{x_i} \phi_i(x_i + U_i(k, \mathbf{x})) dx_i \leq \sum_{j=1}^{n} \int_0^{x_i} \phi_i(x_i) dx_i.$$

Then the zero solution of system (6.5) is globally stable.

Proof. Let us construct the following Lyapunov function:

$$V(\mathbf{x}) = \sum_{i=1}^{n} \int_{0}^{x_i} \phi_i(x_i) dx_i. \tag{6.6}$$

We then have

$$D^+V(\mathbf{x}) = \begin{cases} \sum_{i=1}^n \phi_i(x_i) \sum_{j=1}^n f_{ij}(x_j), \\ \text{at the continuous points of } \phi_i(x_i),\ i = 1, 2, \cdots, n, \\ \max\left\{\sum_{i=1}^n \phi_i(x_i+0) \sum_{j=1}^n f_{ij}(x_j), \right. \\ \left. \quad \sum_{i=1}^n \phi_i(x_i-0) \sum_{j=1}^n f_{ij}(x_j)\right\}, \text{ otherwise.} \end{cases} \tag{6.7}$$

From assumption (3) we know that

$$D^+V(\mathbf{x}) \le -\alpha(\|\mathbf{x}\|). \tag{6.8}$$

Then from Theorem 2.9 we get the conclusion. □

Theorem 6.2 *Assume that*

(1) $f_{ii}(x_i)x_i < 0$ *for* $x_i \neq 0$, $i = 1, 2, \cdots, n$;

(2)

$$\int_0^{\pm\infty} f_{ii}(x_i) dx_i = -\infty, \qquad i = 1, 2, \cdots, n;$$

(3) for $i, j = 1, 2, \cdots, n$, *let*

$$a_{ij}(\mathbf{x}) \triangleq \begin{cases} -\frac{1}{2}\left(\frac{c_i f_{ij}(x_j)}{f_{jj}(x_j)} + \frac{c_j f_{ji}(x_i)}{f_{ii}(x_i)}\right), & x_i x_j \neq 0, \\ 0, & x_i x_j = 0, \end{cases}$$

where constants $c_i > 0$, $i = 1, 2, \cdots, n$, $\epsilon > 0$ *such that* $A(a_{ij}(\mathbf{x}))_{n\times n} + \epsilon I$ *is negative semidefinite and*

$$-\sum_{i=1}^{n} \epsilon f_{ii}^2(x_i) \le -\alpha(\|\mathbf{x}\|), \qquad \alpha \in \mathcal{K};$$

(4)

$$-\sum_{i=1}^{n} \int_0^{x_i} c_i f_{ii}(x_i + U_i(k, \mathbf{x})) dx_i \le -\sum_{i=1}^{n} \int_0^{x_i} c_i f_{ii}(x_i) dx_i.$$

Then the zero solution of system (6.5) is globally stable.

Proof. Construct the following Lyapunov function:

$$V(\mathbf{x}) = -\sum_{i=1}^{n} \int_0^{x_i} c_i f_{ii}(x_i) dx_i. \tag{6.9}$$

We then have

$$\begin{aligned}
D^+V(\mathbf{x}) &= -\sum_{i=1}^{n} c_i f_{ii}(x_i) \sum_{j=1}^{n} f_{ij}(x_j) \\
&= -\frac{1}{2}\sum_{i=1}^{n}\sum_{j=1}^{n}[c_i f_{ii}(x_i) f_{ij}(x_j) + c_j f_{jj}(x_j) f_{ji}(x_i)] \\
&= -\sum_{i=1}^{n} c_i f_{ii}^2(x_i) \\
&\quad -\sum_{\substack{i,j=1 \\ i\neq j}}^{n} \frac{1}{2}\left(\frac{c_i f_{ij}(x_j)}{f_{jj}(x_j)} + \frac{c_j f_{ji}(x_i)}{f_{ii}(x_i)}\right) f_{ii}(x_i) f_{jj}(x_j) \\
&= \sum_{i=1}^{n} a_{ii}(\mathbf{x}) f_{ii}^2(x_i) + \sum_{i=1}^{n} \epsilon f_{ii}^2(x_i) \\
&\quad + \sum_{\substack{i,j=1 \\ i\neq j}}^{n} a_{ij}(\mathbf{x}) f_{ii}(x_i) f_{jj}(x_j) - \sum_{i=1}^{n} \epsilon f_{ii}^2(x_i) \\
&\leq -\sum_{i=1}^{n} \epsilon f_{ii}^2(x_i) \leq -\alpha(\|\mathbf{x}\|).
\end{aligned} \tag{6.10}$$

Then from Theorem 2.9 we get the conclusion. □

Theorem 6.3 *Assume that*

(1) $f_{ii}(x_i)x_i < 0$ *for* $x_i \neq 0$, $i = 1, 2, \cdots, n$;

(2) there are functions $c_i(x_i)$, $i = 1, 2, \cdots, n$, *which are continuous or have only finite discontinuous points of the first or third kind, and satisfy* $c_i(x_i)x_i > 0$ *for* $x_i \neq 0$ *and* $\int_0^{\pm\infty} c_i(x_i) dx_i = +\infty$, $i = 1, 2, \cdots, n$;

(3) there exist functions $\epsilon_i(x_i) > 0$, $i = 1, 2, \cdots, n$, *such that*

$$A(a_{ij}(\mathbf{x}))_{n\times n} + \text{diag}(\epsilon_1(x_1), \epsilon_2(x_2), \cdots, \epsilon_n(x_n))$$

is negative semidefinite, where

$$a_{ij}(\mathbf{x}) \triangleq \begin{cases} \frac{1}{2}\left[\frac{c_i(x_i)f_{ij}(x_j)}{\sqrt{|f_{ii}(x_i)f_{jj}(x_j)|}} + \frac{c_j(x_j)f_{ji}(x_i)}{\sqrt{|f_{jj}(x_j)f_{ii}(x_i)|}}\right], & x_i x_j \neq 0, \\ 0, & x_i x_j = 0, \end{cases}$$

$i, j = 1, 2, \cdots, n$, *and*

$$-\sum_{i=1}^{n} \epsilon_i(x_i)|f_{ii}(x_i)| \leq -\alpha(\|\mathbf{x}\|), \qquad \alpha \in \mathcal{K};$$

(4)

$$\sum_{i=1}^{n} \int_0^{x_i} c_i(x_i + U_i(k, x))dx_i \leq \sum_{i=1}^{n} \int_0^{x_i} c_i(x_i)dx_i.$$

Then the zero solution of system (6.5) is globally stable.

Proof.

Construct the following Lyapunov function:

$$V(\mathbf{x}) = \sum_{i=1}^{n} \int_0^{x_i} c_i(x_i)dx_i. \tag{6.11}$$

$$\begin{aligned} D^+V(\mathbf{x}) &= \sum_{i=1}^{n} c_i(x_i) \sum_{j=1}^{n} f_{ij}(x_j) \\ &= \frac{1}{2}\sum_{i,j=1}^{n} [c_i(x_i)f_{ij}(x_j) + c_j(x_j)f_{ji}(x_i)] \\ &= \sum_{i,j=1}^{n} \frac{1}{2}\left(\frac{c_i(x_i)f_{ij}(x_j)}{\sqrt{|f_{ii}(x_i)f_{jj}(x_j)|}} + \frac{c_j(x_j)f_{ji}(x_i)}{\sqrt{|f_{ii}(x_i)f_{jj}(x_j)|}}\right)\sqrt{|f_{ii}(x_i)f_{jj}(x_j)|} \\ &= \sum_{i,j=1}^{n} \frac{1}{2}\left(\frac{c_i(x_i)f_{ij}(x_j)}{\sqrt{|f_{ii}(x_i)f_{jj}(x_j)|}} + \frac{c_j(x_j)f_{ji}(x_i)}{\sqrt{|f_{ii}(x_i)f_{jj}(x_j)|}}\right)\sqrt{|f_{ii}(x_i)f_{jj}(x_j)|} \\ &\quad + \sum_{i=1}^{n} \epsilon_i(x_i)\sqrt{|f_{ii}(x_i)f_{ii}(x_i)|} - \sum_{i=1}^{n} \epsilon_i(x_i)\sqrt{|f_{ii}(x_i)f_{ii}(x_i)|} \\ &\leq -\sum_{i=1}^{n} \epsilon_i(x_i)\sqrt{|f_{ii}(x_i)f_{ii}(x_i)|} \leq -\alpha(\|\mathbf{x}\|). \end{aligned} \tag{6.12}$$

Then from Theorem 2.9 we get the conclusion. □

Theorem 6.4 *Assume that*

(1) $f_{ii}(x_i)x_i < 0$ *for* $x_i \neq 0$, $i = 1, 2, \cdots, n$*;*
(2) there exist constants $c_i > 0, i = 1, 2, \cdots, n$*, such that*

$$\sum_{i=1}^{n}\left(-c_i|f_{ii}(x_i)| + \sum_{\substack{j=1\\ j\neq i}}^{n} c_i|f_{ij}(x_j)|\right) \leq -\alpha\left(\sum_{i=1}^{n} c_i|x_i|\right),$$

$$\alpha \in \mathcal{K}; \tag{6.13}$$

(3)

$$\sum_{i=1}^{n} c_i|x_i + U_i(k,x)| \leq \sum_{i=1}^{n} c_i|x_i|.$$

Then the zero solution of system (6.5) is globally stable.

Proof. Construct the following Lyapunov function:

$$V(\mathbf{x}) = \sum_{i=1}^{n} c_i|x_i|. \tag{6.14}$$

We then have

$$\begin{aligned} D^+V(\mathbf{x}) &= \sum_{i=1}^{n} c_i \mathrm{sgn}(x_i) \sum_{j=1}^{n} f_{ij}(x_j) \\ &= \sum_{i=1}^{n}\left(-c_i|f_{ii}(x_i)| + \sum_{\substack{j=1\\ j\neq i}}^{n} c_i \mathrm{sgn}(x_i) f_{ij}(x_j)\right) \\ &\leq \sum_{i=1}^{n}\left(-c_i|f_{ii}(x_i)| + \sum_{\substack{j=1\\ j\neq i}}^{n} c_i|f_{ij}(x_j)|\right) \\ &\leq -\alpha(V(\mathbf{x})) \end{aligned} \tag{6.15}$$

Then from Corollary 2.1 we get the conclusion. □

Theorem 6.5 *Assume that*

(1) $f_{ii}(x_i)x_i < 0$ *for* $x_i \neq 0$, $i = 1, 2, \cdots, n$*;*

(2) there exist constants $c_i > 0, i = 1, 2, \cdots, n$, such that

$$\sum_{i=1}^{n}\left(-c_i|f_{ii}(x_i)| + \sum_{\substack{j=1\\ j\neq i}}^{n} c_i|f_{ij}(x_j)|\right) \leq -\alpha(\|\mathbf{x}\|),$$
$$\alpha \in \mathcal{K}; \tag{6.16}$$

(3)

$$\sum_{i=1}^{n} c_i|x_i + U_i(k,x)| \leq \sum_{i=1}^{n} c_i|x_i|.$$

Then the zero solution of system (6.5) is globally stable.

Proof. Construct the following Lyapunov function:

$$V(\mathbf{x}) = \sum_{i=1}^{n} c_i|x_i|. \tag{6.17}$$

Using the similar process of the proof of Theorem 6.4, we have

$$\begin{aligned} D^+V(\mathbf{x}) &\leq \sum_{i=1}^{n}\left(-c_i|f_{ii}(x_i)| + \sum_{\substack{j=1\\ j\neq i}}^{n} c_i|f_{ij}(x_j)|\right) \\ &\leq -\alpha(\|\mathbf{x}\|) \end{aligned} \tag{6.18}$$

Then from Theorem 2.9 we get the conclusion. □

We then consider the global stability of the following more specific system:

$$\begin{aligned} \dot{\mathbf{x}} &= \mathrm{col}\left(\sum_{j=1}^{n} a_{1j}f_j(x_j), \sum_{j=1}^{n} a_{2j}f_j(x_j), \cdots, \sum_{j=1}^{n} a_{nj}f_j(x_j)\right), \quad t \neq \tau_k, \\ \Delta\mathbf{x} &= U(k,\mathbf{x}), \quad t = \tau_k, \\ \mathbf{x}(t_0^+) &= \mathbf{x}_0, \quad t_0 \geq 0, \ k = 1, 2, \cdots, \end{aligned} \tag{6.19}$$

where $f_j \in C[\mathbb{R}, \mathbb{R}]$, $f_j(0) = 0$, $j = 1, 2, \cdots, n$, $U(k, \mathbf{x})$ is continuous and $U(k, 0) = 0$, $k \in \mathbb{N}$.

Theorem 6.6 *Assume that*

(1) $f_i(x_i)x_i > 0$ *for* $x_i \neq 0$,

$$\int_0^{\pm\infty} f_i(x_i)dx_i = +\infty, a_{ii} < 0,$$

$i = 1, 2, \cdots, n$;

(2) there exist constants $c_i > 0, i = 1, 2, \cdots, n$, $\epsilon > 0$, *such that* $B(b_{ij})_{n\times n} + \epsilon I$ *is negative semidefinite, where*

$$b_{ij} \triangleq \begin{cases} c_i a_{ii}, & i = j, \quad i, j = 1, 2, \cdots, n \\ \frac{1}{2}(c_i a_{ij} + c_j a_{ji}), & i \neq j, \quad i, j = 1, 2, \cdots, n, \end{cases} \tag{6.20}$$

and

$$-\sum_{i=1}^{n} \epsilon f_i^2(x_i) \leq -\alpha(\|\mathbf{x}\|);$$

(3)

$$\sum_{i=1}^{n} \int_0^{x_i} c_i f_i(x_i + U_i(k, x))dx_i \leq \sum_{i=1}^{n} \int_0^{x_i} c_i f_i(x_i)dx_i.$$

Then the zero solution of system (6.19) is globally stable.

Proof. Construct the following Lyapunov function:

$$V(\mathbf{x}) = \sum_{i=1}^{n} c_i \int_0^{x_i} f_i(x_i)dx_i. \tag{6.21}$$

$$\begin{aligned} D^+V(\mathbf{x}) &= \sum_{i=1}^{n} c_i f_i(x_i) \sum_{j=1}^{n} a_{ij} f_j(x_j) \\ &= \frac{1}{2} \sum_{i,j=1}^{n} [c_i f_i(x_i) a_{ij} f_j(x_j) + c_j f_j(x_j) a_{ji} f_i(x_i)] \\ &= \sum_{i=1}^{n} c_i a_{ii} f_i^2(x_i) + \sum_{\substack{i,j=1 \\ i\neq j}}^{n} \frac{1}{2}(c_i a_{ij} + c_j a_{ji}) f_i(x_i) f_j(x_j) \\ &\quad + \sum_{i=1}^{n} \epsilon f_i^2(x_i) - \sum_{i=1}^{n} \epsilon f_i^2(x_i) \\ &\leq -\sum_{i=1}^{n} \epsilon f_i^2(x_i) \leq -\alpha(\|\mathbf{x}\|). \end{aligned} \tag{6.22}$$

Then from Theorem 2.9 we get the conclusion. □

Theorem 6.7 *Assume that*

(1) $f_i(x_i)x_i < 0$ *for* $x_i \neq 0$, $a_{ii} > 0$, $i = 1, 2, \cdots, n$;
(2) *there are* n *functions* $c_i(x_i)$, $i = 1, 2, \cdots, n$, *which are continuous or have only finite discontinuous points of the first or third kind, and satisfy* $c_i(x_i)x_i > 0$ *for* $x_i \neq 0$ *and* $\int_0^{\pm\infty} c_i(x_i)dx_i = +\infty$, $i = 1, 2, \cdots, n$;
(3) *there exist functions* $\epsilon_i(x_i) > 0$, $i = 1, 2, \cdots, n$, *such that*

$$A(a_{ij}(\mathbf{x}))_{n\times n} + \mathrm{diag}(\epsilon_1(x_1), \epsilon_2(x_2), \cdots, \epsilon_n(x_n))$$

is negative semidefinite, where

$$a_{ij}(\mathbf{x}) \triangleq \begin{cases} \frac{1}{2}\left[\frac{c_i(x_i)a_{ij}f_j(x_j)}{\sqrt{|f_i(x_i)f_j(x_j)|}} + \frac{c_j(x_j)a_{ji}f_i(x_i)}{\sqrt{|f_j(x_j)f_i(x_i)|}}\right], & x_ix_j \neq 0, \\ 0, & x_ix_j = 0, \end{cases}$$

$i, j = 1, 2, \cdots, n$, *and*

$$-\sum_{i=1}^{n} \epsilon_i(x_i)|f_i(x_i)| \leq -\alpha(||\mathbf{x}||), \qquad \alpha \in \mathcal{K};$$

(4)

$$\sum_{i=1}^{n} \int_0^{x_i} c_i(x_i + U_i(k, x))dx_i \leq \sum_{i=1}^{n} \int_0^{x_i} c_i(x_i)dx_i.$$

Then the zero solution of system (6.19) is globally stable.

Proof. Construct the following Lyapunov function:

$$V(\mathbf{x}) = \sum_{i=1}^{n} \int_0^{x_i} c_i(x_i)dx_i. \tag{6.23}$$

$$\begin{aligned} D^+V(\mathbf{x}) &= \sum_{i=1}^n c_i(x_i)\sum_{j=1}^n a_{ij}f_j(x_j) \\ &= \frac{1}{2}\sum_{i,j=1}^n [c_i(x_i)a_{ij}f_j(x_j) + c_j(x_j)a_{ji}f_i(x_i)] \\ &= \sum_{i,j=1}^n \frac{1}{2}\left(\frac{c_i(x_i)a_{ij}f_j(x_j)}{\sqrt{|f_i(x_i)f_j(x_j)|}} + \frac{c_j(x_j)a_{ji}f_i(x_i)}{\sqrt{|f_j(x_j)f_i(x_i)|}}\right)\sqrt{|f_i(x_i)f_j(x_j)|} \\ &\quad + \sum_{i=1}^n \epsilon_i(x_i)|f_i(x_i)| - \sum_{i=1}^n \epsilon_i(x_i)|f_i(x_i)| \\ &\leq -\sum_{i=1}^n \epsilon_i(x_i)|f_i(x_i)| \leq -\alpha(\|\mathbf{x}\|). \end{aligned} \tag{6.24}$$

Then from Theorem 2.9 we get the conclusion. □

Theorem 6.8 *Assume that*

(1) $f_i(x_i)x_i > 0$ *for* $x_i \neq 0$, $a_{ii} < 0$, $i = 1, 2, \cdots, n$;
(2) there exist constants $c_i > 0, i = 1, 2, \cdots, n$, *such that*

$$\sum_{i=1}^n \left(-c_i|a_{ii}f_i(x_i)| + \sum_{\substack{j=1\\ j\neq i}}^n c_i|a_{ij}f_j(x_j)|\right) \leq -\alpha\left(\sum_{i=1}^n c_i|x_i|\right),$$

$$\alpha \in \mathcal{K}; \tag{6.25}$$

(3)

$$\sum_{i=1}^n c_i|x_i + U_i(k,x)| \leq \sum_{i=1}^n c_i|x_i|.$$

Then the zero solution of system (6.19) is globally stable.

Proof. Construct the following Lyapunov function:

$$V(\mathbf{x}) = \sum_{i=1}^n c_i|x_i|. \tag{6.26}$$

$$
\begin{aligned}
D^+V(\mathbf{x}) &= \sum_{i=1}^{n} c_i \mathrm{sgn}(x_i) \sum_{i=1}^{n} a_{ij} f_j(x_j) \\
&= \sum_{i=1}^{n} \left(-c_i |a_{ii} f_i(x_i)| + c_i \mathrm{sgn}(x_i) \sum_{\substack{j=1 \\ j \neq i}}^{n} a_{ij} f_j(x_j) \right) \\
&\leq \sum_{i=1}^{n} \left(-c_i |a_{ii} f_i(x_i)| + c_i \sum_{\substack{j=1 \\ j \neq i}}^{n} |a_{ij} f_j(x_j)| \right) \\
&\leq -\alpha \left(\sum_{i=1}^{n} c_i |x_i| \right) = -\alpha(V(\mathbf{x})).
\end{aligned} \tag{6.27}
$$

Then from Corollary 2.1 we get the conclusion. □

6.2 Plants with generalized separable variables

In this section we consider the global stability of the following generalized separable variable system:

$$
\begin{aligned}
\dot{\mathbf{x}} &= \mathrm{col}\left(\sum_{j=1}^{n} F_{1j}(\mathbf{x}) f_{1j}(x_j), \sum_{j=1}^{n} F_{2j}(\mathbf{x}) f_{2j}(x_j), \cdots, \right. \\
&\qquad \left. \sum_{j=1}^{n} F_{nj}(\mathbf{x}) f_{nj}(x_j) \right), \qquad t \neq \tau_k, \\
\Delta \mathbf{x} &= U(k, \mathbf{x}), \qquad t = \tau_k, \\
\mathbf{x}(t_0^+) &= \mathbf{x}_0, \qquad t_0 \geq 0, \quad k = 1, 2, \cdots,
\end{aligned} \tag{6.28}
$$

where $F_{ij} \in C[\mathbb{R}^n, \mathbb{R}]$, $f_{ij} \in C[\mathbb{R}, \mathbb{R}]$, $f_j(0) = 0$, $i, j = 1, 2, \cdots, n$, $U(k, \mathbf{x})$ is continuous, $U(k, 0) = 0$, $k \in \mathbb{N}$.

Theorem 6.9 *Assume that*

(1) $f_{ii}(x_i)x_i > 0$ *for* $x_i \neq 0$, $i = 1, 2, \cdots, n$;

(2)

$$\int_0^{\pm\infty} f_{ii}(x_i)dx_i = +\infty, \qquad i = 1, 2, \cdots, n;$$

(3) for $i, j = 1, 2, \cdots, n$, *let*

$$a_{ij}(\mathbf{x}) \triangleq \begin{cases} \frac{1}{2}\left(\frac{F_{ij}(\mathbf{x})f_{ij}(x_j)}{f_{jj}(x_j)} + \frac{F_{ji}(\mathbf{x})f_{ji}(x_i)}{f_{ii}(x_i)}\right), & x_i x_j \neq 0, \\ 0, & x_i x_j = 0, \end{cases}$$

and $\epsilon > 0$ *such that* $A(a_{ij}(\mathbf{x}))_{n\times n} + \epsilon I$ *is negative semidefinite and*

$$-\sum_{i=1}^{n} \epsilon f_{ii}^2(x_i) \leq -\alpha(\|\mathbf{x}\|), \qquad \alpha \in \mathcal{K};$$

(4)

$$\sum_{i=1}^{n} \int_0^{x_i} f_{ii}(x_i + U_i(k, \mathbf{x}))dx_i \leq \sum_{i=1}^{n} \int_0^{x_i} f_{ii}(x_i)dx_i.$$

Then the zero solution of system (6.28) is globally stable.

Proof. Construct the following Lyapunov function:

$$V(\mathbf{x}) = \sum_{i=1}^{n} \int_0^{x_i} f_{ii}(x_i)dx_i. \tag{6.29}$$

We then have

$$\begin{aligned}
D^+V(\mathbf{x}) &= \sum_{i=1}^{n} f_{ii}(x_i) \sum_{j=1}^{n} F_{ij}(\mathbf{x}) f_{ij}(x_j) \\
&= \sum_{i=1}^{n} F_{ii}(\mathbf{x}) f_{ii}^2(x_i) \\
&\quad + \frac{1}{2} \sum_{\substack{i,j=1 \\ i \neq j}}^{n} [f_{ii}(x_i) F_{ij}(\mathbf{x}) f_{ij}(x_j) + f_{jj}(x_j) F_{ji}(\mathbf{x}) f_{ji}(x_i)] \\
&= \sum_{i=1}^{n} a_{ii}(\mathbf{x}) f_{ii}^2(x_i) \\
&\quad + \sum_{\substack{i,j=1 \\ i \neq j}}^{n} \frac{1}{2} \left(\frac{F_{ij}(\mathbf{x}) f_{ij}(x_j)}{f_{jj}(x_j)} + \frac{F_{ji}(\mathbf{x}) f_{ji}(x_i)}{f_{ii}(x_i)} \right) f_{ii}(x_i) f_{jj}(x_j) \\
&= \sum_{i=1}^{n} a_{ii}(\mathbf{x}) f_{ii}^2(x_i) + \sum_{i=1}^{n} \epsilon f_{ii}^2(x_i) \\
&\quad + \sum_{\substack{i,j=1 \\ i \neq j}}^{n} a_{ij}(\mathbf{x}) f_{ii}(x_i) f_{jj}(x_j) - \sum_{i=1}^{n} \epsilon f_{ii}^2(x_i) \\
&\leq -\sum_{i=1}^{n} \epsilon f_{ii}^2(x_i) \leq -\alpha(\|\mathbf{x}\|).
\end{aligned} \tag{6.30}$$

Then from Theorem 2.9 we get the conclusion. □

Theorem 6.10 *Assume that*

(1) $F_{ii}(\mathbf{x}) f_{ii}(x_i) x_i < 0$ *for* $x_i \neq 0,\ i = 1, 2, \cdots, n$;
(2) there exist constants $c_i > 0, i = 1, 2, \cdots, n$, *such that*

$$\begin{aligned}
&\sum_{i=1}^{n} \left(-c_i |F_{ii}(\mathbf{x}) f_{ii}(x_i)| + \sum_{\substack{j=1 \\ j \neq i}}^{n} c_i |F_{ij}(\mathbf{x}) f_{ij}(x_j)| \right) \\
\leq\ & -\alpha \left(\sum_{i=1}^{n} c_i |x_i| \right), \\
& \alpha \in \mathcal{K}, j = 1, 2, \cdots, n;
\end{aligned} \tag{6.31}$$

(3)

$$\sum_{i=1}^{n} c_i|x_i + U_i(k,x)| \leq \sum_{i=1}^{n} c_i|x_i|.$$

Then the zero solution of system (6.28) is globally stable.

Proof. Construct the following Lyapunov function:

$$V(\mathbf{x}) = \sum_{i=1}^{n} c_i|x_i|. \tag{6.32}$$

$$\begin{aligned} D^+V(\mathbf{x}) &= \sum_{i=1}^{n} c_i \mathrm{sgn}(x_i) \sum_{j=1}^{n} F_{ij}(\mathbf{x}) f_{ij}(x_j) \\ &= \sum_{i=1}^{n} \left(-c_i|F_{ii}(\mathbf{x}) f_{ii}(x_i)| + c_i \mathrm{sgn}(x_i) \sum_{\substack{j=1 \\ j\neq i}}^{n} F_{ij}(\mathbf{x}) f_{ij}(x_j) \right) \\ &\leq \sum_{i=1}^{n} \left(-c_i|F_{ii}(\mathbf{x}) f_{ii}(x_i)| + \sum_{\substack{j=1 \\ j\neq i}}^{n} c_i|F_{ij}(\mathbf{x}) f_{ij}(x_j)| \right) \\ &\leq -\alpha\left(\sum_{i=1}^{n} c_i|x_i|\right) = -\alpha(V(\mathbf{x})). \end{aligned} \tag{6.33}$$

Then from Corollary 2.1 we get the conclusion. □

Similarly, we have

Theorem 6.11 *Assume that*

(1) $F_{ii}(\mathbf{x}) f_{ii}(x_i) x_i < 0$ *for* $x_i \neq 0$, $i = 1, 2, \cdots, n$;
(2) there exist constants $c_i > 0, i = 1, 2, \cdots, n$, *such that*

$$\begin{aligned} &\sum_{i=1}^{n} \left(-c_i|F_{ii}(\mathbf{x}) f_{ii}(x_i)| + \sum_{\substack{j=1 \\ j\neq i}}^{n} c_i|F_{ij}(\mathbf{x}) f_{ij}(x_j)| \right) \\ \leq\ & -\alpha(\|\mathbf{x}\|), \\ & \alpha \in \mathcal{K}, j = 1, 2, \cdots, n; \end{aligned} \tag{6.34}$$

(3)

$$\sum_{i=1}^{n} c_i|x_i + U_i(k,x)| \leq \sum_{i=1}^{n} c_i|x_i|.$$

Then the zero solution of system (6.28) is globally stable.

6.3 Nonautonomous plants with separable variables

In this section we study the following nonautonomous impulsive control system with separable variables:

$$\begin{aligned}
\dot{\mathbf{x}} &= \operatorname{col}\left(\sum_{j=1}^{n} f_{1j}(t,x_j), \sum_{j=1}^{n} f_{2j}(t,x_j), \cdots, \sum_{j=1}^{n} f_{nj}(t,x_j)\right), \quad t \neq \tau_k, \\
\Delta\mathbf{x} &= U(k,\mathbf{x}), \quad t = \tau_k, \\
\mathbf{x}(t_0^+) &= \mathbf{x}_0, \quad t_0 \geq 0, \quad k = 1, 2, \cdots, \qquad (6.35)
\end{aligned}$$

where $f_{ij} \in C[\mathbb{R}_+ \times \mathbb{R}, \mathbb{R}]$, $f_{ij}(t,0) = 0$, $i,j = 1,2,\cdots,n$, $U(k,\mathbf{x})$ is continuous and $U(k,0) = 0$, $k \in \mathbb{N}$.

Theorem 6.12 *If there exist functions $\phi_i(x_i)$ on $(-\infty,+\infty)$ ($i = 1,2,\cdots,n$), which are continuous or have only finite discontinuous points of the first or third kind such that*

(1) $\phi_i(x_i)x_i > 0$ *for* $x_i \neq 0$, $i = 1,2,\cdots,n$;
(2)

$$\int_0^{\pm\infty} \phi_i(x_i)dx_i = +\infty, \qquad i = 1,2,\cdots,n;$$

(3)

$$\sum_{i=1}^{n} \phi_i(x_i \pm 0) \sum_{j=1}^{n} f_{ij}(t,x_j) \leq -\alpha(\|\mathbf{x}\|), \qquad \alpha \in \mathcal{K};$$

(4)

$$\sum_{i=1}^{n} \int_0^{x_i} \phi_i(x_i + U_i(k,\mathbf{x}))dx_i \leq \sum_{i=1}^{n} \int_0^{x_i} \phi_i(x_i)dx_i.$$

Then the zero solution of system (6.35) is globally stable.

Proof. Let us construct the following Lyapunov function:

$$V(\mathbf{x}) = \sum_{i=1}^{n} \int_0^{x_i} \phi_i(x_i)dx_i. \tag{6.36}$$

We then have

$$D^+V(\mathbf{x}) = \begin{cases} \sum_{i=1}^n \phi_i(x_i) \sum_{j=1}^n f_{ij}(t,x_j), \\ \quad \text{at the continuous points of } \phi_i(x_i),\ i=1,2,\cdots,n, \\ \max\left\{\sum_{i=1}^n \phi_i(x_i+0) \sum_{j=1}^n f_{ij}(t,x_j), \right. \\ \quad \left. \sum_{i=1}^n \phi_i(x_i-0) \sum_{j=1}^n f_{ij}(t,x_j)\right\}, \text{ otherwise.} \end{cases} \tag{6.37}$$

From assumption (3) we know that

$$D^+V(\mathbf{x}) \le -\alpha(\|\mathbf{x}\|). \tag{6.38}$$

Then from Theorem 2.9 we get the conclusion. □

Theorem 6.13 *Assume that*

(1) $f_{ii}(t,x_i)x_i < 0$ *for* $x_i \neq 0$, $i=1,2,\cdots,n$;
(2) there are functions $F_{ii}(x_i)$ *on* $(-\infty,\infty)$, $i=1,2,\cdots,n$, *which are continuous or have only finite discontinuous points of the first or third kind such that* $F_{ii}(x_i)x_i > 0$ *for* $x_i \neq 0$, $\int_0^{\pm\infty} F_{ii}(x_i)dx_i = +\infty$, $|F_{ii}(x_i)| \le |f_{ii}(t,x_i)|$, $i=1,2,\cdots,n$;
(3) the matrix $A(a_{ij}(t,\mathbf{x}))_{n\times n} + \epsilon I$ *is negative semi-definite with* $0 < \epsilon \ll 1$ *and for* $i,j=1,2,\cdots,n$, *we have*

$$a_{ij}(t,\mathbf{x}) = \begin{cases} -1, & i=j, \\ \frac{1}{2}\left(\frac{f_{ij}(t,x_j)}{F_{jj}(x_j)} + \frac{f_{ji}(t,x_i)}{F_{ii}(x_i)}\right), & i\neq j,\quad x_ix_j \neq 0, \\ 0, & i \neq j, \quad x_ix_j = 0, \end{cases}$$

and

$$-\sum_{i=1}^n \epsilon F_{ii}^2(x_i) \le -\alpha(\|\mathbf{x}\|), \qquad \alpha \in \mathcal{K};$$

(4)

$$\sum_{i=1}^n \int_0^{x_i} F_{ii}(x_i + U_i(k,\mathbf{x}))dx_i \le \sum_{i=1}^n \int_0^{x_i} F_{ii}(x_i)dx_i.$$

Then the zero solution of system (6.35) is globally stable.

Proof. Let us construct the following Lyapunov function:

$$V(\mathbf{x}) = \sum_{i=1}^{n} \int_{0}^{x_i} F_{ii}(x_i)dx_i. \tag{6.39}$$

We then have

$$\begin{aligned} D^+V(\mathbf{x}) &= \sum_{i=1}^{n} F_{ii}(x_i) \sum_{j=1}^{n} f_{ij}(t, x_j) \\ &\leq \sum_{i=1}^{n} a_{ii}(t, \mathbf{x}) F_{ii}^2(x_i) + \sum_{i=1}^{n} \epsilon F_{ii}^2(x_i) \\ &\quad + \sum_{\substack{i,j=1 \\ i \neq j}}^{n} a_{ij}(t, \mathbf{x}) F_{ii}(x_i) F_{jj}(x_j) - \sum_{i=1}^{n} \epsilon F_{ii}^2(x_i) \\ &\leq -\sum_{i=1}^{n} \epsilon F_{ii}^2(x_i) \leq -\alpha(||\mathbf{x}||). \end{aligned} \tag{6.40}$$

Then from Theorem 2.9 we get the conclusion. □

Theorem 6.14 *Assume that*

(1) $f_{ii}(t, x_i)x_i < 0$ *for* $x_i \neq 0$, $i = 1, 2, \cdots, n$;
(2) there are constants $c_i > 0$ *such that*

$$\sum_{i=1}^{n} \left(-c_i |f_{ii}(t, x_i)| + \sum_{\substack{j=1 \\ j \neq i}}^{n} c_i |f_{ij}(t, x_j)| \right) \leq -\alpha \left(\sum_{i=1}^{n} c_i |x_i| \right);$$

(3)

$$\sum_{i=1}^{n} c_i |x_i + U_i(k, \mathbf{x})| \leq \sum_{i=1}^{n} c_i |x_i|.$$

Then the zero solution of system (6.35) is globally stable.

Proof. Let us construct the following Lyapunov function:

$$V(\mathbf{x}) = \sum_{i=1}^{n} c_i |x_i|. \tag{6.41}$$

We then have

$$\begin{aligned}
D^+V(\mathbf{x}) &= \sum_{i=1}^{n} c_i \operatorname{sgn}(x_i) \sum_{j=1}^{n} f_{ij}(t, x_j) \\
&= \sum_{i=1}^{n} \left(-c_i |f_{ii}(t, x_i)| + c_i \operatorname{sgn}(x_i) \sum_{\substack{j=1 \\ j \neq i}}^{n} f_{ij}(t, x_j) \right) \\
&\leq \sum_{i=1}^{n} \left(-c_i |f_{ii}(t, x_i)| + \sum_{\substack{j=1 \\ j \neq i}}^{n} c_i |f_{ij}(t, x_j)| \right) \\
&\leq -\alpha \left(\sum_{i=1}^{n} c_i |x_i| \right) = -\alpha(V(\mathbf{x})). \qquad (6.42)
\end{aligned}$$

Then from Corollary 2.1 we get the conclusion. □

Similarly, we have the following theorem:

Theorem 6.15 *Assume that*

(1) $f_{ii}(t, x_i)x_i < 0$ *for* $x_i \neq 0$, $i = 1, 2, \cdots, n$;

(2) there are constants $c_i > 0$ *such that*

$$\sum_{i=1}^{n} \left(-c_i |f_{ii}(t, x_i)| + \sum_{\substack{j=1 \\ j \neq i}}^{n} c_i |f_{ij}(t, x_j)| \right) \leq -\alpha(\|\mathbf{x}\|);$$

(3)

$$\sum_{i=1}^{n} c_i |x_i + U_i(k, \mathbf{x})| \leq \sum_{i=1}^{n} c_i |x_i|.$$

Then the zero solution of system (6.35) is globally stable.

Theorem 6.16 *Assume that*

(1) there exist functions $\phi_i(x_i)$ *on* $(-\infty, +\infty)$ *(* $i = 1, 2, \cdots, n$*), which are continuous or have only finite discontinuous points of the first or third kind such that* $\phi_i(x_i)x_i > 0$ *for* $x_i \neq 0$ *and*

$$\int_0^{\pm\infty} \phi_i(x_i) dx_i = +\infty, \qquad i = 1, 2, \cdots, n;$$

(2) there are functions $a_i(x_i) > 0$ *for* $x_i \neq 0$, $i = 1, 2, \cdots, n$, *such that*

$$\sum_{i=1}^{n} \phi_i(x_i) f_{ij}(t, x_j) \leq -a_j(x_j), \qquad j = 1, 2, \cdots, n,$$

and

$$-\sum_{j=1}^{n} a_j(x_j) \leq -\alpha(\|\mathbf{x}\|), \qquad \alpha \in \mathcal{K};$$

(3)

$$\sum_{i=1}^{n} \int_0^{x_i} \phi_i(x_i + U_i(k, \mathbf{x})) dx_i \leq \sum_{i=1}^{n} \int_0^{x_i} \phi_i(x_i) dx_i.$$

Then the zero solution of system (6.35) is globally stable.

Proof. Let us construct the following Lyapunov function:

$$V(\mathbf{x}) = \sum_{i=1}^{n} \int_0^{x_i} \phi_i(x_i) dx_i. \tag{6.43}$$

Then we have

$$\begin{aligned} D^+V(\mathbf{x}) &\leq \sum_{i=1}^{n} \phi_i(x_i) \sum_{j=1}^{n} f_{ij}(t, x_j) \\ &\leq -\sum_{j=1}^{n} a_j(x_j) \leq -\alpha(\|\mathbf{x}\|) \end{aligned} \tag{6.44}$$

Then from Theorem 2.9 we get the conclusion. □

Chapter 7

Absolute Stability of Impulsive Control

In this chapter we study the absolute stability of impulsive Lur'e control systems given by

$$\begin{aligned}
\dot{\mathbf{x}} &= A\mathbf{x} + \mathbf{b}f(\sigma), \quad \sigma = \mathbf{c}^\top \mathbf{x} = \sum_{i=1}^{n} c_i x_i, \qquad t \neq \tau_k \\
\Delta \mathbf{x} &= U(k, \mathbf{x}), \qquad t = \tau_k, \\
\mathbf{x}(t_0^+) &= \mathbf{x}_0, \qquad t_0 \geq 0, \quad k = 1, 2, \cdots,
\end{aligned} \tag{7.1}$$

where $\mathbf{x} = \mathrm{col}(x_1, x_2, \cdots, x_n)$, $\mathbf{b} = \mathrm{col}(b_1, b_2, \cdots, b_n)$ and $\mathbf{c} = \mathrm{col}(c_1, c_2, \cdots, c_n)$ are n-dimensional real vectors, and $A = (a_{ij})_{n\times n}$ is a real matrix. $U(k, \mathbf{x})$ is continuous and $U(k, 0) = 0$, $k \in \mathbb{N}$.

We use the following notations:

$$\begin{aligned}
\mathcal{F} &\triangleq \{f |\ f(0) = 0, \sigma f(\sigma) > 0 \text{ for } \sigma \neq 0, f \in C[(-\infty, +\infty), \mathbb{R}]\}, \\
\mathcal{F}_\kappa &\triangleq \{f |\ f(0) = 0, 0 \leq f(\sigma)/\sigma \leq \kappa < +\infty \\
&\qquad \text{for } \sigma \neq 0, f \in C[(-\infty, +\infty), \mathbb{R}]\}, \\
\mathcal{F}_{\kappa_1,\kappa_2} &\triangleq \{f |\ f(0) = 0, 0 \leq \kappa_1 \leq f(\sigma)/\sigma \leq \kappa_2 < +\infty \\
&\qquad \text{for } \sigma \neq 0, f \in C[(-\infty, +\infty), \mathbb{R}]\}.
\end{aligned} \tag{7.2}$$

Definition 7.1 The zero solution of system (7.1) is said to be ***absolutely stable*** if for any $f \in \mathcal{F}$, the zero solution of system (7.1) is globally stable. The zero solution of system (7.1) is said to be ***absolutely stable in the Hurwitz angle*** $[0, \kappa]$ (***absolutely stable in*** $[0, \kappa]$ for short) if for any $f \in \mathcal{F}_\kappa$, the zero solution of system (7.1) is globally stable. The zero solution of system

(7.1) is said to be *absolutely stable in* $[\kappa_1, \kappa_2]$ if for any $f \in \mathcal{F}_{\kappa_1,\kappa_2}$, the zero solution of system (7.1) is globally stable.

7.1 General theory

In this section we study the following system:

$$\begin{aligned} \dot{\mathbf{x}} &= A\mathbf{x} + \mathbf{b}f(\sigma), \quad \sigma = \mathbf{c}^\top \mathbf{x} = \sum_{i=1}^{n} c_i x_i, \qquad t \neq \tau_k \\ \Delta \mathbf{x} &= U(k, \mathbf{x}), \qquad t = \tau_k, \\ \mathbf{x}(t_0^+) &= \mathbf{x}_0, \qquad t_0 \geq 0, \quad k = 1, 2, \cdots, \end{aligned} \tag{7.3}$$

where $f \in \mathcal{F}$, or $f \in \mathcal{F}_\kappa$. $U(k, \mathbf{x})$ is continuous and $U(k, 0) = 0$, $k \in \mathbb{N}$.

Definition 7.2 The zero solution of system (7.3) is said to be *absolutely stable for the set* $\Omega \triangleq \{\mathbf{x}|\ \sigma = 0\}$ (resp. *absolutely stable for* Ω *in* $[0, \kappa]$) if for every $\epsilon > 0$, there exists $\delta(\epsilon) > 0$ such that for any $f(\sigma) \in \mathcal{F}$ (resp., for any $f(\sigma) \in \mathcal{F}_\kappa$) and $\|\mathbf{x}_0\| < \delta(\epsilon)$ the solution $\mathbf{x}(t) = \mathbf{x}(t, t_0, \mathbf{x}_0)$ of system (7.3) satisfies $|\sigma(t, t_0, \mathbf{x}_0)| = |\mathbf{c}^\top \mathbf{x}(t, t_0, \mathbf{x}_0)| < \epsilon$ for all $t \geq t_0$. And for any $\mathbf{x}_0 \in \mathbb{R}^n$ we have

$$\lim_{t \to +\infty} \sigma(t, t_0, \mathbf{x}_0) = \lim_{t \to +\infty} \mathbf{c}^\top \mathbf{x}(t, t_0, \mathbf{x}_0) = 0.$$

Let $\Psi(t, s)$ be the state transition matrix of the following reference impulsive system of (7.3):

$$\begin{aligned} \dot{\mathbf{x}} &= (A + \mathbf{b}\mathbf{c}^\top \theta)\mathbf{x}, \qquad t \neq \tau_k \\ \Delta \mathbf{x} &= U(k, \mathbf{x}), \qquad t = \tau_k, \\ \mathbf{x}(t_0^+) &= \mathbf{x}_0, \qquad t_0 \geq 0, \quad k = 1, 2, \cdots, \end{aligned} \tag{7.4}$$

where $\theta = 0$ or $\theta = 1$(resp., $\theta = 0$ or $\theta = \kappa$) if we study the absolute stability (resp., absolute stability in $[0, \kappa]$) of system (7.3).

Theorem 7.1 *Assume that*

(1) there exist constants $M \geq 1$ and $a > 0$ such that $\Psi(t, s) \leq Me^{-a(t-s)}$;
(2) the zero solution of system (7.4) is asymptotically stable;
(3) the zero solution of system (7.3) is absolutely stable(resp. absolutely stable on $[0, \kappa]$) for Ω.

Then the zero solution of (7.3) is absolutely stable (resp. absolutely stable on $[0, \kappa]$).

Proof. For any $f \in \mathcal{F}$(resp. $f \in \mathcal{F}_\kappa$), the solution of system (7.3) is given by

$$\mathbf{x}(t) = \Psi(t,t_0)\mathbf{x}_0 + \int_{t_0}^{t} \Psi(t,s)[\mathbf{b}f(\sigma(s)) - \theta\mathbf{b}\sigma(s)]ds + \sum_{t_0<\tau_k<t} \Psi(t,t_k^+)U(k,\mathbf{x}(\tau_k)).$$

From assumption (3) we know that $\sigma(t) \to 0$ as $t \to +\infty$, and since $\sigma(t)$ continuously depends on $\mathbf{x}_0$, and $f(\sigma(t))$ is a continuous function of $\mathbf{x}_0$ and $f(\sigma(t)) \to 0$ as $t \to +\infty$, then for any $\epsilon > 0$, there exist $\delta(\epsilon) > 0$ and $t_1 > t_0$ such that if $\|\mathbf{x}_0\| < \delta(\epsilon)$ we have

$$\int_{t_0}^{t_1} Me^{-a(t-s)}[\|\mathbf{b}f(\sigma(s))\| + \|\theta\mathbf{b}\sigma(s)\|]ds < \frac{\epsilon}{4},$$

$$\int_{t_1}^{t} Me^{-a(t-s)}[\|\mathbf{b}f(\sigma(s))\| + \|\theta\mathbf{b}\sigma(s)\|]ds < \frac{\epsilon}{4} \quad \text{for all } t \geq t_1,$$

From assumption (2) we have

$$\|\Psi(t,t_0)\mathbf{x}_0\| \leq Me^{-a(t-t_0)}\|\mathbf{x}_0\| < \frac{\epsilon}{4} \quad \text{for all } t \geq t_0,$$

$$\left\| \sum_{t_0<\tau_k<t} \Psi(t,t_k^+)U(k,\mathbf{x}(\tau_k)) \right\| < \frac{\epsilon}{4} \quad \text{for all } t \geq t_0.$$

Therefore, it follows

$$\begin{aligned}
\|\mathbf{x}(t)\| \quad &\leq \quad \|\Psi(t,t_0)\mathbf{x}_0\| + \int_{t_0}^{t_1} Me^{-a(t-s)}[\|\mathbf{b}f(\sigma(s))\| + \|\theta\mathbf{b}\sigma(s)\|]ds \\
&\quad + \int_{t_1}^{t} Me^{-a(t-s)}[\|\mathbf{b}f(\sigma(s))\| + \|\theta\mathbf{b}\sigma(s)\|]ds \\
&\quad + \left\| \sum_{t_0<\tau_k<t} \Psi(t,t_k^+)U(k,\mathbf{x}(\tau_k)) \right\| \\
&< \quad \frac{\epsilon}{4} + \frac{\epsilon}{4} + \frac{\epsilon}{4} + \frac{\epsilon}{4} = \epsilon, \quad \text{for all } t \geq t_0. \qquad (7.5)
\end{aligned}$$

From assumptions (1) and (2) we know

$$\lim_{t\to+\infty}\left\|\sum_{t_0<\tau_k<t}\Psi(t,t_k^+)U(k,\mathbf{x}(\tau_k))\right\|=0.$$

For any $\mathbf{x}_0\in\mathbb{R}^n$, we have

$$\begin{aligned}0 &\le \lim_{t\to+\infty}\|\mathbf{x}(t)\|\\ &\le \lim_{t\to+\infty} Me^{-a(t-t_0)}\|\mathbf{x}_0\|\\ &\quad+\lim_{t\to+\infty}\int_{t_0}^{t} Me^{-a(t-s)}[\|\mathbf{b}f(\sigma(s))\|+\|\theta\mathbf{b}\sigma(s)\|]ds\\ &\quad+\lim_{t\to+\infty}\left\|\sum_{t_0<\tau_k<t}\Psi(t,t_k^+)U(k,\mathbf{x}(\tau_k))\right\|=0.\end{aligned}\tag{7.6}$$

Thus, the zero solution of system (7.3) is absolutely stable(resp. absolutely stable in $[0,\kappa]$). □

Theorem 7.2 *Assume that*

(1) there exist constants $M\ge 1$ and $a>0$ such that $\Psi(t,s)\le Me^{-a(t-s)}$;
(2) the zero solution of system (7.4) is asymptotically stable;
(3) there exists a differentiable function $V(\mathbf{x})\in C[\mathbb{R}^n,\mathbb{R}]$ such that $V(0)=0$,

$$V(\mathbf{x})\ge\alpha(|\sigma|),\quad \alpha\in\mathcal{KR}\tag{7.7}$$

and

$$\left.\frac{dV}{dt}\right|_{(7.3)}\le-\beta(|\sigma|),\quad \beta\in\mathcal{K}.\tag{7.8}$$

Then the zero solution of (7.3) is absolutely stable (resp. absolutely stable on $[0,\kappa]$).

Proof. It suffices to prove that assumption (3) implies that the zero solution of system (7.3) is absolutely stable(resp. absolutely stable on $[0,\kappa]$) for Ω.

Since $V(0)=0\in\Omega$ and $V(\mathbf{x})$ is continuous, for any $\epsilon>0$, there exists $\delta(\epsilon)>0$ such that

$$V(\mathbf{x}_0)<\alpha(\epsilon)\quad\text{for } \|\mathbf{x}_0\|<\delta(\epsilon).$$

From Eqs. (7.7) and (7.8) we have

$$\alpha(|\sigma(t)|) \le V(\mathbf{x}) \le V(\mathbf{x}_0) < \alpha(\epsilon).$$

Thus we know that $|\sigma(t)| < \epsilon$ which means that the zero solution of system (7.3) is stable for Ω.

We then need to show that for any $\mathbf{x}_0 \in \mathbb{R}$, $\lim_{t\to+\infty}\sigma(t) = 0$. It follows from Eq. (7.8) that for any $\mathbf{x}_0 \in \mathbb{R}^n$, $V(\mathbf{x})$ is monotone decreasing and has a lower bound. Thus there exists

$$\inf_{t\ge t_0} V(\mathbf{x}(t)) = \lim_{t\to+\infty} V(\mathbf{x}(t)) \triangleq a \ge 0$$

and a can only be reached on Ω. If a is reached outside Ω, then there must exist a constant $b > 0$ such that $|\sigma(t)| \ge b > 0$ for all $t \ge t_0$. Otherwise, there is a sequence $\{t_k\}$ with $t_k \to +\infty$ as $k \to +\infty$ such that $\lim_{t_k\to+\infty}\sigma(t_k) = 0$, therefore we have

$$a = \inf_{t\ge t_0} V(\mathbf{x}(t)) = \lim_{t\to+\infty} V(\mathbf{x}(t)) = \lim_{t_k\to+\infty} V(\mathbf{x}(t_k)) = \lim_{\sigma(t_k)\to 0} V(\mathbf{x}(t_k)),$$

which means that a is reached on Ω. This is a contradiction.

For any $\mathbf{x}_0 \in \mathbb{R}^n$, it follows from Eq. (7.8) that

$$|\sigma(t)| \le |\mathbf{c}^\top \mathbf{x}(t_0)| \triangleq h < +\infty.$$

If $\lim_{t\to+\infty}\sigma(t) \ne 0$, then from the uniform continuity of $\sigma(t)$, there are constants $b > 0$, $\xi > 0$ and a sequence $\{t_j\}$ such that

$$|\sigma(t)| \ge b \;\text{ for } t \in [t_j - \xi, t_j + \xi] \;\text{ and } \xi < \inf_{t_j} \frac{1}{2}(t_{j+1} - t_j).$$

Let us fix $r = \inf_{b\le|\sigma|\le h} \beta(\sigma) > 0$, we have

$$\begin{aligned} V(\mathbf{x}(t)) &= V(\mathbf{x}(t_0)) + \int_{t_0}^{t} \frac{dV}{dt}dt \\ &\le V(\mathbf{x}(t_0)) - \int_{t_0}^{t} \beta(|\sigma(s)|)ds \\ &\le V(\mathbf{x}(t_0)) - \sum_{j=1}^{n} \int_{t_j-\xi}^{t_j+\xi} \beta(|\sigma(s)|)ds \\ &\le V(\mathbf{x}(t_0)) - 2\xi n r \to -\infty \;\text{ as } n \to +\infty, \end{aligned} \tag{7.9}$$

which is in contradiction with $V(\mathbf{x}(t)) \geq 0$. Therefore we have $\lim_{t\to+\infty} \sigma(t) = 0$. Thus, we prove that the zero solution of system (7.3) is absolutely stable(resp. absolutely stable on $[0, \kappa]$) for Ω. From Theorem 7.1 we get the conclusion. □

Theorem 7.3 *Assume that*

(1) there exist constants $M \geq 1$ and $a > 0$ such that $\Psi(t,s) \leq Me^{-a(t-s)}$;
(2) the zero solution of system (7.4) is asymptotically stable;
(3) $\theta = 0$ or $\theta = 1$ and there exists a symmetric matrix $P_{n\times n}$ such that

$$\begin{aligned}
&\mathbf{x}^\top P\mathbf{x} \geq \alpha\sigma^2, \qquad \alpha > 0,\\
&\mathbf{x}^\top(PA + A^\top P)\mathbf{x} + (2P\mathbf{b} + \beta A^\top \mathbf{c})^\top \mathbf{x} f(\sigma) + \beta\mathbf{c}^\top \mathbf{b} f^2(\sigma) \leq -\epsilon\mu,\\
&\beta \geq 0.
\end{aligned} \tag{7.10}$$

or

$$\begin{aligned}
&\mathbf{x}^\top P\mathbf{x} \geq 0, \quad \int_0^{\pm\infty} f(\sigma)d\sigma = +\infty,\\
&\mathbf{x}^\top(PA + A^\top P)\mathbf{x} + (2P\mathbf{b} + \beta A^\top \mathbf{c})^\top \mathbf{x} f(\sigma) + \beta\mathbf{c}^\top \mathbf{b} f^2(\sigma) \leq -\epsilon\mu,\\
&\beta > 0.
\end{aligned} \tag{7.11}$$

where $\mu \in \{\sigma^2, \sigma f(\sigma), f^2(\sigma)\}$, α, β, ϵ are constants and $0 < \epsilon \ll 1$.

Then the zero solution of (7.3) is absolutely stable.

Proof. It suffices to prove that assumption (3) implies assumption (3) of Theorem 7.2. Let us construct

$$V(\mathbf{x}) = \mathbf{x}^\top P\mathbf{x} + \beta \int_0^\sigma f(\sigma)d\sigma,$$

from assumption (3) we have

$$V(\mathbf{x}) \geq \alpha\sigma^2 \geq \psi(|\sigma|), \qquad \psi \in \mathcal{KR},$$

or

$$V(\mathbf{x}) \geq \beta \int_0^\sigma f(\sigma)d\sigma \geq \phi(|\sigma|), \qquad \phi \in \mathcal{KR},$$

and

$$\begin{aligned}
\left.\frac{dV}{dt}\right|_{(7.3)} &= \mathbf{x}^\top(PA + A^\top P)\mathbf{x} + (2P\mathbf{b} + \beta A^\top \mathbf{c})^\top \mathbf{x} f(\sigma) + \beta\mathbf{c}^\top \mathbf{b} f^2(\sigma)\\
&\leq -\epsilon\mu \leq -\gamma(|\sigma|), \qquad \gamma \in \mathcal{K}.
\end{aligned} \tag{7.12}$$

Then from Theorem 7.2 we get the conclusion. □

Theorem 7.4 *Assume that*

(1) there exist constants $M \geq 1$ and $a > 0$ such that $\Psi(t,s) \leq Me^{-a(t-s)}$;
(2) the zero solution of system (7.4) is asymptotically stable;
(3) $\theta = 0$ or $\theta = 1$ and there exist a constant $\beta \geq 0$ and a symmetric positive definite matrix $P_{n\times n}$ such that

$$\mathbf{x}^\top (PA + A^\top P)\mathbf{x} + 2\left(P\mathbf{b} + \frac{1}{2}\beta A^\top \mathbf{c}\right)^\top \mathbf{x} f(\sigma) + \beta \mathbf{c}^\top \mathbf{b} f^2(\sigma)$$

is negative definite.

Then the zero solution of (7.3) is absolutely stable.

Proof. It suffices to prove that assumption (3) implies assumption (3) of Theorem 7.2. Let us construct the following Lyapunov function:

$$V(\mathbf{x}) = \mathbf{x}^\top P\mathbf{x} + \beta \int_0^\sigma f(\sigma)d\sigma.$$

It is easy to see that $V(0) = 0$ and let $\lambda_{\min}$ be the smallest eigenvalue of P, we have

$$\begin{aligned} V(\mathbf{x}) &\geq \lambda_{\min}\mathbf{x}^\top\mathbf{x} \geq \lambda_{\min}\frac{n\sum_{i=1}^n |c_i x_i|^2}{n\max_{i=1}^n |c_i|^2} \geq \lambda_{\min}\frac{\left(\sum_{i=1}^n c_i x_i\right)^2}{n\max_{i=1}^n |c_i|^2} \\ &= \lambda_{\min}\frac{\sigma^2}{n\max_{i=1}^n |c_i|^2} \triangleq \psi(|\sigma|) \in \mathcal{KR}. \end{aligned} \tag{7.13}$$

Therefore $V(\mathbf{x})$ is radially unbounded positive definite for Ω. We then have

$$\begin{aligned} \left.\frac{dV}{dt}\right|_{(7.3)} &= \mathbf{x}^\top (PA + A^\top P)\mathbf{x} + 2\left(P\mathbf{b} + \frac{1}{2}\beta A^\top \mathbf{c}\right)^\top \mathbf{x} f(\sigma) + \beta \mathbf{c}^\top \mathbf{b} f^2(\sigma) \\ &\leq -\gamma(\|\mathbf{x}\|) \leq -\gamma\left(\frac{1}{\max_{i=1}^n |c_i|}\sum_{i=1}^n |c_i x_i|\right) \\ &= -\gamma\left(\frac{1}{\max_{i=1}^n |c_i|}|\sigma|\right) \triangleq -\gamma_1(|\sigma|) \end{aligned} \tag{7.14}$$

where $\gamma, \gamma_1 \in \mathcal{K}$. From Theorem 7.2 we get the conclusion. □

7.2 Examples

Example 7.1 Let us study the stability of the following impulsive control system:

$$\left.\begin{array}{l}\dot{x}_1 = x_2 - f(x_1 - x_2) \\ \dot{x}_2 = -x_1 + f(x_1 - x_2)\end{array}\right\}, t \neq \tau_k,$$
$$\left.\begin{array}{l}\Delta x_1 = -0.2x_1 \\ \Delta x_2 = -0.6x_2\end{array}\right\}, t = \tau_k, k \in \mathbb{N} \tag{7.15}$$

where $f \in \mathcal{F}$. Let $\theta = 1$ then the reference system is given by

$$\left.\begin{array}{l}\dot{x}_1 = -x_1 + 2x_2 \\ \dot{x}_2 = -x_2\end{array}\right\}, t \neq \tau_k,$$
$$\left.\begin{array}{l}\Delta x_1 = -0.2x_1 \\ \Delta x_2 = -0.6x_2\end{array}\right\}, t = \tau_k, k \in \mathbb{N} \tag{7.16}$$

It is easy to verify that assumptions (1) and (2) of Theorem 7.3 are satisfied. Let us construct the following Lyapunov function:

$$V(\mathbf{x}) = \frac{1}{2}(x_1^2 + x_2^2) \tag{7.17}$$

then we know

$$\begin{aligned}\left.\frac{dV}{dt}\right|_{(7.15)} &= -x_1 f(x_1 - x_2) + x_2 f(x_1 - x_2) \\ &= -(x_1 - x_2) f(x_1 - x_2)\end{aligned} \tag{7.18}$$

is negative definite for $\Omega = \{\mathbf{x} |\ \sigma = x_1 - x_2 = 0\}$. Therefore, the third assumption of Theorem 7.3 is also satisfied. Thus, the zero solution of system (7.15) is absolutely stable. Figure 7.1 shows the simulation results with $f(x_1 - x_2) = (x_1 - x_2)^3$. In this simulation, we set $t_0 = 0$ and $\tau_k = 0.09k, k \in \mathbb{N}$.

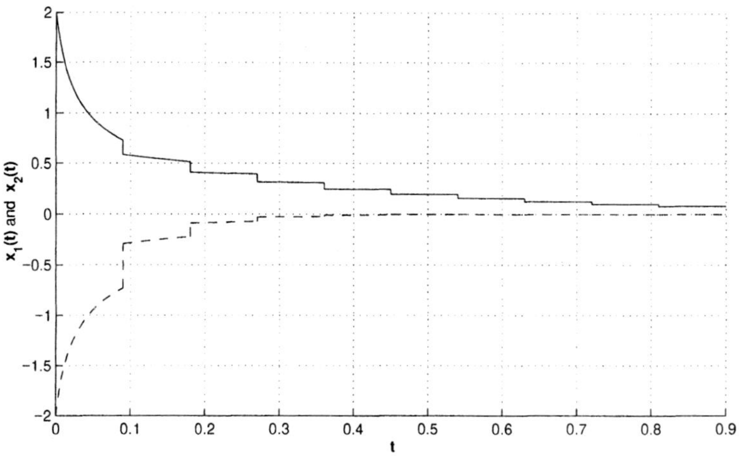

Fig. 7.1 Simulation results of impulsive control system (7.15). Initial condition is $(x_1(0), x_2(0)) = (2, -2)$. The solid and dashed waveforms represent $x_1(t)$ and $x_2(t)$, respectively.

Example 7.2 Let us study the stability of the following impulsive control system:

$$\left.\begin{aligned} \dot{x}_1 &= -3x_1 + x_2 + x_3 - f(x_1 + 2x_2 + x_3) \\ \dot{x}_2 &= x_1 - 2x_2 + x_3 + f(x_1 + 2x_2 + x_3) \\ \dot{x}_3 &= x_1 + 3x_2 - 3x_3 - 2f(x_1 + 2x_2 + x_3) \end{aligned}\right\}, \quad t \neq \tau_k,$$
$$\left.\begin{aligned} \Delta x_1 &= -0.3x_1 \\ \Delta x_2 &= -0.2x_2 \\ \Delta x_3 &= -0.1x_3 \end{aligned}\right\}, \quad t = \tau_k, \quad k \in \mathbb{N} \tag{7.19}$$

where $f \in \mathcal{F}$. Let $\theta = 0$ then the reference system is given by

$$\left.\begin{aligned} \dot{x}_1 &= -3x_1 + x_2 + x_3 \\ \dot{x}_2 &= x_1 - 2x_2 + x_3 \\ \dot{x}_3 &= x_1 + 3x_2 - 3x_3 \end{aligned}\right\}, \quad t \neq \tau_k,$$
$$\left.\begin{aligned} \Delta x_1 &= -0.3x_1 \\ \Delta x_2 &= -0.2x_2 \\ \Delta x_3 &= -0.1x_3 \end{aligned}\right\}, \quad t = \tau_k, \quad k \in \mathbb{N} \tag{7.20}$$

It is easy to verify that assumptions (1) and (2) of Theorem 7.3 are satisfied. Let us construct the following Lyapunov function:

$$V(\mathbf{x}) = (x_1 + 2x_2 + x_3)^2 \tag{7.21}$$

then we know

$$\begin{aligned} \left.\frac{dV}{dt}\right|_{(7.19)} &= 2(x_1 + 2x_2 + x_3)(\dot{x}_1 + 2\dot{x}_2 + \dot{x}_3) \\ &= -2(x_1 + 2x_2 + x_3)f(x_1 + 2x_2 + x_3) \end{aligned} \tag{7.22}$$

is negative definite for $\Omega = \{\mathbf{x} | \ \sigma = x_1 + 2x_2 + x_3 = 0\}$. Therefore, the third assumption of Theorem 7.3 is also satisfied. Thus, the zero solution of system (7.19) is absolutely stable. Figure 7.2 shows the simulation results with $f(x_1 + 2x_2 + x_3) = (x_1 + 2x_2 + x_3)^3$. In this simulation, we set $t_0 = 0$ and $\tau_k = 0.09k, k \in \mathbb{N}$.

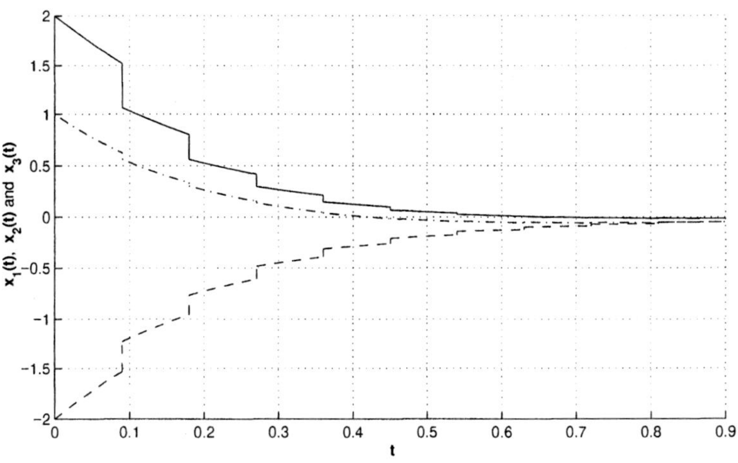

Fig. 7.2 Simulation results of impulsive control system (7.19). Initial condition is $(x_1(0), x_2(0), x_3(0)) = (2, -2, 1)$. The solid, dashed and dash-dotted waveforms represent $x_1(t)$, $x_2(t)$ and $x_3(t)$, respectively.

7.3 Methods based on a kind of Lyapunov function

To prove that the zero solution of system (7.3) is absolutely stable for Ω, we can construct the following Lyapunov function:

$$V(\mathbf{x}) = \mathbf{x}^\top P\mathbf{x} + \beta \int_0^\sigma f(\sigma)d\sigma \tag{7.23}$$

where $\beta \geq 0$ is a constant and P is the solution of the Lyapunov matrix equation

$$PA + A^\top P = -R$$

where R is a given symmetric positive definite matrix. We then have

$$\begin{aligned}
\left.\frac{dV}{dt}\right|_{(7.3)} &= \mathbf{x}^\top(PA + A^\top P)\mathbf{x} + (2P\mathbf{b} + \beta A^\top \mathbf{c})^\top \mathbf{x} f(\sigma) + \beta \mathbf{c}^\top \mathbf{b} f^2(\sigma) \\
&= \mathbf{x}^\top(PA + A^\top P)\mathbf{x} + (2P\mathbf{b} + \beta A^\top \mathbf{c})^\top \mathbf{x} f(\sigma) + \beta \mathbf{c}^\top \mathbf{b} f^2(\sigma) \\
&\quad + \alpha f(\sigma)\left(\sigma - \frac{f(\sigma)}{\kappa}\right) - \alpha f(\sigma)\left(\sigma - \frac{f(\sigma)}{\kappa}\right) \\
&= -S(\mathbf{x},\sigma) - \alpha f(\sigma)\left(\sigma - \frac{f(\sigma)}{\kappa}\right),
\end{aligned} \tag{7.24}$$

where $\alpha \geq 0$ is a constant and

$$\begin{aligned}
S(\mathbf{x},\sigma) &= \mathbf{x}^\top R\mathbf{x} + 2\mathbf{d}^\top \mathbf{x} f(\sigma) + r f^2(\sigma), \\
\mathbf{d} &= -\left[P\mathbf{b} + \frac{1}{2}(\alpha \mathbf{c} + \beta A^\top \mathbf{c})\right], \\
r &= -\beta \mathbf{c}^\top \mathbf{b} + \frac{\alpha}{\kappa}.
\end{aligned} \tag{7.25}$$

If $S(\mathbf{x},\sigma)$ is positive definite in $\mathbf{x}$ and σ, then $\frac{dV}{dt}\big|_{(7.3)}$ is negative definite.

Theorem 7.5 *Assume that*

(1) *there exist constants $M \geq 1$ and $a > 0$ such that $\Psi(t,s) \leq Me^{-a(t-s)}$;*
(2) *the zero solution of system (7.4) is asymptotically stable;*
(3) *there exist constants $\alpha > 0$, $\beta \geq 0$ and a real symmetric positive definite matrix $P_{n\times n}$ such that*

$$r > 0, \qquad R - \frac{1}{r}\mathbf{d}\mathbf{d}^\top > 0 \tag{7.26}$$

or

$$R > 0, \qquad r - \mathbf{d}^\top R^{-1}\mathbf{d} > 0. \tag{7.27}$$

Then the zero solution of (7.3) is absolutely stable.

Proof. It suffices to prove that assumption (3) implies that (7.24) is negative definite. To do this, we need to prove that $S(\mathbf{x}, \sigma)$ is positive definite; namely,

$$\det \begin{vmatrix} R & \mathbf{d} \\ \mathbf{d}^\top & r \end{vmatrix} > 0.$$

We can reach the conclusion from the following two relations:

$$\begin{pmatrix} I & -\mathbf{d}/r \\ 0 & 1 \end{pmatrix} \begin{pmatrix} R & \mathbf{d} \\ \mathbf{d}^\top & r \end{pmatrix} \begin{pmatrix} I & 0 \\ -\mathbf{d}^\top/r & 1 \end{pmatrix} = \begin{pmatrix} R - \mathbf{d}\mathbf{d}^\top/r & 0 \\ 0 & r \end{pmatrix},$$

and

$$\begin{pmatrix} I & 0 \\ -\mathbf{d}^\top R^{-1} & 1 \end{pmatrix} \begin{pmatrix} R & \mathbf{d} \\ \mathbf{d}^\top & r \end{pmatrix} \begin{pmatrix} I & -R^{-1}\mathbf{d} \\ 0 & 1 \end{pmatrix} = \begin{pmatrix} R & 0 \\ 0 & r - \mathbf{d}^\top R^{-1}\mathbf{d} \end{pmatrix}.$$

□

Theorem 7.6 *Assume that*

(1) there exist constants $M \geq 1$ and $a > 0$ such that $\Psi(t,s) \leq Me^{-a(t-s)}$;
(2) the zero solution of system (7.4) is asymptotically stable;
(3) there exist constants $\alpha > 0$, $\beta \geq 0$ and a real symmetric positive definite matrix $P_{n\times n}$ such that

$$r > 0, \qquad R - \frac{1}{r}\mathbf{d}\mathbf{d}^\top \geq 0 \tag{7.28}$$

or

$$R > 0, \qquad r - \mathbf{d}^\top R^{-1}\mathbf{d} \geq 0, \tag{7.29}$$

and $0 \leq f(\sigma)/\sigma \leq \kappa \leq +\infty$.

Then the zero solution of (7.3) is absolutely stable.

Proof. Let us construct the Lyapunov function as that in Eq. (7.23). When $\kappa = +\infty$, the condition (7.28) or (7.29) implies that $S(\mathbf{x}, \sigma) \geq 0$. Therefore

$$\left.\frac{dV}{dt}\right|_{(7.3)} \leq -\alpha f(\sigma)\sigma$$

is negative definite for Ω. □

Corollary 7.1 *Assume that*

(1) there exist constants $M \geq 1$ *and* $a > 0$ *such that* $\Psi(t,s) \leq Me^{-a(t-s)}$;
(2) the zero solution of system (7.4) is asymptotically stable;
(3) $\kappa = +\infty$ *and one of the following conditions holds:*

(a)

$$\left(\frac{\beta A^\top \mathbf{c}}{2} + P\mathbf{b}\right)^\top R^{-1}\left(\frac{\beta A^\top \mathbf{c}}{2} + P\mathbf{b}\right) + \beta\mathbf{c}^\top\mathbf{b} < 0$$

(b)

$$\left(\frac{\beta A^\top \mathbf{c}}{2} + P\mathbf{b}\right)^\top R^{-1}\mathbf{c} < 0$$

and

$$\left[\left(\frac{\beta A^\top \mathbf{c}}{2} + P\mathbf{b}\right)^\top R^{-1}\mathbf{c}\right]^2 - (\mathbf{c}^\top R^{-1}\mathbf{c})\left[\left(\frac{\beta A^\top \mathbf{c}}{2} + P\mathbf{b}\right)^\top R^{-1}\left(\frac{\beta A^\top \mathbf{c}}{2} + P\mathbf{b}\right) + \beta\mathbf{c}^\top\mathbf{b}\right] \geq 0;$$

(c)

$$\left(\frac{\beta A^\top \mathbf{c}}{2} + P\mathbf{b}\right)^\top R^{-1}\mathbf{c} \leq 0$$

and

$$\left[\left(\frac{\beta A^\top \mathbf{c}}{2} + P\mathbf{b}\right)^\top R^{-1}\mathbf{c}\right]^2 - (\mathbf{c}^\top R^{-1}\mathbf{c})\left[\left(\frac{\beta A^\top \mathbf{c}}{2} + P\mathbf{b}\right)^\top R^{-1}\left(\frac{\beta A^\top \mathbf{c}}{2} + P\mathbf{b}\right) + \beta\mathbf{c}^\top\mathbf{b}\right] > 0.$$

Then there exists $\alpha > 0$ *such that* $S(\mathbf{x}, \sigma) \geq 0$.

Then the zero solution of (7.3) is absolutely stable.

Proof. It suffices to prove that assumption (3) of Theorem 7.6 holds. We have $R > 0$. Let $\mathbf{d}^\top R^{-1}\mathbf{d} - r = 0$; namely,

$$\left(\frac{\beta A^\top \mathbf{c}}{2} + P\mathbf{b} + \frac{\alpha \mathbf{c}}{2}\right)^\top R^{-1}\left(\frac{\beta A^\top \mathbf{c}}{2} + P\mathbf{b} + \frac{\alpha \mathbf{c}}{2}\right) + \beta \mathbf{c}^\top \mathbf{b} = 0, \quad (7.30)$$

or

$$\frac{\mathbf{c}^\top}{2} R^{-1} \frac{\mathbf{c}}{2}\alpha^2 + \left[\left(\frac{\beta A^\top \mathbf{c}}{2} + P\mathbf{b}\right)^\top R^{-1}\frac{\mathbf{c}}{2} + \frac{\mathbf{c}^\top}{2} R^{-1}\left(\frac{\beta A^\top \mathbf{c}}{2} + P\mathbf{b}\right)\right]\alpha$$
$$+ \left(\frac{\beta A^\top \mathbf{c}}{2} + P\mathbf{b}\right)^\top R^{-1}\left(\frac{\beta A^\top \mathbf{c}}{2} + P\mathbf{b}\right) + \beta \mathbf{c}^\top \mathbf{b} = 0, \quad (7.31)$$

which has the positive solution α, if and only if one of the conditions (a), (b) and (c) holds. Then assumption (3) of Theorem 7.6 holds. □

Corollary 7.2 *Assume that*

(1) there exist constants $M \geq 1$ and $a > 0$ such that $\Psi(t, s) \leq Me^{-a(t-s)}$;
(2) the zero solution of system (7.4) is asymptotically stable;
(3) if $\kappa < +\infty$ and the following conditions hold:

$$\frac{1}{\kappa} - \mathbf{c}^\top R^{-1}\mathbf{d} > 0,$$

$$\left(\frac{1}{\kappa} - \mathbf{c}^\top R^{-1}\mathbf{d}\right)^2 - \mathbf{c}^\top R^{-1}\mathbf{c}(\mathbf{d}^\top R^{-1}\mathbf{d} + \beta \mathbf{c}^\top \mathbf{b}) > 0.$$

Then the zero solution of (7.3) is absolutely stable in $[0, \kappa]$.

Proof. It suffices to prove that assumption (3) implies that the derivative of $V(\mathbf{x})$ in (7.23) along the solution of (7.3) is negative definite for Ω. Assumption (3) implies that there exists ϵ with $0 < \epsilon \ll 1$ such that

$$\frac{1}{\kappa + \epsilon} - \mathbf{c}^\top R^{-1}\mathbf{d} > 0, \quad (7.32)$$

$$\left(\frac{1}{\kappa + \epsilon} - \mathbf{c}^\top R^{-1}\mathbf{d}\right)^2 - \mathbf{c}^\top R^{-1}\mathbf{c}(\mathbf{d}^\top R^{-1}\mathbf{d} + \beta \mathbf{c}^\top \mathbf{b}) > 0, \quad (7.33)$$

We have

$$-\left.\frac{dV}{dt}\right|_{(7.3)} = \mathbf{x}^\top R\mathbf{x} + 2\mathbf{d}^\top \mathbf{x} f(\sigma) + \tilde{r} f^2(\sigma) + \alpha f(\sigma)\left(\sigma - \frac{1}{\kappa+\epsilon} f(\sigma)\right)$$
$$= \tilde{S}(\mathbf{x},\sigma) + \alpha f(\sigma)\left(\sigma - \frac{1}{\kappa+\epsilon} f(\sigma)\right), \tag{7.34}$$

where

$$\tilde{r} = -\left(\beta \mathbf{c}^\top \mathbf{b} - \frac{\alpha}{\kappa+\epsilon}\right).$$

It is clear that R is positive definite. Therefore, the conditions in Eqs. (7.32) and (7.33) guarantee that

$$\det\begin{vmatrix} R & \mathbf{d} \\ \mathbf{d}^\top & \tilde{r} \end{vmatrix} = 0$$

has positive solution for α. Thus we have $\tilde{S}(\mathbf{x},\sigma) \geq 0$ and therefore the zero solution of system (7.3) is absolutely stable in $[0,\kappa]$. □

7.4 Detailed results

Theorem 7.7 *Assume that*

(1) there exist constants $M \geq 1$ and $a > 0$ such that $\Psi(t,s) \leq Me^{-a(t-s)}$;
(2) the zero solution of system (7.4) is asymptotically stable;
(3) there exist a symmetric positive definite matrix P and a constant $\beta \geq 0$ such that for $V(\mathbf{x})$ defined in (7.23) the derivative in (7.24) is negative definite with respect to $\mathbf{x}$, in particular, there exists $\alpha > 0$ such that

$$T(\mathbf{x},\sigma) \triangleq \mathbf{x}^\top (PA + A^\top P)\mathbf{x} + 2\left(P\mathbf{b} + \frac{1}{2}\beta A^\top \mathbf{c}\right)^\top \mathbf{x} f(\sigma)$$
$$+\alpha f(\sigma)\left(\sigma - \frac{f(\sigma)}{\kappa}\right) \tag{7.35}$$

is negative definite with respect to $\mathbf{x}$ and σ.

Then the zero solution of (7.3) is absolutely stable(resp. absolutely stable in $[0,\kappa]$).

Proof. For any $f \in \mathcal{F}$(resp. $f \in \mathcal{F}_\kappa$), $V(\mathbf{x})$ is a radially unbounded positive definite function. Since $T(\mathbf{x}, \sigma)$ is negative definite with respect to $\mathbf{x}$ and σ, we know (7.24) is negative definite. Therefore the zero solution of (7.3) is absolutely stable(resp. absolutely stable in $[0, \kappa]$). □

Theorem 7.8 *Assume that*

(1) there exist constants $M \geq 1$ and $a > 0$ such that $\Psi(t, s) \leq Me^{-a(t-s)}$;
(2) the zero solution of system (7.4) is asymptotically stable;
(3) $f \in \mathcal{F}_\kappa$ and the symmetric positive definite matrix P is the solution of

$$A^\top P + PA = -R,$$

and

$$\frac{1}{\kappa} - \mathbf{c}^\top R^{-1}\mathbf{d}_1 > 0 \tag{7.36}$$

where $\mathbf{d}_1 = P\mathbf{b} + \beta A^\top \mathbf{c}$ with a constant $\beta \geq 0$. And

$$\frac{1}{\mathbf{c}^\top R^{-1}\mathbf{c}}\left(\frac{1}{\kappa} - \mathbf{c}^\top R^{-1}\mathbf{d}_1\right)^2 - \mathbf{d}_1^\top R^{-1}\mathbf{d}_1 - 2\beta\mathbf{c}^\top\mathbf{b} > 0. \tag{7.37}$$

Then the zero solution of (7.3) is absolutely stable in $[0, \kappa]$.

Proof. Let us construct the Lyapunov function as

$$V(\mathbf{x}) = \mathbf{x}^\top P\mathbf{x} + 2\beta\int_0^\sigma f(\sigma)d\sigma \tag{7.38}$$

and there exists H such that $R = H^\top H$ and $R^{-1} = H^{-1}(H^{-1})^\top$. We then have

$$\begin{aligned}
-\left.\frac{dV}{dt}\right|_{(7.3)} &= \mathbf{x}^\top R\mathbf{x} - 2\mathbf{x}^\top(P\mathbf{b} + \beta A^\top\mathbf{c})f(\sigma) - 2\beta\mathbf{c}^\top\mathbf{b}f^2(\sigma) \\
&= \mathbf{x}^\top H^\top H\mathbf{x} - 2\mathbf{x}^\top\mathbf{d}_1 f(\sigma) - 2\beta\mathbf{c}^\top\mathbf{b}f^2(\sigma) \\
&= (H\mathbf{x})^\top H\mathbf{x} - 2(H\mathbf{x})^\top(H^{-1})^\top\mathbf{d}_1 f(\sigma) - 2\beta\mathbf{c}^\top\mathbf{b}f^2(\sigma) \\
&= [H\mathbf{x} - (H^{-1})^\top\mathbf{d}_1 f(\sigma)]^\top[H\mathbf{x} - (H^{-1})^\top\mathbf{d}_1 f(\sigma)] \\
&\quad -(\mathbf{d}_1^\top R^{-1}\mathbf{d} + 2\beta\mathbf{c}^\top\mathbf{b})f^2(\sigma) \\
&= \begin{cases} 0, & \text{for } \mathbf{x} = 0, \\ (H\mathbf{x})^\top H\mathbf{x} = \mathbf{x}^\top R\mathbf{x} > 0, & \text{for } \mathbf{x} \neq 0,\ f(\sigma) = 0, \\ \mu f^2(\sigma), & \text{for } f(\sigma) \neq 0, \end{cases}
\end{aligned} \tag{7.39}$$

where

$$\mu = \left[H\frac{\mathbf{x}}{f(\sigma)} - (H^{-1})^\top \mathbf{d}_1\right]^\top \left[H\frac{\mathbf{x}}{f(\sigma)} - (H^{-1})^\top \mathbf{d}_1\right] - (\mathbf{d}_1^\top R^{-1}\mathbf{d}_1 + 2\beta\mathbf{c}^\top\mathbf{b}). \tag{7.40}$$

Therefore, it is enough to show that $\mu > 0$ for any $\mathbf{c}^\top\mathbf{x} \geq 1/\kappa$. Let us use notation $\xi = H\mathbf{x} - (H^{-1})^\top\mathbf{d}_1$ then we have

$$\mu = \xi^\top\xi - \psi, \quad \psi = \mathbf{d}_1^\top R^{-1}\mathbf{d}_1 + 2\beta\mathbf{c}^\top\mathbf{b} \tag{7.41}$$

and the condition $\mathbf{c}^\top\mathbf{x} \geq 1/\kappa$ is equivalent to

$$\mathbf{c}^\top H^{-1}[\xi + (H^{-1})^\top\mathbf{d}_1] \geq 1/\kappa. \tag{7.42}$$

It is easy to see that $\psi \geq 0$. Therefore we have $\mu \leq 0$ for $\xi = 0$. Thus we need to guarantee that $\xi = 0$ does not satisfy (7.42); namely,

$$\mathbf{c}^\top H^{-1}[0 + (H^{-1})^\top\mathbf{d}_1] = \mathbf{c}^\top H^{-1}(H^{-1})^\top\mathbf{d}_1 = \mathbf{c}^\top R^{-1}\mathbf{d}_1 < 1/\kappa \tag{7.43}$$

which is condition (7.36).

Let ξ_1 be the intersection point of the hyperplane

$$\mathbf{c}^\top H^{-1}[\xi + (H^{-1})^\top\mathbf{d}_1] = 1/\kappa$$

and its normal line passing $\xi = 0$. Let us assume $\xi_1 = \lambda(\mathbf{c}^\top H^{-1})^\top$ and λ can be determined by

$$\mathbf{c}^\top H^{-1}[\lambda(\mathbf{c}^\top H^{-1})^\top + (H^{-1})^\top\mathbf{d}_1] = 1/\kappa$$

which leads to

$$\lambda = \frac{1/\kappa - \mathbf{c}^\top R^{-1}\mathbf{d}_1}{\mathbf{c}^\top R^{-1}\mathbf{c}}.$$

Since μ with constraint (7.42) has its minimum at $\xi = \xi_1$, we need to guarantee that the following expression greater than 0:

$$\begin{aligned}\mu(\xi_1) &= [\lambda(\mathbf{c}^\top H^{-1})^\top]^\top[\lambda(\mathbf{c}^\top H^{-1})^\top] - \psi \\ &= \lambda^2\mathbf{c}^\top R^{-1}\mathbf{c} - (\mathbf{d}_1^\top R^{-1}\mathbf{d}_1 + 2\beta\mathbf{c}^\top\mathbf{b}) \\ &= \frac{1}{\mathbf{c}^\top R^{-1}\mathbf{c}}\left(\frac{1}{\kappa} - \mathbf{c}^\top R^{-1}\mathbf{d}_1\right)^2 - \mathbf{d}_1^\top R^{-1}\mathbf{d}_1 - 2\beta\mathbf{c}^\top\mathbf{b} \\ &> 0 \end{aligned} \tag{7.44}$$

which is condition (7.37). □

By using the similar process of the proof of Theorem 7.8, we can prove the following two corollaries.

Corollary 7.3 *Assume that*

(1) there exist constants $M \geq 1$ *and* $a > 0$ *such that* $\Psi(t, s) \leq Me^{-a(t-s)}$*;*
(2) the zero solution of system (7.4) is asymptotically stable;
(3) $f \in \mathcal{F}_\kappa$ *and*

$$0 \leq f(\sigma)\sigma < \kappa\sigma^2$$

and the symmetric positive definite matrix P *is the solution of*

$$A^\top P + PA = -R,$$

and

$$\frac{1}{\kappa} - \mathbf{c}^\top R^{-1}\mathbf{d}_1 \geq 0 \tag{7.45}$$

where $\mathbf{d}_1 = P\mathbf{b} + \beta A^\top \mathbf{c}$ *with a constant* $\beta \geq 0$*. And*

$$\frac{1}{\mathbf{c}^\top R^{-1}\mathbf{c}}\left(\frac{1}{\kappa} - \mathbf{c}^\top R^{-1}\mathbf{d}_1\right)^2 - \mathbf{d}_1^\top R^{-1}\mathbf{d}_1 - 2\beta\mathbf{c}^\top\mathbf{b} \geq 0. \tag{7.46}$$

Then the zero solution of (7.3) is absolutely stable in $[0, \kappa)$*.*

Corollary 7.4 *Assume that*

(1) there exist constants $M \geq 1$ *and* $a > 0$ *such that* $\Psi(t, s) \leq Me^{-a(t-s)}$*;*
(2) the zero solution of system (7.4) is asymptotically stable;
(3) $f \in \mathcal{F}$ *and the symmetric positive definite matrix* P *is the solution of*

$$A^\top P + PA = -R,$$

and

$$\mathbf{c}^\top R^{-1}\mathbf{d}_1 \leq 0 \tag{7.47}$$

where $\mathbf{d}_1 = P\mathbf{b} + \beta A^\top \mathbf{c}$ *with a constant* $\beta \geq 0$*. And*

$$\frac{1}{\mathbf{c}^\top R^{-1}\mathbf{c}}\left(\mathbf{c}^\top R^{-1}\mathbf{d}_1\right)^2 - \mathbf{d}_1^\top R^{-1}\mathbf{d}_1 - 2\beta\mathbf{c}^\top\mathbf{b} \geq 0. \tag{7.48}$$

Then the zero solution of (7.3) is absolutely stable.

7.5 Simplified forms

Without loss of generality let us assume that in system (7.3) $c_n \neq 0$. Let the matrix $G = (g_{ij})_{n\times n}$ be

$$g_{ij} = \begin{cases} 1, & i = j = 1, 2, \cdots, n-1, \\ c_j, & j = 1, 2, \cdots, n, \quad i = n \\ 0, & \text{otherwise,} \end{cases}$$

and let

$$\mathbf{y} = G\mathbf{x}$$

then system (7.3) is transformed into

$$\begin{aligned} \dot{\mathbf{y}} &= A\mathbf{x} + \mathbf{b}f(\sigma), \quad \sigma = \mathbf{c}^\top \mathbf{x} = \sum_{i=1}^{n} c_i x_i, \qquad t \neq \tau_k \\ \Delta \mathbf{x} &= U(k, \mathbf{x}), \qquad t = \tau_k, \\ \mathbf{x}(t_0^+) &= \mathbf{x}_0, \qquad t_0 \geq 0, \quad k = 1, 2, \cdots, \end{aligned} \tag{7.49}$$

where $f \in \mathcal{F}$, or $f \in \mathcal{F}_K$. $U(k, \mathbf{x})$ is continuous and $U(k, 0) = 0$, $k \in \mathbb{N}$.

Chapter 8

Optimal Impulsive Control

8.1 Linear impulsive systems: Fixed terminal conditions

In this section we consider the optimal impulsive control of the following time-varying linear system:

$$\dot{\mathbf{x}}(t) = A(t)\mathbf{x}(t) \tag{8.1}$$

where $\mathbf{x} \in \Re^n$ and $A \in C[\Re, \Re^n \times \Re^n]$. The corresponding impulsive controlled system is given by

$$\begin{cases} \dot{\mathbf{x}}(t) = A(t)\mathbf{x}(t), & t \neq \tau_k \\ \Delta \mathbf{x} = B(\tau_k)\mathbf{u}_k, & t = \tau_k, \quad k = 1, 2, ..., \end{cases} \tag{8.2}$$

where $B \in C^1[\Re, \Re^n \times \Re^m]$ and $\mathbf{u}_k \in \Re^m$, $1 \leq m < n$.

The terminal conditions are given by $(t_1, \mathbf{x}_1)$ and $(t_2, \mathbf{x}_2)$, where $\mathbf{x}(t_1) = \mathbf{x}_1$ and $\mathbf{x}(t_2) = \mathbf{x}_2$. We will consider the following two cost functionals:

$$J_1 \triangleq \sum_{k=1}^{p} \|B(\tau_k)\mathbf{u}_k\|, \tag{8.3}$$

and

$$J_2 \triangleq \sum_{k=1}^{p} \|\mathbf{u}_k\|. \tag{8.4}$$

From Section 3.3 we know that

$$\sum_{k=1}^{p} \Phi^{-1}(\tau_k)B(\tau_k)\mathbf{u}_k = \Phi^{-1}(t_2)\mathbf{x}_2 - \Phi^{-1}(t_1)\mathbf{x}_1 \tag{8.5}$$

where $\Phi(t)$ is the fundamental matrix of Eq. (8.1). Let us define

$$\mathbf{v} \triangleq \Phi^{-1}(t_2)\mathbf{x}_2 - \Phi^{-1}(t_1)\mathbf{x}_1. \tag{8.6}$$

Lemma 8.1 *If Eq. (8.5) has a solution, $(\tau_k, \mathbf{u}_k)$, $k = 1, 2, \cdots, p$, then the gradient vectors of the independent equations of (8.5) are linearly independent at any solution of (8.5).*

Proof. Let us denote the n equations of (8.5) as

$$\Gamma_i(\tau_k, \mathbf{u}_k) = \left[\sum_{k=1}^{p} \Phi^{-1}(\tau_k)B(\tau_k)\mathbf{u}_k - \mathbf{v}\right]_i, \quad i = 1, 2, \cdots, n, \tag{8.7}$$

where $[\cdot]_i$ denotes the ith row of the matrix $[\cdot]$. The gradient vectors are given by

$$\nabla\Gamma_i = \left(\frac{\partial\Gamma_i}{\partial\tau_1}, \frac{\partial\Gamma_i}{\partial\tau_2}, \cdots, \frac{\partial\Gamma_i}{\partial\tau_p}, \frac{\partial\Gamma_i}{\partial\mathbf{u}_1}, \frac{\partial\Gamma_i}{\partial\mathbf{u}_2}, \cdots, \frac{\partial\Gamma_i}{\partial\mathbf{u}_p}\right)^{\top}, i = 1, 2, \cdots, n, \tag{8.8}$$

where

$$\frac{\partial\Gamma_i}{\partial\mathbf{u}_k} = \left(\frac{\partial\Gamma_i}{\partial u_{k1}}, \frac{\partial\Gamma_i}{\partial u_{k2}}, \cdots, \frac{\partial\Gamma_i}{\partial u_{km}}\right). \tag{8.9}$$

Because $A(t)$ is continuous we know that $\Phi^{-1}(t)$ is C^1 and because $B(\tau_k)$ is C^1 we know that the above derivatives exist. We then have

$$\frac{\partial\Gamma_i}{\partial\mathbf{u}_k} = [\Phi^{-1}(\tau_k)B(\tau_k)]_i \tag{8.10}$$

then

$$\begin{aligned} & \begin{bmatrix} \frac{\partial\Gamma_1}{\partial\mathbf{u}_1} & \frac{\partial\Gamma_1}{\partial\mathbf{u}_2} & \cdots & \frac{\partial\Gamma_1}{\partial\mathbf{u}_p} \\ \frac{\partial\Gamma_2}{\partial\mathbf{u}_1} & \frac{\partial\Gamma_2}{\partial\mathbf{u}_2} & \cdots & \frac{\partial\Gamma_2}{\partial\mathbf{u}_p} \\ \vdots & \vdots & \vdots & \vdots \\ \frac{\partial\Gamma_n}{\partial\mathbf{u}_1} & \frac{\partial\Gamma_n}{\partial\mathbf{u}_2} & \cdots & \frac{\partial\Gamma_n}{\partial\mathbf{u}_p} \end{bmatrix} \\ = \; & [\Phi^{-1}(\tau_1)B(\tau_1), \;\; \Phi^{-1}(\tau_2)B(\tau_2), \cdots, \Phi^{-1}(\tau_p)B(\tau_p)] \\ \triangleq \; & W. \end{aligned} \tag{8.11}$$

The number of independent equations of (8.5) is equal to Rank$[W, \mathbf{v}]$. At any solution of (8.5) we must have Rank$[W, \mathbf{v}]$ =Rank$[W]$, therefore from Eq. (8.11) we can see that the gradient vectors of the independent equations of (8.5) are linearly independent at any solution of (8.5), □

Theorem 8.1 *Let*

$$\begin{aligned} \alpha_k &\triangleq |B(\tau_k)\mathbf{u}_k|, \\ \nu_k &\triangleq \frac{B(\tau_k)\mathbf{u}_k}{\alpha_k} \quad \text{if} \quad \alpha_k \neq 0, \\ & k = 1, 2, \cdots, p. \end{aligned} \tag{8.12}$$

and

$$K(t) \triangleq B(t)B^{\dagger}(t), \tag{8.13}$$

where $B^{\dagger}(t)$ is the pseudo-inverse of the rectangle matrix $B(t)$. Given the terminal conditions $(t_1, \mathbf{x}_1)$ and $(t_2, \mathbf{x}_2)$ and suppose that $\mathbf{v} \neq \mathbf{0}$, then for each $p = 1, 2, \cdots$, the element $(\tau_1, \tau_2, \cdots, \tau_p, \mathbf{u}_1, \mathbf{u}_2, \cdots, \mathbf{u}_p)$ minimizes J_1 over the set $[t_1, t_2]^p \times \Re^{mp}$ if and only if there exists a nonzero vector $\lambda \in \Re^n$ such that

$$\alpha_k = 0, \quad \text{or} \quad \nu_k = -K(\tau_k)^{\top}\Phi^{-1}(\tau_k)^{\top}\lambda, \quad k = 1, 2, \cdots, p, \tag{8.14}$$

$$\begin{aligned} &\alpha_k = 0, \quad \text{or} \quad \lambda^{\top}\Phi^{-1}(\tau_k)K(\tau_k)K(\tau_k)^{\top}\Phi^{-1}(\tau_k)^{\top}\lambda = 1, \\ &k = 1, 2, \cdots, p, \end{aligned} \tag{8.15}$$

$$\begin{aligned} &\alpha_k = 0, \quad \text{or} \quad \tau_k = t_1, \quad \text{or} \quad \tau_k = t_2, \quad \text{or} \\ &\lambda^{\top}\frac{d(\Phi^{-1}(\tau_k)K(\tau_k)K(\tau_k)^{\top}\Phi^{-1}(\tau_k)^{\top})}{d\tau_k}\lambda = 0, \\ &k = 1, 2, \cdots, p, \end{aligned} \tag{8.16}$$

$$\sum_{k=1}^{p}[\Phi^{-1}(\tau_k)K(\tau_k)K(\tau_k)^{\top}\Phi^{-1}(\tau_k)^{\top})\lambda]\alpha_k = -\mathbf{v}, \tag{8.17}$$

$$\alpha_k \geq 0, \quad k = 1, 2, \cdots, p, \tag{8.18}$$

$$\sum_{k=1}^{p}\alpha_k = -\lambda^{\top}\mathbf{v} > 0, \tag{8.19}$$

$$\begin{aligned} &-\lambda^{\top}\mathbf{v} \textit{ is a minimum on the set} \\ &\{\lambda \in \Re^n | \textit{ Eqs. (8.15)-(8.19) are valid}\}. \end{aligned} \tag{8.20}$$

Proof. Suppose that a solution $(\tau_1, \tau_2, \cdots, \tau_p, \mathbf{u}_1, \mathbf{u}_2, \cdots, \mathbf{u}_p)$ can minimize J_1, then from Lemma 8.1, we know that the constraints (8.5) guarantee the existence of a unique Lagrange multiplier $\lambda \in \Re^n$ such that the first partial derivatives of the function

$$G(\mathbf{u}_k, \tau_k, \lambda) = \sum_{k=1}^{p} \|B(\tau_k)\mathbf{u}_k\| + \lambda^\top \left(\sum_{k=1}^{p} \Phi^{-1}(\tau_k)B(\tau_k)\mathbf{u}_k - \mathbf{v} \right), \quad k = 1, 2, \cdots, p, \tag{8.21}$$

with respect to $\mathbf{u}_k$, τ_k and λ must vanish except possibly at the end points t_1, t_2, or where $G(\cdot, \cdot, \cdot)$ is not differentiable, i.e., at $\alpha_k = 0$. Thus

$$\alpha_k = 0 \quad \text{or} \quad \frac{B(\tau_k)^\top B(\tau_k)\mathbf{u}_k}{\|B(\tau_k)\mathbf{u}_k\|} + B(\tau_k)^\top \Phi^{-1}(\tau_k)^\top \lambda = 0, \quad k = 1, 2, \cdots, p, \tag{8.22}$$

$$\alpha_k = 0 \quad \text{or} \quad \tau_k = t_1 \quad \text{or} \quad \tau_k = t_2 \quad \text{or}$$
$$\frac{\mathbf{u}_k^\top \frac{dB(\tau_k)}{d\tau_k}^\top B(\tau_k)\mathbf{u}_k}{\|B(\tau_k)\mathbf{u}_k\|} + \lambda^\top \frac{d(\Phi^{-1}(\tau_k)B(\tau_k))}{d\tau_k}\mathbf{u}_k = 0, \quad k = 1, 2, \cdots, p, \tag{8.23}$$

and

$$\sum_{k=1}^{p} \Phi^{-1}(\tau_k)B(\tau_k)\mathbf{u}_k = \mathbf{v}. \tag{8.24}$$

We then have

$$\alpha_k = 0 \quad \text{or} \quad B(\tau_k)^\top(\nu_k + \Phi^{-1}(\tau_k)^\top\lambda) = 0, \quad k = 1, 2, \cdots, p. \tag{8.25}$$

Since ν_k is an element of the column space of $B(\tau_k)$, we have

$$C(\tau_k)^\top \nu_k = 0, \quad k = 1, 2, \cdots, p, \tag{8.26}$$

where $C(t)$ is an $n \times (n - \text{Rank}B(t))$ matrix whose columns form an orthonormal basis of the orthogonal complement of the column space of $B(t)$. From Eqs. (8.25) and (8.26) we have

$$\begin{bmatrix} B(\tau_k)^\top \\ C(\tau_k)^\top \end{bmatrix} \nu_k = -\begin{bmatrix} B(\tau_k)^\top \\ \mathbf{0} \end{bmatrix} \Phi^{-1}(\tau_k)^\top \lambda, \quad k = 1, 2, \cdots, p, \tag{8.27}$$

where the matrix on the left-hand side is of order $(m+n-\text{Rank}B(\tau_k)\times n)$ and has rank n. The unique solution of Eq. (8.27) for ν_k is given by

$$\begin{aligned}\nu_k &= -\begin{bmatrix} B(\tau_k)^\top \\ C(\tau_k)^\top \end{bmatrix}^\dagger \begin{bmatrix} B(\tau_k)^\top \\ \mathbf{0} \end{bmatrix} \Phi^{-1}(\tau_k)^\top \lambda \\ &= -[B^\dagger(\tau_k)^\top \mid C(\tau_k)] \begin{bmatrix} B(\tau_k)^\top \\ \mathbf{0} \end{bmatrix} \Phi^{-1}(\tau_k)^\top \lambda \\ &= -B^\dagger(\tau_k)^\top B(\tau_k)^\top \Phi^{-1}(\tau_k)^\top \lambda, \qquad k = 1, 2, \cdots, p. \end{aligned} \tag{8.28}$$

From which we have Eq. (8.14) and since ν_k is a unit vector, we immediately have Eqs. (8.15) followed from (8.14).

The first three relations of (8.23) are identical to those of (8.16). In terms of α_k and ν_k, the last relation in (8.23) becomes

$$\begin{aligned} & \frac{d\alpha_k}{d\tau_k} + \lambda^\top \frac{d(\Phi^{-1}(\tau_k)\alpha_k\nu_k)}{d\tau_k} \\ = & \frac{d\alpha_k}{d\tau_k} + \frac{d\alpha_k}{d\tau_k}\lambda^\top \Phi^{-1}(\tau_k)\nu_k + \alpha_k\lambda^\top \frac{d(\Phi^{-1}(\tau_k)\nu_k)}{d\tau_k} \\ = & \ 0. \end{aligned} \tag{8.29}$$

From Eq. (8.25) we have

$$\nu_k = -\Phi^{-1}(\tau_k)^\top \lambda + a_k, \qquad \text{where } a_k \in [\text{col}\{B(\tau_k)\}]^\perp, \tag{8.30}$$

and since $\nu_k \in \text{col}\{B(\tau_k)\}$, multiplying the transpose of Eq. (8.30) by ν_k gives

$$1 = -\lambda^\top \Phi^{-1}(\tau_k)\nu_k, \tag{8.31}$$

substitute which into Eq. (8.29) gives

$$\alpha_k = 0, \quad \text{or} \quad \lambda^\top \frac{\Phi^{-1}(\tau_k)\nu_k}{d\tau_k} = 0. \tag{8.32}$$

Since $K(t)$ is symmetric and satisfies

$$K(t)K(t)^\top = K(t) \tag{8.33}$$

and from Eq. (8.14) we get the final relation of Eq. (8.16).

Equation (8.17) follows from substituting Eqs. (8.12), (8.33), and (8.14) into Eq. (8.24). Equation (8.18) is a consequence of the definition of α_k, and Eq. (8.19) is established by multiplying Eq. (8.17) with $\lambda^\top$ and using Eq. (8.15) to simplify the result. The right hand side of Eq. (8.19) must be positive since otherwise we would necessarily have α_k all zero and consequently Eq. (8.17) could not be satisfied since $\mathbf{v}$ is nonzero by assumption.

All of Eqs. (8.14)-(8.19) must hold at any local minimizing solution. The relation (8.20) is a necessary condition for the solution to be a global minimum. We have thus established the necessity of relations in Eqs. (8.14)-(8.19). We then prove that they are sufficient.

Let $\lambda, \tau_1, \tau_2, \cdots, \tau_p, \alpha_1, \alpha_2, \cdots, \alpha_p$ satisfy Eqs. (8.15)-(8.20) and whenever $\alpha_k \neq 0$, let ν_k satisfy Eq. (8.14). For $k = 1, 2, \cdots, p$, set

$$B(\tau_k)\mathbf{u}_k = \begin{cases} 0, & \text{if } \alpha_k = 0, \\ \nu_k \alpha_k, & \text{if } \alpha_k \neq 0. \end{cases} \tag{8.34}$$

Since $\nu_k \in \text{col}\{B(\tau_k)\}$, Eq. (8.34) is consistent and has a solution $\mathbf{u}_k$. In the case where $B(\tau_k)$ has a nontrivial kernel the solution, $\mathbf{u}_k$, to Eq. (8.34) will not be unique but rather Eq. (8.34) then defines an equivalent class for $\mathbf{u}_k$. In this case we choose the solution $\mathbf{u}_k$ form the equivalence class that is orthogonal to the kernel of $B(\tau_k)$. Thus we have an element

$$\mu_0 = (\tau_1, \tau_2, \cdots, \tau_p, \mathbf{u}_1, \mathbf{u}_2, \cdots, \mathbf{u}_p)$$

defined by (8.34) that satisfy Eqs. (8.14)-(8.20) through the definitions in Eq. (8.12). In particular, (8.5) holds so that μ_0 will take a solution $\mathbf{x}(t)$ of (8.1) satisfying the terminal conditions.

If we define $U \triangleq [t_1, t_2]^p \times \Re^{mp}$ then we must show that

$$J_1(\mu_0) \leq J_1(\mu), \qquad \forall \mu \in U.$$

Since

$$J_1(\mu_0) = \sum_{k=1}^{p} \|B(\tau_k)\mathbf{u}_k\|$$

is finite, we may choose a real number $M > J_1(\mu_0)$ and define the set

$$U_M \triangleq \{\mu \in U | J_1(\mu) \leq M\}.$$

The set U_M is clearly closed and $\mu_0 \in U_M$ so that U_M is nonempty. It can also be shown that U_M is bounded since the τ_k lie within $[t_1, t_2]$ and the vectors $\mathbf{u}_k$ are necessarily bounded due to the definition of J_1 and the fact that the components of $\mathbf{u}_k$ in the kernel of $B(\tau_k)$ have been chosen as zero. The cost function J_1 is continuous over the compact set U_M hence there is an element μ^* such that $J_1(\mu^*)$ is a minimum of J_1 over U_M. Therefore

$J_1(\mu^*) \leq J_1(\mu_0) < M$ so that μ^* is an interior point of U_M; consequently μ^* is a local minimum and must satisfy Eqs. (8.14)-(8.19), but since μ_0 also satisfies (8.20) we have $J_1(\mu_0) \leq J_1(\mu^*)$, hence $J_1(\mu_0) = J_1(\mu^*)$.

Suppose that there exists a $\mu \in U$ such that $J_1(\mu) = J_1(\mu_0)$; then since $J_1(\mu_0) < M$, we have $\mu \in U_M$, and since μ^* is minimal over U_M we have $J_1(\mu_0) = J_1(\mu^*) \leq J_1(\mu)$, a contradiction. □

Theorem 8.2 *Let*

$$\begin{aligned} \tilde{\alpha}_k &\stackrel{\triangle}{=} \|\mathbf{u}_k\|, \\ \tilde{\nu}_k &\stackrel{\triangle}{=} \frac{\mathbf{u}_k}{\tilde{\alpha}_k} \quad \text{if} \quad \tilde{\alpha}_k \neq 0, \\ & \quad k = 1, 2, \cdots, p. \end{aligned} \tag{8.35}$$

Given the terminal conditions $(t_1, \mathbf{x}_1)$ and $(t_2, \mathbf{x}_2)$ and suppose that $\mathbf{v} \neq \mathbf{0}$, then for each $p = 1, 2, \cdots$, the element $(\tau_1, \tau_2, \cdots, \tau_p, \mathbf{u}_1, \mathbf{u}_2, \cdots, \mathbf{u}_p)$ minimizes J_2 over the set $[t_1, t_2]^p \times \Re^{mp}$ if and only if there exists a nonzero $\lambda \in \Re^n$ such that

$$\tilde{\alpha}_k = 0, \quad \text{or} \quad \tilde{\nu}_k = -B(\tau_k)^\top \Phi^{-1}(\tau_k)^\top \lambda, \quad k = 1, 2, \cdots, p, \tag{8.36}$$

$$\begin{aligned} &\tilde{\alpha}_k = 0, \quad \text{or} \quad \lambda^\top \Phi^{-1}(\tau_k) B(\tau_k) B(\tau_k)^\top \Phi^{-1}(\tau_k)^\top \lambda = 1, \\ &k = 1, 2, \cdots, p, \end{aligned} \tag{8.37}$$

$$\begin{aligned} &\tilde{\alpha}_k = 0, \quad \text{or} \quad \tau_k = t_1, \quad \text{or} \quad \tau_k = t_2, \quad \text{or} \\ &\lambda^\top \frac{d(\Phi^{-1}(\tau_k) B(\tau_k) B(\tau_k)^\top \Phi^{-1}(\tau_k)^\top)}{d\tau_k} \lambda = 0, \\ &k = 1, 2, \cdots, p, \end{aligned} \tag{8.38}$$

$$\sum_{k=1}^{p} [\Phi^{-1}(\tau_k) B(\tau_k) B(\tau_k)^\top \Phi^{-1}(\tau_k)^\top \lambda] \tilde{\alpha}_k = -\mathbf{v}, \tag{8.39}$$

$$\tilde{\alpha}_k \geq 0, \quad k = 1, 2, \cdots, p, \tag{8.40}$$

$$\sum_{k=1}^{p} \tilde{\alpha}_k = -\lambda^\top \mathbf{v} > 0, \tag{8.41}$$

$$-\lambda^\top \mathbf{v} \text{ is a minimum on the set}$$
$$\{\lambda \in \Re^n | \text{ Eqs. (8.37)-(8.41) are valid}\}. \tag{8.42}$$

Proof. The Lagrange function is give by

$$\tilde{G}(\mathbf{u}_k, \tau_k, \lambda) = \sum_{k=1}^{p} ||\mathbf{u}_k|| + \lambda^\top \left(\sum_{k=1}^{p} \Phi^{-1}(\tau_k) B(\tau_k) \mathbf{u}_k - \mathbf{v} \right),$$
$$k = 1, 2, \cdots, p, \tag{8.43}$$

and the necessary conditions for the constrained extremum are

$$\tilde{\alpha}_k = 0, \quad \text{or} \quad \frac{\mathbf{u}_k}{||\mathbf{u}_k||} + B(\tau_k)^\top \Phi^{-1}(\tau_k)^\top \lambda = 0,$$
$$k = 1, 2, \cdots, p, \tag{8.44}$$

$$\tilde{\alpha}_k = 0, \quad \text{or} \quad \tau_k = t_1, \quad \text{or} \quad \tau_k = t_2, \quad \text{or}$$
$$\lambda^\top \frac{d(\Phi^{-1}(\tau_k) B(\tau_k))}{d\tau_k} \mathbf{u}_k = 0,$$
$$k = 1, 2, \cdots, p, \tag{8.45}$$

and

$$\sum_{k=1}^{p} \Phi^{-1}(\tau_k) B(\tau_k) \mathbf{u}_k = \mathbf{v}. \tag{8.46}$$

From Eqs. (8.35) and (8.44) we have Eq. (8.36) and the other conclusions can be obtained following the similar process of the proof of Theorem 8.1. □

8.2 Example 1: Fuel optimal trajectories

Consider the following impulsive control system[Car91]:

$$\begin{pmatrix} \dot{x}_1 \\ \dot{x}_2 \end{pmatrix} = \begin{bmatrix} 0 & 1 \\ -1 & 0 \end{bmatrix} \begin{pmatrix} x_1 \\ x_2 \end{pmatrix}, \quad t \neq \tau_k,$$
$$\Delta \begin{pmatrix} x_1 \\ x_2 \end{pmatrix} = \begin{bmatrix} 0 \\ 1 \end{bmatrix} u_k, \quad t = \tau_k, \quad k = 1, \cdots, p. \tag{8.47}$$

We will find the fuel optimal solution of this system with respect to cost J_2. The terminal times are given by $t_1 = 0$, $t_2 = 2\pi$, $\mathbf{x}_1 = (x_1(t_1), x_2(t_1))^\top$ and $\mathbf{x}_2 = (x_1(t_2), x_2(t_2))^\top$.

Followed Eq. (8.35), we have

$$\tilde{\alpha}_k = |u_k|, \qquad \tilde{\nu}_k = \frac{u_k}{|u_k|} = \operatorname{sgn}(u_k). \tag{8.48}$$

We have

$$\Phi(t) = \begin{bmatrix} \cos(t) & \sin(t) \\ -\sin(t) & \cos(t) \end{bmatrix}, \quad \Phi^{-1}(t) = \begin{bmatrix} \cos(t) & -\sin(t) \\ \sin(t) & \cos(t) \end{bmatrix}. \tag{8.49}$$

Then the $\mathbf{v}$ defined in Eq. (8.6) is given by

$$\mathbf{v} = \begin{pmatrix} v_1 \\ v_2 \end{pmatrix} = \begin{pmatrix} x_1(t_2) - x_1(t_1) \\ x_2(t_2) - x_2(t_1) \end{pmatrix}. \tag{8.50}$$

We know that

$$B(\tau_k) = \begin{pmatrix} 0 \\ 1 \end{pmatrix} \tag{8.51}$$

and we define $Q(\tau_k)$ as

$$Q(\tau_k) \triangleq \Phi^{-1}(\tau_k)B(\tau_k) = \begin{pmatrix} -\sin(t) \\ \cos(t) \end{pmatrix} \tag{8.52}$$

Then from Eqs. (8.36) and (8.48) we have

$$\begin{aligned} u_k &= 0, \ \text{ or} \\ \operatorname{sgn}(u_k) &= -Q(\tau_k)^\top \lambda \\ &= (\sin(\tau_k) \quad -\cos(\tau_k)) \begin{pmatrix} \lambda_1 \\ \lambda_2 \end{pmatrix} \\ &= \sin(\tau_k)\lambda_1 - \cos(\tau_k)\lambda_2. \end{aligned} \tag{8.53}$$

And from Eqs. (8.37) and (8.48) we have

$$u_k = 0, \quad \text{or} \quad [-\sin(\tau_k)\lambda_1 + \cos(\tau_k)\lambda_2]^2 = 1. \tag{8.54}$$

From Eq. (8.38) we have

$$\begin{aligned} &u_k = 0, \quad \text{or} \quad \tau_k = t_1, \quad \text{or} \quad \tau_k = t_2, \quad \text{or} \\ &\sin(\tau_k)\cos(\tau_k)(\lambda_1^2 - \lambda_2^2) + (\sin^2(\tau_k) - \cos^2(\tau_k))\lambda_1\lambda_2 = 0. \end{aligned} \tag{8.55}$$

The last relation in Eq (8.55) can be rewritten as

$$[\cos(\tau_k)\lambda_1 + \sin(\tau_k)\lambda_2][\sin(\tau_k)\lambda_1 - \cos(\tau_k)\lambda_2] = 0 \tag{8.56}$$

Followed Eq. (8.53) we have

$$[\cos(\tau_k)\lambda_1 + \sin(\tau_k)\lambda_2]\text{sgn}(u_k) = 0 \tag{8.57}$$

from which we have

$$\cos(\tau_k)\lambda_1 + \sin(\tau_k)\lambda_2 = 0, \quad \text{if } u_k \neq 0. \tag{8.58}$$

From Eq. (8.39) we have

$$\sum_{k=1}^{p} \begin{pmatrix} (\sin^2(\tau_k)\lambda_1 - \sin(\tau_k)\cos(\tau_k)\lambda_2)|u_k| \\ (-\sin(\tau_k)\cos(\tau_k)\lambda_1 + \cos^2(\tau_k)\lambda_2)|u_k| \end{pmatrix} = -\begin{pmatrix} v_1 \\ v_2 \end{pmatrix} \tag{8.59}$$

from which we have

$$\sum_{k=1}^{p} \begin{pmatrix} \sin(\tau_k)(\sin(\tau_k)\lambda_1 - \cos(\tau_k)\lambda_2)|u_k| \\ \cos(\tau_k)(-\sin(\tau_k)\lambda_1 + \cos(\tau_k)\lambda_2)|u_k| \end{pmatrix} = -\begin{pmatrix} v_1 \\ v_2 \end{pmatrix} \tag{8.60}$$

by using the second relation in Eq. (8.53) we have

$$\sum_{k=1}^{p} \begin{pmatrix} \sin(\tau_k)(\text{sgn}(u_k))|u_k| \\ \cos(\tau_k)(-\text{sgn}(u_k))|u_k| \end{pmatrix} = -\begin{pmatrix} v_1 \\ v_2 \end{pmatrix} \tag{8.61}$$

which leads to

$$\sum_{k=1}^{p} \sin(\tau_k)u_k = -v_1, \quad \sum_{k=1}^{p} \cos(\tau_k)u_k = v_2. \tag{8.62}$$

From Eqs. (8.53) and (8.58) we know that if $u_k \neq 0$ at an interior point $\tau_k \in (0, 2\pi)$, then

$$\lambda_1 = \text{sgn}(u_k)\sin(\tau_k), \qquad \lambda_2 = -\text{sgn}(u_k)\cos(\tau_k) \tag{8.63}$$

from which we have

$$\theta_{k+1} = \theta_k + \pi, \qquad \text{if } 0 < \theta_k < \pi. \tag{8.64}$$

This means that there can be at most two interior impulses. In this section, we set $p = 4$. If there is a nonzero impulse at an endpoint, then from Eq. (8.53) we know $\lambda_2 = -\text{sgn}(u(0))$ or $\lambda_2 = -\text{sgn}(u(2\pi))$. In this case, from Eq. (8.63) we know that a nonzero interior impulse can only appear at π. We then have the following two cases.

8.2.1 *Case 1*

There are at most two nonzero impulse at $\tau_1 \in (0, \pi)$ and $\tau_2 = \tau_1 + \pi$. In this case, from Eq. (8.62) we have

$$-\sin(\tau_1)(u_1 - u_2) = v_1, \quad \cos(\tau_1)(u_1 - u_2) = v_2, \tag{8.65}$$

from which we have

$$(u_1 - u_2)^2 = v_1^2 + v_2^2 \tag{8.66}$$

It follows from Eqs. (8.63) and (8.41)that

$$|u_1| + |u_2| = |u_1 - u_2|, \tag{8.67}$$

so that u_1 and u_2 must have opposite signs, and

$$J_2 = |u_1| + |u_2| = ||\mathbf{v}||. \tag{8.68}$$

The condition (8.42) shows that Eq. (8.68) is an absolute minimum. The point τ_1 is determined from

$$\tan(\tau_1) = \frac{\sin\tau_1}{\cos\tau_1} = -\frac{v_1}{v_2} \tag{8.69}$$

and $\tau_2 = \tau_1 + \pi$. Since $\tau_1 \in (1, \pi)$, we have $v_1 \neq 0$. Then we have

$$u_1 = -\xi\text{sgn}(v_1), \qquad u_2 = (||\mathbf{v}|| - \xi)\text{sgn}(v_1), \qquad \xi \in [0, ||\mathbf{v}||]. \tag{8.70}$$

8.2.1.1 *Simulation results*

Let $(x_1(0), x_2(0)) = (2, 4)$ and $(x_1(2\pi), x_2(2\pi)) = (6, -10)$, then from Eq. (8.50) we know that

$$v_1 = x_1(2\pi) - x_1(0) = 4, \quad v_2 = x_2(2\pi) - x_2(0) = -14. \tag{8.71}$$

Then from Eq. (8.69) we know that

$$\tau_1 = \tan^{-1}\left(-\frac{v_1}{v_2}\right) = \tan^{-1}(-14/4) = 0.2783 \tag{8.72}$$

and

$$\tau_2 = \tau_2 + \pi = 3.4199. \qquad (8.73)$$

From Eqs. (8.70) and (8.71) we have

$$u_1 = -\xi, \qquad u_2 = (14.5602 - \xi), \qquad \xi \in [0, 14.5602]. \qquad (8.74)$$

Thus, the cost function in this example is $J_2 = |u_1| + |u_2| = 14.5602$ regardless of the values of ξ's. Figure 8.1(a) shows the simulation results with $\xi = 3$. In this case, $u_1 = -3$ and $u_2 = 11.5602$. Figure 8.1(b) shows the simulation results with $\xi = 5.5$. In this case, $u_1 = -5.5$ and $u_2 = 9.0602$.

8.2.2 *Case 2*

Nonzero impulses can appear only at 0, π, and 2π. In this case from Eqs. (8.36) and (8.39) we have

$$u_1 - u_2 + u_3 = v_2. \qquad (8.75)$$

Apply Eq. (8.63) to at the interior point π, we have that $\lambda_1 = 0$ and $\lambda_2 = \pm 1$. Then from Eq. (8.41) we have

$$J_2 = |u_1| + |u_2| + |u_3| = |v_2|. \qquad (8.76)$$

Thus, u_2 has an opposite sign from u_1 and u_3, then the solution of the impulsive optimal control problem is given by

$$u_1 = \xi \text{sgn}(v_2), \quad u_2 = -\zeta \text{sgn}(v_2), \quad u_3 = (|v_2| - \xi - \zeta)\text{sgn}(v_2),$$
$$\xi + \zeta \in [0, |v_2|], \quad \xi \geq 0, \quad \zeta \geq 0. \qquad (8.77)$$

8.2.2.1 *Simulation results*

Let $(x_1(0), x_2(0)) = (2, 2)$ and $(x_1(2\pi), x_2(2\pi)) = (2, 6)$, then from Eq. (8.50) we know that

$$v_1 = x_1(2\pi) - x_1(0) = 0, \quad v_2 = x_2(2\pi) - x_2(0) = 4. \qquad (8.78)$$

From Eqs. (8.77) and (8.78) we have

$$u_1 = \xi, \quad u_2 = -\zeta, \quad u_3 = (4 - \xi - \zeta),$$
$$\xi + \zeta \in [0, 4], \quad \xi \geq 0, \quad \zeta \geq 0. \qquad (8.79)$$

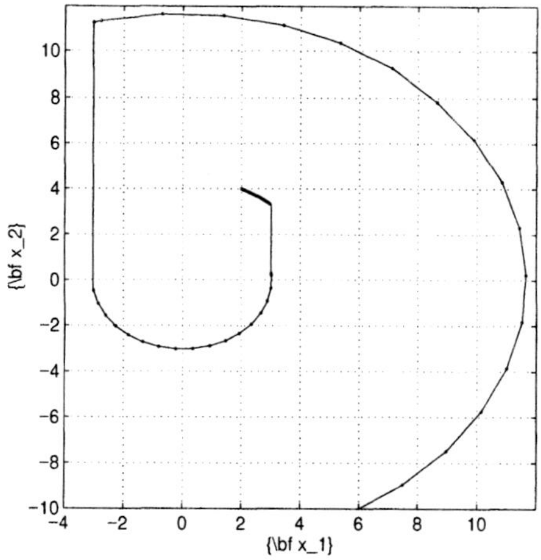

(a)

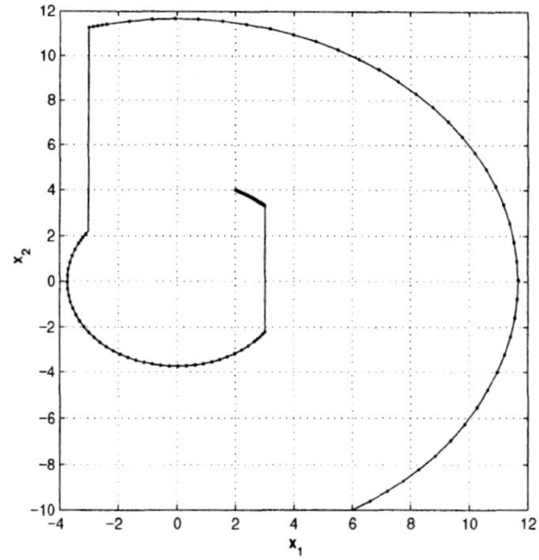

(b)

Fig. 8.1 Optimal impulsive control with respect to J_2. (a) $\xi = 3$. (b) $\xi = 5.5$.

Thus, the cost function in this example is $J_2 = |u_1| + |u_2| + |u_3| = 4$ regardless of the values of ξ's and ζ's. Figure 8.2(a) shows the simulation results with $\xi = 0.5$ and $\zeta = 1$. In this case, $u_1 = 0.5$, $u_2 = -1$ and $u_2 = 2.5$. Figure 8.1(b) shows the simulation results with $\xi = 1$ and $\zeta = 1$. In this case, $u_1 = 1$, $u_2 = -1$ and $u_2 = 2$.

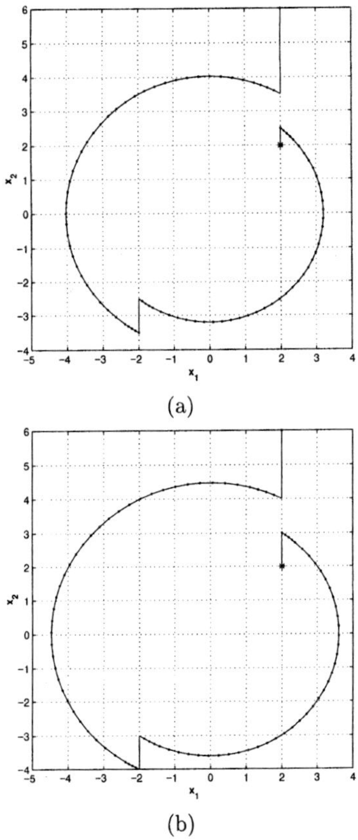

Fig. 8.2 Optimal impulsive control with respect to J_2. (a) $\xi = 0.5$ and $\zeta = 1$. (b) $\xi = 1$ and $\zeta = 1$.

8.3 Example 2

Consider the following Clohessy-Wiltshire model of impulsive maneuver if a spacecraft in near circular orbit[Car91; Liu95]:

$$
\begin{aligned}
\begin{pmatrix} \dot{x}_1 \\ \dot{x}_2 \\ \dot{x}_3 \\ \dot{x}_4 \end{pmatrix} &= \begin{bmatrix} 0 & 0 & 1 & 0 \\ 0 & 0 & 0 & 1 \\ 0 & 0 & 0 & 2 \\ 0 & 3 & -2 & 0 \end{bmatrix} \begin{pmatrix} x_1 \\ x_2 \\ x_3 \\ x_4 \end{pmatrix}, \qquad t \neq \tau_k, \\
\Delta \begin{pmatrix} x_1 \\ x_2 \\ x_3 \\ x_4 \end{pmatrix} &= \begin{bmatrix} 0 & 0 \\ 0 & 0 \\ 1 & 0 \\ 0 & 1 \end{bmatrix} \begin{pmatrix} u_1(\tau_k) \\ u_2(\tau_k) \end{pmatrix}, \qquad t = \tau_k,
\end{aligned} \tag{8.80}
$$

whose fundamental matrix is given by

$$\Phi(t) = \begin{bmatrix} 1 & -3t & 2\cos(t) & 2\sin(t) \\ 0 & -2 & -\sin(t) & \cos(t) \\ 0 & -3 & -2\sin(t) & 2\cos(t) \\ 0 & 0 & \cos(t) & -\sin(t) \end{bmatrix} \tag{8.81}$$

and we have

$$\Phi^{-1}(t) = \begin{bmatrix} 1 & -6t & 3t & 2 \\ 0 & -2 & 1 & 0 \\ 0 & 3\sin(t) & -2\sin(t) & -\cos(t) \\ 0 & -3\cos(t) & 2\cos(t) & -\sin(t) \end{bmatrix}. \tag{8.82}$$

In this example, the terminal condition is given by $t_1 = 0$ and $t_2 = 2\pi$. We also fix the impulse moments at $\tau_1 = 0$, $\tau_2 = \pi/2$, $\tau_3 = 3\pi/2$ and $\tau_4 = 2\pi$. In this case we have $p = 4$ impulses. We will optimize impulsive control of the spacecraft with respect to J_2. To avoid clutter, we define

$$Q(\tau_k) \triangleq \Phi^{-1}(\tau_k)B(\tau_k) = \begin{bmatrix} 3\tau_k & 2 \\ 1 & 0 \\ -2\sin(\tau_k) & -\cos(\tau_k) \\ 2\cos(\tau_k) & -\sin(\tau_k) \end{bmatrix}. \tag{8.83}$$

Then from Eq. (8.36) we have

$$\begin{aligned} u_i(\tau_k) &= 0, \quad i = 1, 2, \quad \text{or} \\ \tilde{\nu}_k &= -Q(\tau_k)^\top \lambda \\ &= \begin{pmatrix} -3\tau_k\lambda_1 - \lambda_2 + 2\sin(\tau_k)\lambda_3 - 2\cos(\tau_k)\lambda_4 \\ -2\lambda_1 + \cos(\tau_k)\lambda_3 + \sin(\tau_k)\lambda_4 \end{pmatrix}, \\ & k = 1, 2, 3, 4. \end{aligned} \tag{8.84}$$

From Eq. (8.37) we have

$$\begin{aligned} u_i(\tau_k) &= 0, \quad i = 1, 2, \quad \text{or} \\ 1 &= (3\tau_k\lambda_1 + \lambda_2 - 2\sin(\tau_k)\lambda_3 + 2\cos(\tau_k)\lambda_4)^2 \\ &\quad +(2\lambda_1 - \cos(\tau_k)\lambda_3 - \sin(\tau_k)\lambda_4)^2, \\ & k = 1, 2, 3, 4. \end{aligned} \tag{8.85}$$

Since $\tau_1 = 0$, $\tau_2 = \pi/2$, $\tau_3 = 3\pi/2$ and $\tau_4 = 2\pi$, Eq. (8.85) gives the following 16 solutions of λ:

$$\begin{pmatrix} -0.065987657 \\ 0.62191901 \\ 0.65110625 \\ 0 \end{pmatrix}, \begin{pmatrix} 0.065987657 \\ -0.62191901 \\ -0.65110625 \\ 0 \end{pmatrix}, \begin{pmatrix} 0.10606582 \\ -0.99964676 \\ 0.23870886 \\ 0 \end{pmatrix},$$
$$\begin{pmatrix} -0.10606582 \\ 0.99964676 \\ -0.23870886 \\ 0 \end{pmatrix}, \begin{pmatrix} -0.098201378 \\ 1.7981234 \\ 0.18228086 \\ -0.43629862 \end{pmatrix}, \begin{pmatrix} 0.098201378 \\ -1.7981234 \\ -0.18228086 \\ 0.43629862 \end{pmatrix},$$
$$\begin{pmatrix} -0.098201378 \\ 0.052928938 \\ 0.18228086 \\ 0.43629862 \end{pmatrix}, \begin{pmatrix} 0.098201378 \\ -0.052928938 \\ -0.18228086 \\ -0.43629862 \end{pmatrix}, \begin{pmatrix} 0 \\ 0 \\ -1/\sqrt(5) \\ -1/\sqrt(5) \end{pmatrix},$$
$$\begin{pmatrix} 0 \\ 0 \\ 1/\sqrt(5) \\ -1/\sqrt(5) \end{pmatrix}, \begin{pmatrix} 0 \\ 0 \\ -1/\sqrt(5) \\ 1/\sqrt(5) \end{pmatrix}, \begin{pmatrix} 0 \\ 0 \\ 1/\sqrt(5) \\ 1/\sqrt(5) \end{pmatrix},$$
$$\begin{pmatrix} 0 \\ 1 \\ 0 \\ 0 \end{pmatrix}, \begin{pmatrix} 0 \\ -1 \\ 0 \\ 0 \end{pmatrix}, \begin{pmatrix} 0 \\ 3/5 \\ 0 \\ -4/5 \end{pmatrix}, \begin{pmatrix} 0 \\ -3/5 \\ 0 \\ 4/5 \end{pmatrix}. \tag{8.86}$$

From Eq. (8.38) we know that an interior impulse should satisfy the following relation:

$$\begin{aligned} u_i(\tau_k) &= 0, \quad i = 1, 2, \quad \text{or} \\ 0 &= 3\tau_k a\lambda_1 + 2b\lambda_1 + b\lambda_2 - 2\sin(\tau_k)a\lambda_3 - \cos(\tau_k)b\lambda_3 \\ &\quad +2\cos(\tau_k)a\lambda_4 - \sin(\tau_k)b\lambda_4, \\ a &= 3\lambda_1 - 2\cos(\tau_k)\lambda_3 - 2\sin(\tau_k)\lambda_4, \\ b &= \sin(\tau_k)\lambda_3 - \cos(\tau_k)\lambda_4, \\ &\quad k = 1, 2, 3, 4. \end{aligned} \tag{8.87}$$

which can be simplified as

$$
\begin{aligned}
u_i(\tau_k) &= 0, \quad i = 1, 2, \quad \text{or} \\
0 &= 9\tau_k \lambda_1^2 - (4\sin(\tau_k) + 6\tau_k \cos(\tau_k))\lambda_1\lambda_3 \\
&\quad +(4\cos(\tau_k) - 6\tau_k \sin(\tau_k))\lambda_1\lambda_4 \\
&\quad + \sin(\tau_k)\lambda_2\lambda_3 - \cos(\tau_k)\lambda_2\lambda_4 \\
&\quad +3\sin(\tau_k)\cos(\tau_k)\lambda_3^2 \\
&\quad +3(\sin^2(\tau_k) - \cos^2(\tau_k))\lambda_3\lambda_4 \\
&\quad -3\sin(\tau_k)\cos(\tau_k)\lambda_4^2, \\
&\quad k = 1, 2, 3, 4.
\end{aligned} \tag{8.88}
$$

Notes

The theory of linear optimal impulsive control was presented in [Car91; Liu95].

Chapter 9

Applications to Nanoelectronics

In this chapter we investigate the nonlinear dynamics of circuits made of single-electron tunneling junctions (SETJ) driven by a sinusoidal pump and biased by a DC voltage source. The mathematical model of an isolated SETJ circuit is a first-order non-autonomous *impulsive* differential equation. The tunneling effect of each SETJ can be realistically modeled by the impulsive effect of the junction voltage, which we choose to be the state variable of the circuit model. Based on this model we present theoretical results on the stability of the periodic and almost periodic solutions of driven SETJs. Theoretical results show there are two *phase states* in each isolated SETJ circuit, which corresponds to two phase-shifted periodic solutions. We present theoretical and numerical results of return maps of an isolated SETJ circuit. The results also show that if the tunneling events are equi-potentially almost periodic then the attractors generated by two-coupled SETJs will be confined to the vicinity of some periodic orbits. This result provides a foundation for implementing robust logic operations in nanoelectronics.

The driven single-electron tunneling junction (SETJ) structure shown in Fig. 9.1 has been investigated for implementing logic operations via "tunnel phase logic" (TPL) in [OK96; LAK99; KO95; YC00b]. This TPL operation has many potential applications in digital technology[OK96; LAK99; KO95]. The driven SETJ circuit in Fig. 9.1 can be modeled by a non-autonomous impulsive differential equation. The general form of a non-

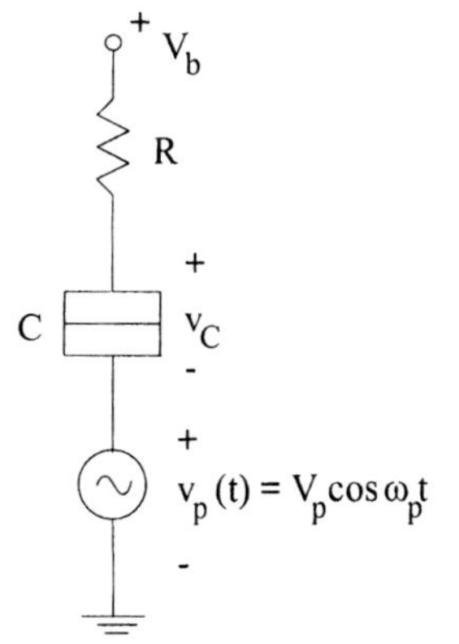

Fig. 9.1 An isolated driven single-electron tunneling junction circuit.

autonomous impulsive differential equation is given by

$$\begin{aligned} \frac{d\mathbf{x}}{dt} &= \mathbf{f}(t,\mathbf{x}) + \mathbf{g}(t), \quad t \neq \tau_i(\mathbf{x}), \\ \Delta\mathbf{x}(\tau_i) &\triangleq \mathbf{x}(\tau_i^+) - \mathbf{x}(\tau_i) = \mathbf{I}_i(\mathbf{x}) + \mathbf{h}_i, \quad t = \tau_i(\mathbf{x}), \\ \mathbf{x}(t_0^+) &= \mathbf{x}_0, t_0 \geq 0, \\ & i = 1, 2, \cdots \end{aligned} \tag{9.1}$$

where $\mathbf{x} \in \mathbb{R}^n$, $\mathbf{f} : \mathbb{R}_+ \times \mathbb{R}^n \to \mathbb{R}^n$ is continuous, $\mathbf{g} : \mathbb{R}_+ \to \mathbb{R}^n$ is a continuous n-vector function of time, and $\mathbf{I}_i : \mathbb{R}^n \to \mathbb{R}^n$ is a continuous n-vector function of state variables. The function $\tau_i : \mathbb{R}^n \to \mathbb{R}_+$ gives the time moment of the ith impulse τ_i; $\mathbf{h}_i \in \mathbb{R}^n$ is a constant n-vector. To avoid clutter, we use a set $\{\tau_i\}$ to denote the time moments of the impulses $\tau_1, \tau_2, \cdots, \tau_i, \cdots$, where $0 < \tau_1 < \tau_2 < \cdots < \tau_i < \tau_{i+1} < \cdots$, $\tau_i \to \infty$ as $i \to \infty$.

9.1 Basic structure and model of isolated SETJ circuit

The block diagram of an isolated SETJ circuit biased by a DC voltage source V_b and driven by a sinusoidal pump $v_p(t)$ is shown in Fig. 9.1. The small box denotes a SETJ with junction capacitance C. The voltage across the SETJ is denoted by v_C and the sinusoidal voltage source (pump) is defined by

$$v_p(t) = V_p \cos(\omega_p t).$$

The DC voltage source V_b is in series with a resistor R.

9.1.1 *Circuit model and its dimensionless form*

The dynamics of the SETJ circuit in Fig. 9.1 is governed by the following non-autonomous impulsive differential equation:

$$\begin{aligned} \frac{dv_C(t)}{dt} &= \frac{1}{RC}(V_b - v_C(t) - V_p \cos(\omega_p t)), \quad \text{if } |v_C(t)| < V_T, \\ \Delta v_C(t_i) &\triangleq v_c(t_i^+) - v_c(t_i) \\ &= \begin{cases} -2V_T, & v_c(t_i) \geq V_T, \\ 2V_T, & v_c(t_i) \leq -V_T, \end{cases} \end{aligned} \tag{9.2}$$

where $\{t_1, t_2, \cdots, t_i, \cdots\}$ are discrete times satisfying $v_C(t_i) = \pm V_T$. Here $v_C(t)$ is the junction voltage, and

$$V_T = \frac{e}{2C}$$

is the quantum-mechanical *tunneling voltage*, where e is the electron charge and C is the junction capacitance. By simple algebra we can normalize (9.2) into the following dimensionless form:

$$\begin{aligned} \frac{d\theta}{d\tau} &= \gamma\left(a - b\cos\tau - \frac{\theta}{\pi}\right), \quad \text{if } |\theta(t)| < \pi, \\ \Delta\theta(\tau_i) &\triangleq \theta(\tau_i^+) - \theta(\tau_i) \\ &= \begin{cases} -2\pi, & \theta(\tau_i) \geq \pi, \\ 2\pi, & \theta(\tau_i) \leq -\pi, \end{cases} \end{aligned} \tag{9.3}$$

where

$$a = \frac{V_b}{V_T}, \quad b = \frac{V_p}{V_T}, \quad \theta = \frac{2\pi C v_C}{e}, \quad \tau = \omega_p t, \quad \gamma = \frac{\pi}{RC\omega_p}.$$

Without loss of generality Eq. (9.3) can be simplified as:

$$\begin{aligned}\frac{d\theta}{d\tau} &= \gamma\left(a - b\cos\tau - \frac{\theta}{\pi}\right), \qquad \text{if } \theta(t) < \pi, \\ \Delta\theta(\tau_i) &= -2\pi, \qquad \text{if } \theta(\tau_i) = \pi, \\ &\quad |\theta(0)| < \pi. \end{aligned} \tag{9.4}$$

Observe that since Eq. (9.4) is not written in the standard mathematical form for impulsive differential equations as that defined in Eq. (9.1), we need to write out explicitly the set $\{\tau_i\}$ as follows:

$$\{\tau_i\} \triangleq \{\tau_i |\ \theta(\tau_i) = \pi\}. \tag{9.5}$$

For convenience, we use t to denote the normalized time τ in Eq. (9.4) and rewrite (9.4) into the following more compact form of non-autonomous impulsive differential equations:

$$\begin{aligned}\frac{d\theta}{dt} &= -\frac{\gamma}{\pi}\theta + \gamma(a - b\cos t), \qquad t \neq \tau_i, \\ \Delta\theta(\tau_i) &= -2\pi, \qquad t = \tau_i, \\ |\theta(0)| &< \pi. \end{aligned} \tag{9.6}$$

where $\{\tau_1, \tau_2, \cdots, \tau_i, \cdots\}$ are discrete times when $\theta(t) = \pi$.

9.1.2 *T-periodic solutions of an isolated SETJ circuit*

In this section we give the conditions under which an isolated SETJ circuit has T-periodic solutions, where T is a positive real number. Since the pump source $b\cos(t)$ is 2π-periodic and the tunneling events for a T-periodic solution should also be T-periodic, we have

$$T = 2n\pi, \qquad \tau_{i+p} = \tau_i + T, \tag{9.7}$$

where n and p are positive integers. The following theorem gives the analytical form of the T-periodic solutions of an isolated SETJ.

Theorem 9.1 *There exists a $T = 2n\pi$ (n is some positive integer) such that the dimensionless SETJ equation (9.6) has n distinct T-periodic solu-*

tions given by

$$\begin{aligned}\theta^*(t) &= e^{-\frac{\gamma}{\pi}t}\frac{1}{1-e^{-\frac{\gamma}{\pi}T}}\left[\int_0^T e^{-\frac{\gamma}{\pi}(T-\tau)}\gamma(a-b\cos(\tau+2q\pi))d\tau \right.\\ &\quad \left. -\sum_{i=1}^{p} 2\pi e^{-\frac{\gamma}{\pi}(T-\tau_i)}\right]\\ &\quad +\int_0^t e^{-\frac{\gamma}{\pi}(t-\tau)}\gamma(a-b\cos(\tau+2q\pi))d\tau - \sum_{0<\tau_i<t} 2\pi e^{-\frac{\gamma}{\pi}(t-\tau_i)},\\ &\quad q=0,1,\cdots,n-1. \end{aligned} \tag{9.8}$$

where $\tau_{i+p} = \tau_i + T$.

Proof. Let the initial condition be $\theta(0,\theta_0)=\theta_0$. Any solution $\theta(t,\theta_0)$ of Eq. (9.6) is given by

$$\theta(t,\theta_0) = e^{-\frac{\gamma}{\pi}t}\theta_0 + \int_0^t e^{-\frac{\gamma}{\pi}(t-\tau)}\gamma(a-b\cos\tau)d\tau - \sum_{0<\tau_i<t} 2\pi e^{-\frac{\gamma}{\pi}(t-\tau_i)}. \tag{9.9}$$

Since $\gamma(a-b\cos\tau)$ is a 2π-periodic function, a 2π phase shift will result in a new solution sharing the same initial condition θ_0. Thus, Eq. (9.9) can be rewritten as

$$\begin{aligned}\theta(t,\theta_0) &= e^{-\frac{\gamma}{\pi}t}\theta_0 + \int_0^t e^{-\frac{\gamma}{\pi}(t-\tau)}\gamma(a-b\cos(\tau+2q\pi))d\tau\\ &\quad - \sum_{0<\tau_i<t} 2\pi e^{-\frac{\gamma}{\pi}(t-\tau_i)}, \quad q=0,1,2,\cdots. \end{aligned} \tag{9.10}$$

For T-periodic solutions that has θ_0 as its initial state, we have the following condition:

$$\begin{aligned}(1-e^{-\frac{\gamma}{\pi}T})\theta_0 &= \int_0^T e^{-\frac{\gamma}{\pi}(T-\tau)}\gamma(a-b\cos(\tau+2q\pi))d\tau\\ &\quad -\sum_{i=1}^{p} 2\pi e^{-\frac{\gamma}{\pi}(T-\tau_i)}, \quad q=0,1,2,\cdots,n-1. \end{aligned} \tag{9.11}$$

Since $(1 - e^{-\frac{\gamma}{\pi}T}) \neq 0$, it follows that

$$\theta_0 = \frac{1}{1 - e^{-\frac{\gamma}{\pi}T}} \left[\int_0^T e^{-\frac{\gamma}{\pi}(T-\tau)} \gamma(a - b\cos(\tau + 2q\pi)) d\tau - \sum_{i=1}^{p} 2\pi e^{-\frac{\gamma}{\pi}(T-\tau_i)} \right], \quad q = 0, 1, 2, \cdots, n-1. \tag{9.12}$$

Substituting (9.12) into (9.9) we obtain the following n distinct T-periodic solutions:

$$\theta^*(t) = e^{-\frac{\gamma}{\pi}t} \frac{1}{1 - e^{-\frac{\gamma}{\pi}T}} \left[\int_0^T e^{-\frac{\gamma}{\pi}(T-\tau)} \gamma(a - b\cos(\tau + 2q\pi)) d\tau - \sum_{i=1}^{p} 2\pi e^{-\frac{\gamma}{\pi}(T-\tau_i)} \right] + \int_0^t e^{-\frac{\gamma}{\pi}(t-\tau)} \gamma(a - b\cos(\tau + 2q\pi)) d\tau - \sum_{0<\tau_i<t} 2\pi e^{-\frac{\gamma}{\pi}(t-\tau_i)}, \quad q = 0, 1, 2, \cdots, n-1. \tag{9.13}$$

□

Remark: Observe that Theorem 9.1 asserts that the SETJ circuit in Fig. 9.1 has at least one T-periodic solution. So far, a $T = 4n\pi$ solution has been found in [OK96]. This theorem shows that for the same periodic orbit (limit cycle) associated with an isolated driven SETJ, the phases of the tunneling events (i.e., the moments of impulsive effects) can be shifted by 2π without changing the initial condition which gives rise to a T-periodic solution. It also shows that a driven SETJ can have n distinct $2n\pi$-periodic solutions. Thus, we immediately have the following corollary.

Corollary 9.1 *If the SETJ equation (9.6) has a $2n\pi$-periodic solution, then there are $(n-1)$ additional distinct $2n\pi$-periodic solutions with 2π phase shifts between consecutive solution pairs.*

Remark: The existence of n distinct $2n\pi$-periodic solutions in an isolated driven SETJ circuit can be used to build robust digital memories as suggested in [OK96]. This is because the different phase shifts can be used to represent different memory states. For example, to build a binary digital memory, we must choose SETJs that have 4π-periodic solutions. The coexistence of two 4π-periodic solutions has been reported in [OK96].

9.1.3 *Simulation results of an isolated SETJ cell*

To demonstrate that the SETJ cell in Fig. 9.1, henceforth called an isolated *SETJ cell*, has two distinct phase states and that the SETJ model in the form of a non-autonomous impulsive differential equation is equivalent to that presented in [OK96], we numerically solve the SETJ model in Eq. (9.3) with parameters $a = 1.77$, $b = 2.0$, and $\gamma = 1/3$. The simulation results with two different initial conditions $\theta(0) = -0.2$ and $\theta(0) = 1.8$ are shown as shadowed and dark curves in Fig. 9.2(a), respectively. From this figure we can see that an isolated SETJ cell can have two stable 4π-periodic solutions. Comparing this simulation result with that presented in Fig. 3 of [OK96], we can see that they are indeed identical. This verifies that the impulsive differential model is equivalent to that presented in [OK96]. In Fig. 9.2(a), the sudden jumps in the shadowed and dark curves occur when electrons tunnel through the SETJ. From Theorem 9.1 we know that these two co-existing 4π-periodic solutions in fact converge to the same limit cycle. This is shown in Fig. 9.2(b) and (c) where the limit cycle is shown in light curves. Observe that these two solutions converge asymptotically to the same limit cycle.

9.2 Return maps of isolated SETJ cells

Since the isolated SETJ cell in Eq. (9.6) is driven by a periodic signal, it is convenient to construct return maps based on the period of the pump signal. The *pth-time return map* can be defined by $\Pi_p : \theta_n \mapsto \theta_{n+p}$, where $\theta_n = \theta(2n\pi)$. In [OK96], the authors presented some numerical results on Π_2. In this section, we will present both theoretical and numerical results on Π_p, $p = 1, 2, \cdots$. In this section, we will use extensively the following formula:

$$\int e^{ax}\cos(bx)dx = \frac{1}{a^2+b^2}e^{ax}[a\cos(bx) + b\sin(bx)] + c \tag{9.14}$$

where c is a constant.

9.2.1 *Theory of* $\Pi_1 : \theta_n \mapsto \theta_{n+1}$

To study the return map $\Pi_1 : \theta_n \mapsto \theta_{n+1}$, we can choose $\theta_n \in [-\pi, \pi)$ as initial conditions at $t = 0$ and calculate θ_{n+1} at time $t = 2\pi$. Using

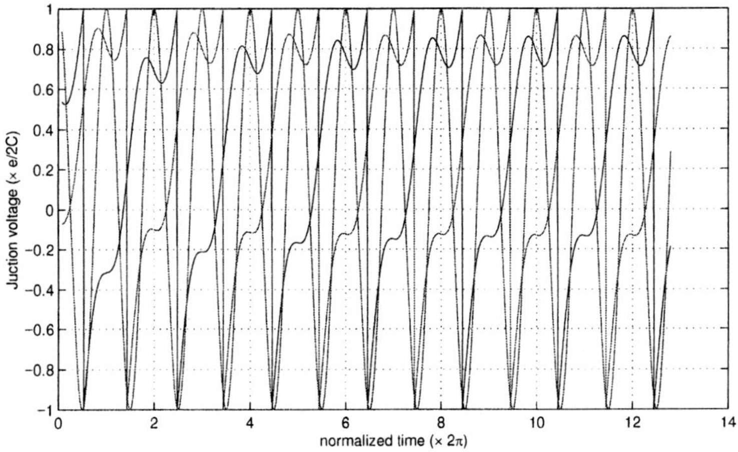

(a)

Fig. 9.2 Simulation results of the two phase states generated by an isolated SETJ cell with two different initial conditions. (a) The waveforms with initial conditions $\theta(0) = -0.2$(shadowed) and $\theta(0) = 1.8$(dark). The pump signal is shown in light curve. (b) The trajectory with $\theta(0) = 1.8$ (dark) converges asymptotically to the limit cycle (shown in light curves). (c) The trajectory with $\theta(0) = -0.2$ (shadowed) converges asymptotically to the limit cycle (shown in light curves).

Eq. (9.9) with initial condition $\theta_0 = \theta_n$, Π_1 is given by the solution of the following two equations:

$$\begin{aligned}
\theta_{n+1} &= e^{-\frac{\gamma}{\pi}(2\pi)}\theta_n + \int_0^{2\pi} e^{-\frac{\gamma}{\pi}(2\pi-\tau)}\gamma(a - b\cos\tau)d\tau \\
&\quad - \sum_{0<\tau_i<2\pi} 2\pi e^{-\frac{\gamma}{\pi}(2\pi-\tau_i)} \\
&= e^{-2\gamma}\theta_n + a\pi - a\pi e^{-2\gamma} - b\gamma e^{-2\gamma}\int_0^{2\pi} e^{\frac{\gamma}{\pi}\tau}\cos\tau d\tau \\
&\quad - \sum_{0<\tau_i(\theta_n)<2\pi} 2\pi e^{-\frac{\gamma}{\pi}(2\pi-\tau_i(\theta_n))} \\
&= e^{-2\gamma}\theta_n + (1 - e^{-2\gamma})\left(a\pi - \frac{b\gamma^2\pi}{\pi^2+\gamma^2}\right) \\
&\quad - \sum_{0<\tau_i(\theta_n)<2\pi} 2\pi e^{-\frac{\gamma}{\pi}(2\pi-\tau_i(\theta_n))}, \\
\theta(\tau_i(\theta_n), \theta_n) &= \pi, \quad \tau_i(\theta_n) \in (0, 2\pi).
\end{aligned} \tag{9.15}$$

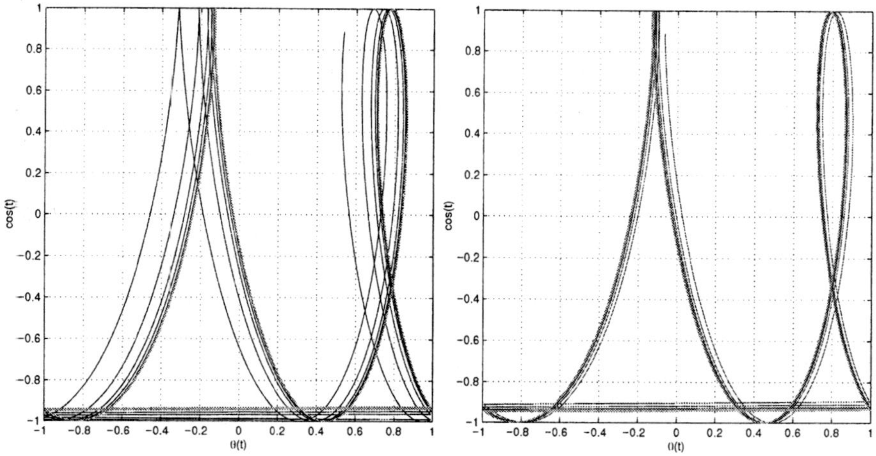

(b) (c)

Fig. 9.2 (*Continued*).

Observe that the only uncertainty comes from impulsive events. Let us consider first the case when there is no impulsive event for a given θ_n:

$$\begin{aligned}\theta(t,\theta_n) &= e^{-\frac{\gamma}{\pi}t}\theta_n + \int_0^t e^{-\frac{\gamma}{\pi}(t-\tau)}\gamma(a - b\cos\tau)d\tau \\ &= e^{-\frac{\gamma}{\pi}t}\theta_n + a\pi(1-e^{-\frac{\gamma}{\pi}t}) \\ &\quad -\frac{b\gamma\pi^2}{\gamma^2+\pi^2}\left(\sin(t)+\frac{\gamma}{\pi}\cos(t)-\frac{\gamma}{\pi}e^{-\frac{\gamma}{\pi}t}\right) \\ &< \pi, \quad \text{for a given } \theta_n \in [-\pi,\pi) \text{ and } t \in (0,2\pi]. \end{aligned} \tag{9.16}$$

By combining the results in Eqs. (9.15) and (9.16) we can derive a closed form expression for Π_1 when there is no impulsive event as follows:

$$\begin{aligned}\theta_{n+1} &= e^{-2\gamma}\theta_n + (1-e^{-2\gamma})\left(a\pi - \frac{b\gamma^2\pi}{\pi^2+\gamma^2}\right), \\ \theta(t,\theta_n) &= e^{-\frac{\gamma}{\pi}t}\theta_n + a\pi(1-e^{-\frac{\gamma}{\pi}t}) \\ &\quad -\frac{b\gamma\pi^2}{\gamma^2+\pi^2}\left(\sin(t)+\frac{\gamma}{\pi}\cos(t)-\frac{\gamma}{\pi}e^{-\frac{\gamma}{\pi}t}\right) \\ &< \pi, \quad \text{for a given } \theta_n \in [-\pi,\pi) \text{ and } t \in (0,2\pi]. \end{aligned} \tag{9.17}$$

This result shows that if there is no impulsive event in an isolated SETJ cell, then Π_1 is a straight line with a slope $e^{-2\gamma}$ and a shift given by

$$\alpha \triangleq (1 - e^{-2\gamma}) \left(a\pi - \frac{b\gamma^2 \pi}{\pi^2 + \gamma^2} \right)$$

along the θ_{n+1}-axis. Observe that the slope $e^{-2\gamma} < 1$ is always satisfied because of physical settings. Hence, if $\alpha \in [-\pi, \pi)$, we can guarantee that the 2π-periodic solutions are asymptotically stable.

Let us consider next the situation when there is an impulsive event within any 2π time period. The interest is to find the initial condition θ^* for a 2π-periodic solution which has only one impulsive event within each period. In this case, $\theta_n = \theta_{n+1} \triangleq \theta^*$ should hold and if $\tau_1^* \in (0, 2\pi]$ is the moment of the impulsive event, we can write

$$\theta^* = e^{-2\gamma}\theta^* + \left(a\pi - \frac{b\gamma^2 \pi}{\pi^2 + \gamma^2} \right)(1 - e^{-2\gamma}) - 2\pi e^{-\frac{\gamma}{\pi}(2\pi - \tau_1^*)}. \tag{9.18}$$

In this equation, there are two unknown parameters θ^* and τ_1^*. We need to construct one more equation in order to find both unknown parameters. One condition we know is that $\theta(\tau_1^*, \theta^*) = \pi$ which leads to the following equation:

$$\begin{aligned} \pi &= e^{-\frac{\gamma}{\pi}\tau_1^*}\theta^* + a\pi(1 - e^{-\frac{\gamma}{\pi}\tau_1^*}) \\ &\quad - \frac{b\gamma\pi^2}{\gamma^2 + \pi^2} \left(\sin(\tau_1^*) + \frac{\gamma}{\pi}\cos(\tau_1^*) - \frac{\gamma}{\pi} e^{-\frac{\gamma}{\pi}\tau_1^*} \right). \end{aligned} \tag{9.19}$$

Since $1 - e^{-2\gamma} \neq 0$ is guaranteed by physical settings, from Eq. (9.18) we have

$$\begin{aligned} \theta^* &= \frac{1}{1 - e^{-2\gamma}} \left[\left(a\pi - \frac{b\gamma^2\pi}{\pi^2 + \gamma^2} \right)(1 - e^{-2\gamma}) - 2\pi e^{-\frac{\gamma}{\pi}(2\pi - \tau_1^*)} \right] \\ &= a\pi - \frac{b\gamma^2\pi}{\pi^2 + \gamma^2} - \frac{2\pi e^{-\frac{\gamma}{\pi}(2\pi - \tau_1^*)}}{1 - e^{-2\gamma}}. \end{aligned} \tag{9.20}$$

Substituting Eq. (9.20) into Eq. (9.19) we have

$$\begin{aligned} \pi &= e^{-\frac{\gamma}{\pi}\tau_1^*} \left(a\pi - \frac{b\gamma^2\pi}{\pi^2 + \gamma^2} - \frac{2\pi e^{-\frac{\gamma}{\pi}(2\pi - \tau_1^*)}}{1 - e^{-2\gamma}} \right) + a\pi(1 - e^{-\frac{\gamma}{\pi}\tau_1^*}) \\ &\quad - \frac{b\gamma\pi^2}{\gamma^2 + \pi^2} \left(\sin(\tau_1^*) + \frac{\gamma}{\pi}\cos(\tau_1^*) - \frac{\gamma}{\pi} e^{-\frac{\gamma}{\pi}\tau_1^*} \right). \end{aligned} \tag{9.21}$$

The solutions of this equation gives τ_1^*. We can solve this equation by numerical methods. The reasonable solution of τ_1^* should be in interval

$(0, 2\pi]$. If there is no solution in interval $(0, 2\pi]$, then it means that there is not 2π-periodic solution with this set of parameters.

As for other initial condition $\theta_n \neq \theta^*$, the moment of the impulsive event satisfies $\tau_1 \neq \tau_1^*$. The return map is then given by the solution of the following equations:

$$\begin{aligned}
\theta_{n+1} &= e^{-2\gamma}\theta_n + (1 - e^{-2\gamma})\left(a\pi - \frac{b\gamma^2\pi}{\pi^2 + \gamma^2}\right) \\
&\quad -2\pi e^{-\frac{\gamma}{\pi}(2\pi - \tau_1(\theta_n))}, \\
\theta(\tau_1(\theta_n), \theta_n) &= e^{-\frac{\gamma}{\pi}\tau_1(\theta_n)}\theta_n + a\pi(1 - e^{-\frac{\gamma}{\pi}\tau_1(\theta_n)}) \\
&\quad -\frac{b\gamma\pi^2}{\gamma^2 + \pi^2}\left(\sin(\tau_1(\theta_n)) + \frac{\gamma}{\pi}\cos(\tau_1(\theta_n)) - \frac{\gamma}{\pi}e^{-\frac{\gamma}{\pi}\tau_1(\theta_n)}\right) \\
&= \pi, \quad \tau_1(\theta_n) \in (0, 2\pi).
\end{aligned} \tag{9.22}$$

Remark: Using the same method presented in this section we can derive many other return maps. Observe that whenever an impulsive event occurs in a return map, there will be an exponential effect on the straight line segments in the return map. Π_p is the solution of the following equations:

$$\begin{aligned}
\theta_{n+p} &= e^{-\frac{\gamma}{\pi}(2p\pi)}\theta_n + \int_0^{2p\pi} e^{-\frac{\gamma}{\pi}(2p\pi - \tau)}\gamma(a - b\cos\tau)d\tau \\
&\quad - \sum_{0<\tau_i<2p\pi} 2\pi e^{-\frac{\gamma}{\pi}(2p\pi - \tau_i)} \\
&= e^{-2p\gamma}\theta_n + a\pi - a\pi e^{-2p\gamma} - b\gamma e^{-2p\gamma}\int_0^{2p\pi} e^{\frac{\gamma}{\pi}\tau}\cos\tau d\tau \\
&\quad - \sum_{0<\tau_i(\theta_n)<2p\pi} 2\pi e^{-\frac{\gamma}{\pi}(2p\pi - \tau_i(\theta_n))} \\
&= e^{-2p\gamma}\theta_n + (1 - e^{-2p\gamma})\left(a\pi - \frac{b\gamma^2\pi}{\pi^2 + \gamma^2}\right) \\
&\quad - \sum_{0<\tau_i(\theta_n)<2p\pi} 2\pi e^{-\frac{\gamma}{\pi}(2p\pi - \tau_i(\theta_n))}, \\
\theta(\tau_i^-(\theta_n), \theta_n) &= \pi, \quad \tau_i(\theta_n) \in (0, 2p\pi).
\end{aligned} \tag{9.23}$$

9.2.2 *Numerical results*

In this section, we present numerical results to verify our theoretical conclusions. The fixed parameters are given by: $b = 2.0$ and $\gamma = 1/3$. The return maps and the corresponding stable waveforms are shown in Fig. 9.3. Figure 9.3(a) shows Π_1 when $a = 0$. In this case, there is no tunneling event in the isolated SETJ cell. Based on the theoretical prediction, the slope of this line is 0.51* and it has a −0.03 vertical shift†. Figure 9.3(b) shows the only 2π-periodic solution which is asymptotically stable.

Figure 9.4(a) shows the return map Π_1 when $a = 1.77$. In this case, the isolated SETJ cell has two asymptotically stable 4π-periodic solutions as shown in Fig. 9.2. This can be clearly seen from Π_2 in Fig. 9.4(b) where Π_2 intersects the $\theta_{n+2} = \theta_n$ line twice.

Figure 9.5(a) shows Π_1 when $a = 1.0$. The sudden jump in Π_1 is because of the impulsive event. Observe that before the jumping occurs, the return map is a straight line. After the jump, the return map becomes an exponential process converging to a straight line. This has been predicted by the theories. In this case, within any 2π time interval, there is no more than one impulsive event. Observe that there is no stable 2π-periodic solution in this case because Π_1 does not intersect the $\theta_{n+1} = \theta_n$ line. However, there is an 8π-periodic solution, as shown in Fig. 9.5(b). As predicted by Corollary 9.1, there should be four distinct 8π-periodic solutions. This theoretical claim is verified by Π_4 shown in Figure 9.5(c). In Fig. 9.5(c) we can see that Π_4 consists of four discontinuous segments each with a slope less than one and intersecting the $\theta_n = \theta_{n+4}$ line at one point. These four intersection points correspond to four stable 8π-periodic solutions. Observe that the basin of attraction of these four 8π-periodic solutions are big enough to represent four logic states.‡

Figure 9.6(a) shows Π_1 when $a = 5$. Observe that the lower part of the return map has many regions where the slopes are bigger than 1 while the upper part has a slope smaller than 1. In this case, the SETJ cell has qusi-periodic solutions. The waveform is shown in Figure 9.6(b) and the corresponding attractor is shown in Figure 9.6(c).

*The simulation result is 0.50. The error is contributed by the numerical integration algorithm.

†Due to the resolution of the printer, this small shift can not be seen clearly. The simulated result is around −0.031.

‡Since in an isolated SETJ cell, we use different periodic solutions to represent logic states, the size of the basin of attraction of a periodic solution decides how robust a logic state can be. This is different from SET transistors because a SET transistor uses the value of charge to encode operations and logic states, and hence any noise will reduce the operating range.

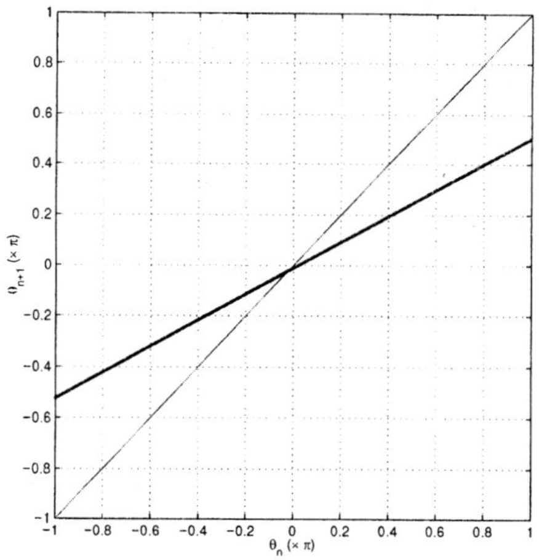

(a)

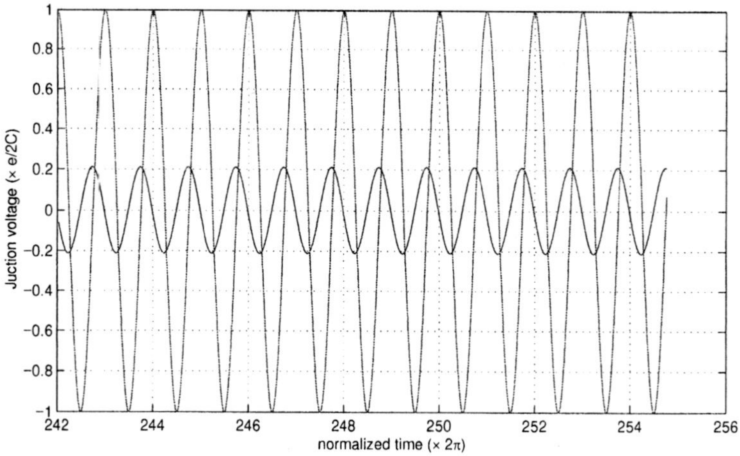

(b)

Fig. 9.3 Return map Π_1 and $\theta(t)$ with $a = 0$. (a) Π_1(think curve). (b) Waveform of $\theta(t)$ (dark) and reference pump signal (shown in light curves).

From the circuit model, we can see that if a increases, then the smallest initial condition that can cause an impulsive event decreases. This in turn shifts the jumping points of the return map towards the left side. In Fig. 9.7 we show an example of this effect. Figure 9.7(a)-(d) show Π_4 with $a = 0.9$, 0.95, 1.05 and 1.1, respectively. Observe that the return maps shift towards

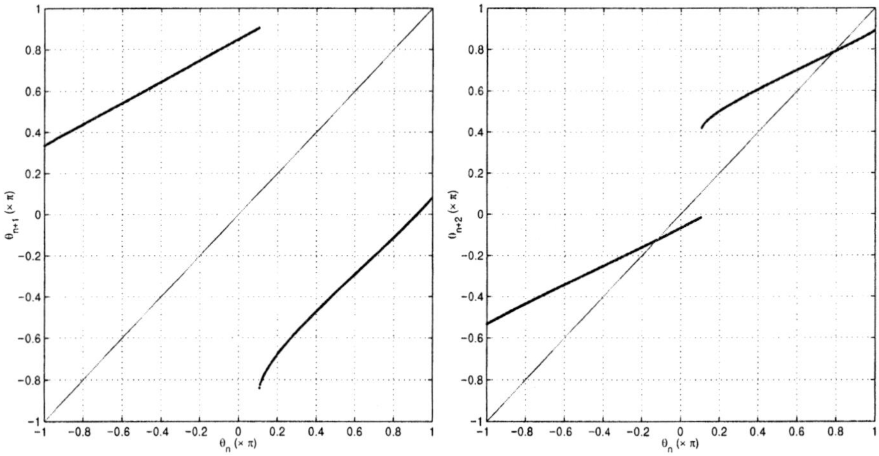

(a) (b)

Fig. 9.4 Return maps (shown in think curves) with $a = 1.77$. (a) Π_1. (b) Π_2.

the right side as a increases. When $a = 0.95$, the four segments of the return maps intersect the shadowed line ($\theta_{n+4} = \theta_n$) giving rise to four stable 8π-periodic solutions. As a becomes larger and larger, the return maps move further and further towards the left side until all four segments have no intersections with the shadowed line (shown in Fig. 9.7(c) and (d)), in this case, the four 8π-periodic solutions lose their stability.

9.3 Basic structure and model of two-coupled SETJ circuits

Let us consider two SETJs coupled through a capacitor C_{in} which is connected in series with a resistor R_{in}, as shown in Fig. 9.8. The dynamics of this circuit model is governed by the following non-autonomous impulsive

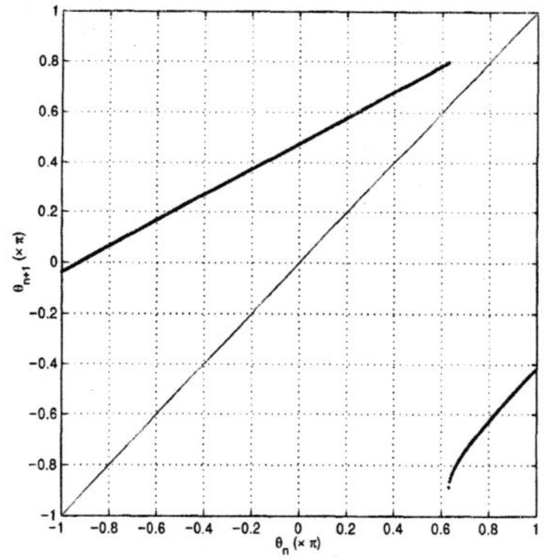

(a)

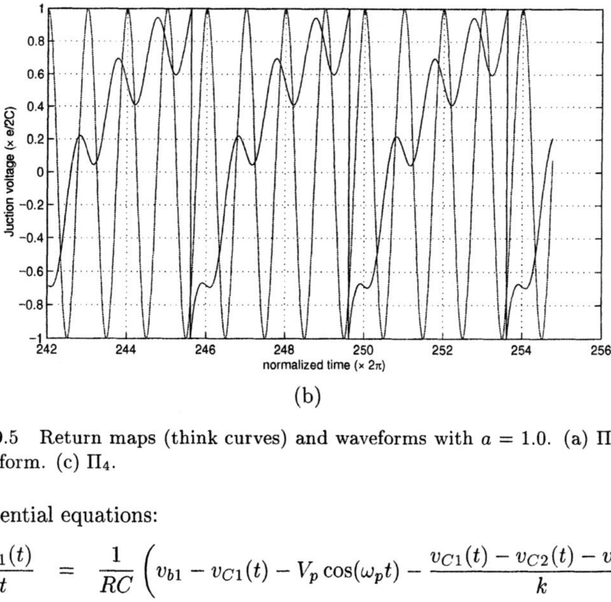

(b)

Fig. 9.5 Return maps (think curves) and waveforms with $a = 1.0$. (a) Π_1. (b) Waveform. (c) Π_4.

differential equations:

$$
\begin{aligned}
\frac{dv_{C1}(t)}{dt} &= \frac{1}{RC}\left(v_{b1} - v_{C1}(t) - V_p\cos(\omega_p t) - \frac{v_{C1}(t) - v_{C2}(t) - v_{\text{in}}(t)}{k}\right), \\
&\quad \text{if } |v_{C1}(t)| < V_T, \\
\frac{dv_{C2}(t)}{dt} &= \frac{1}{RC}\left(v_{b2} - v_{C2}(t) - V_p\cos(\omega_p t) + \frac{v_{C1}(t) - v_{C2}(t) - v_{\text{in}}(t)}{k}\right), \\
&\quad \text{if } |v_{C2}(t)| < V_T, \\
\frac{dv_{\text{in}}(t)}{dt} &= \frac{1}{RC\epsilon k}(v_{C1}(t) - v_{C2}(t) - v_{\text{in}}(t)), \\
\Delta v_{C1}(t_i) &= \begin{cases} -2V_T, & v_{C1}(t_i) \geq V_T, \\ 2V_T, & v_{C1}(t_i) \leq -V_T, \end{cases} \\
\Delta v_{C2}(t_i) &= \begin{cases} -2V_T, & v_{C2}(t_i) \geq V_T, \\ 2V_T, & v_{C2}(t_i) \leq -V_T, \end{cases}
\end{aligned} \tag{9.24}
$$

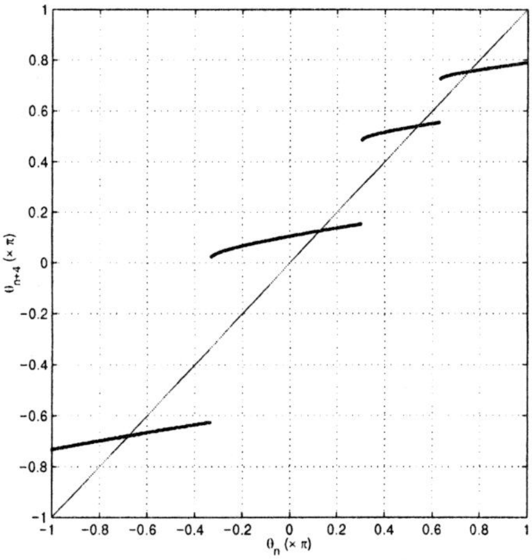

(c)
Fig. 9.5. (*Continued*)

where

$$\epsilon = \frac{C_{\text{in}}}{C}, \quad k = \frac{R_{\text{in}}}{R}.$$

The dimensionless form is given by

$$\begin{aligned}
\frac{d\theta_1}{d\tau} &= \gamma\left(a_1 - b\cos\tau - \frac{(k+1)\theta_1}{k\pi} + \frac{\theta_2}{k\pi} + \frac{\theta_{\text{in}}}{k\pi}\right), \quad \text{if } |\theta_1| < \pi,\\
\frac{d\theta_2}{d\tau} &= \gamma\left(a_2 - b\cos\tau - \frac{(k+1)\theta_2}{k\pi} + \frac{\theta_1}{k\pi} - \frac{\theta_{\text{in}}}{k\pi}\right), \quad \text{if } |\theta_2| < \pi,\\
\frac{d\theta_{\text{in}}}{d\tau} &= \frac{\gamma}{k\epsilon\pi}(\theta_1 - \theta_2 - \theta_{\text{in}}),\\
\Delta\theta_1(\tau_i) &= \begin{cases} -2\pi, & \theta_1(\tau_i) \geq \pi,\\ 2\pi, & \theta_1(\tau_i) \leq -\pi,\end{cases}\\
\Delta\theta_2(\tau_i) &= \begin{cases} -2\pi, & \theta_2(\tau_i) \geq \pi,\\ 2\pi, & \theta_2(\tau_i) \leq -\pi,\end{cases}
\end{aligned} \tag{9.25}$$

where

$$a_1 = \frac{v_{b1}}{V_T}, \quad \theta_1 = \frac{2\pi C v_{C1}}{e}, \quad a_2 = \frac{v_{b2}}{V_T}, \quad \theta_2 = \frac{2\pi C v_{C2}}{e}, \quad \theta_{\text{in}} = \frac{2\pi C v_{\text{in}}}{e}.$$

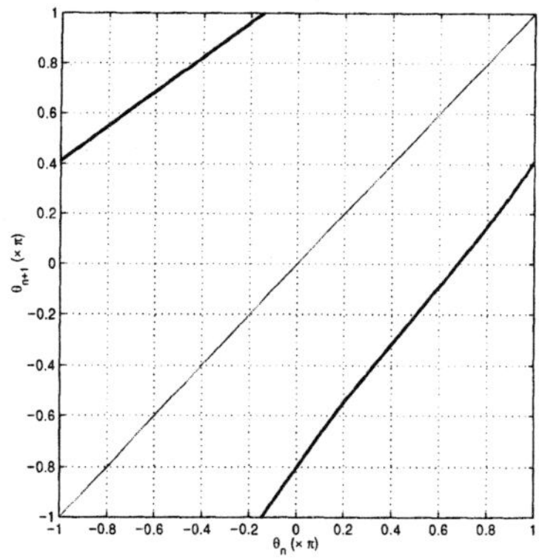

(a)

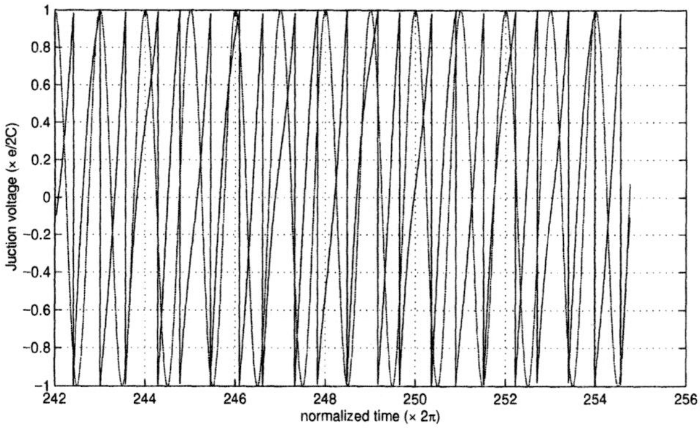

(b)

Fig. 9.6 Return map Π_1, waveform and attractor with $a = 5$. (a) Π_1 (think curves). (b) Waveform. (c) Attractor.

Defining $\mathbf{x} = (\theta_1, \theta_1, \theta_{\text{in}})^\top$ and assuming that $|\theta_1(0)| < \pi$ and $|\theta_2(0)| < \pi$ we can rewrite (9.25) as follow:

$$\begin{aligned}
\frac{d\mathbf{x}}{dt} &= \underbrace{\frac{\gamma}{k\pi}\begin{pmatrix} -k-1 & 1 & 1 \\ 1 & -k-1 & -1 \\ 1/\epsilon & -1/\epsilon & -1/\epsilon \end{pmatrix}}_{\mathbf{A}}\mathbf{x} + \underbrace{\begin{pmatrix} \gamma(a_1 - b\cos t) \\ \gamma(a_2 - b\cos t) \\ 0 \end{pmatrix}}_{\mathbf{g}(t)}, \\
& \quad t \neq \tau_i, \\
\Delta\mathbf{x}(\tau_i^{[1]}) &= \begin{pmatrix} -2\pi \\ 0 \\ 0 \end{pmatrix}, \quad \text{if there is no } j \text{ such that } \tau_i^{[1]} = \tau_j^{[2]}, \\
\Delta\mathbf{x}(\tau_i^{[2]}) &= \begin{pmatrix} 0 \\ -2\pi \\ 0 \end{pmatrix}, \quad \text{if there is no } j \text{ such that } \tau_j^{[1]} = \tau_i^{[2]}, \\
\Delta\mathbf{x}(\tau_i^{[1]}) &= \Delta\mathbf{x}(\tau_j^{[2]}) = \begin{pmatrix} -2\pi \\ -2\pi \\ 0 \end{pmatrix}, \quad \text{if } \tau_i^{[1]} = \tau_j^{[2]}, \\
& \quad |\theta_1(0)| < \pi, \quad |\theta_2(0)| < \pi
\end{aligned} \tag{9.26}$$

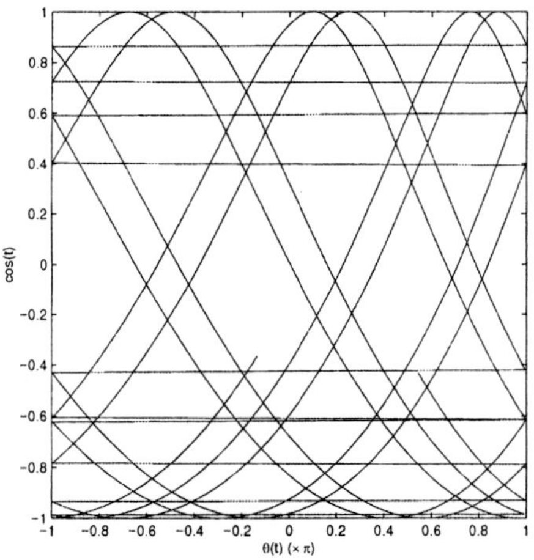

(c)

Fig. 9.6. (*Continued*)

where $\tau_i^{[1]}$ and $\tau_j^{[2]}$ are the ith and jth tunneling moments caused by the first and the second SETJ, respectively, and where

$$\{\tau_i\} = \{\tau_i^{[1]}\} \cup \{\tau_i^{[2]}\} \tag{9.27}$$

denotes the moments of the impulses.

9.3.1 *Theory of two-coupled SETJ circuits*

In many applications, we are concerned with the relative phase differences of the tunneling events occurring in two-coupled SETJ circuits. Let us therefore recast Eq. (9.26) into the following form with respect to the error $r(t) = \theta_1(t) - \theta_2(t)$ and let $a_1 = a_2 = a$:

$$\begin{aligned}
\begin{pmatrix} \frac{dr}{dt} \\ \frac{d\theta_{\text{in}}}{dt} \end{pmatrix} &= \frac{\gamma}{k\pi}\begin{pmatrix} -k-2 & 2 \\ 1/\epsilon & -1/\epsilon \end{pmatrix}\begin{pmatrix} r \\ \theta_{\text{in}} \end{pmatrix}, \quad \text{if } t \neq \tau_i, \\
&\Delta r(\tau_i^{[1]}) = -2\pi, \quad \text{if there is no } j \text{ such that } \tau_i^{[1]} = \tau_j^{[2]}, \\
&\Delta r(\tau_i^{[2]}) = 2\pi, \quad \text{if there is no } j \text{ such that } \tau_j^{[1]} = \tau_i^{[2]}, \\
&\Delta r(\tau_i) = 0, \quad \text{if } \tau_i = \tau_j^{[1]} = \tau_k^{[2]}, \\
&|\theta_1(0)| < \pi, \quad |\theta_2(0)| < \pi,
\end{aligned} \tag{9.28}$$

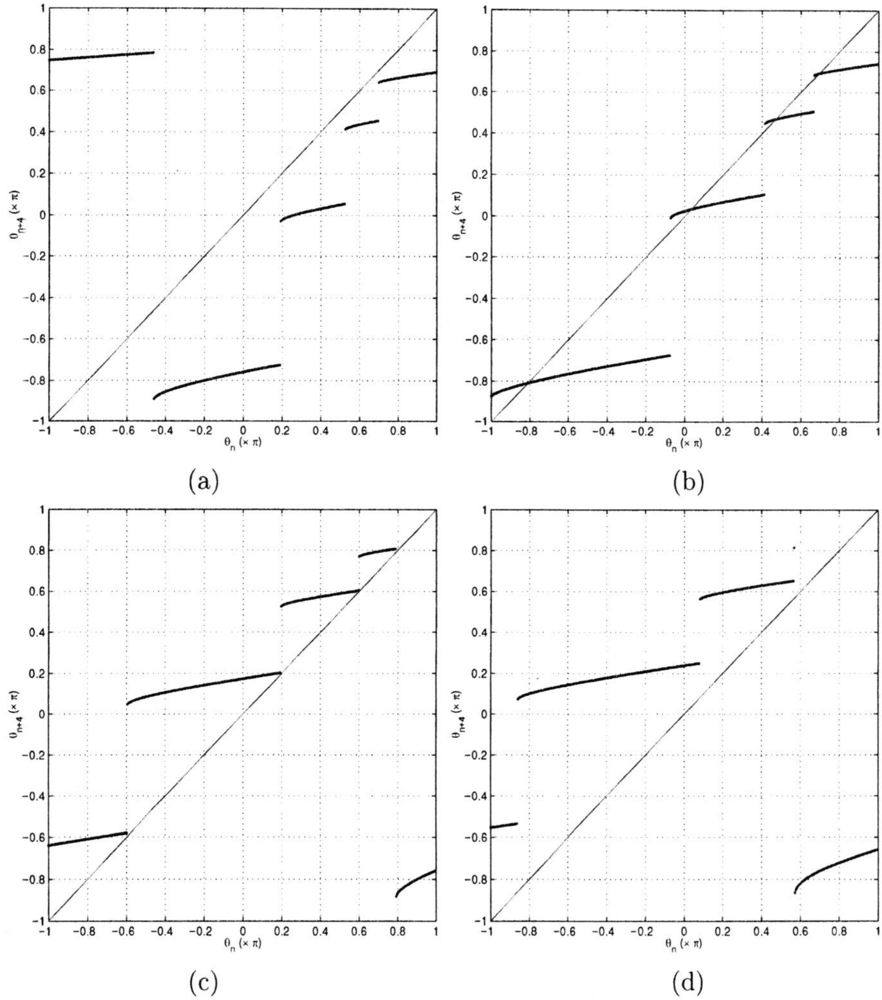

Fig. 9.7 Return maps $\Pi_4 : \theta_n \mapsto \theta_{n+4}$ with different a (shown in think curves). (a) $a = 0.9$. (b) $a = 0.95$. (c) $a = 1.05$. (d) $a = 1.1$.

where $\frac{\gamma}{k\pi} > 0$ is guaranteed by the physical setup in this chapter. Let us choose a and $b\cos(t)$ such that $\theta_1(t) < \pi$ and $\theta_2(t) < \pi$ are always satisfied. Hence, we can guarantee that there is no tunneling events and both SETJ

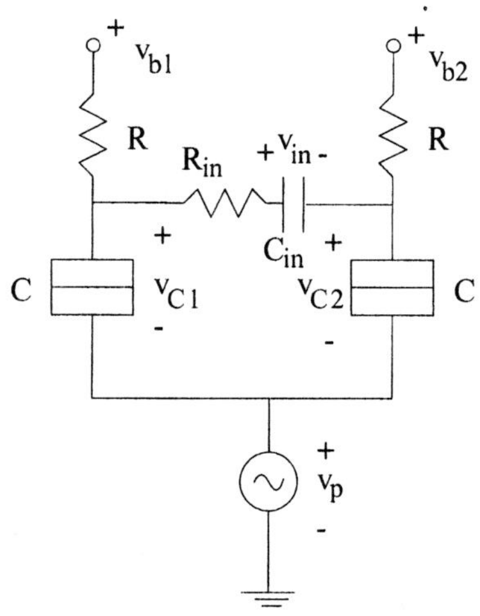

Fig. 9.8 Block diagram of two-coupled SETJ circuits.

circuits will be synchronized to each other if the matrix

$$\mathbf{A}^* = \frac{\gamma}{k\pi}\begin{pmatrix} -k-2 & 2 \\ 1/\epsilon & -1/\epsilon \end{pmatrix}$$

are negative definite. The eigenvalues of the matrix $\mathbf{A}^*$ are the solutions of the characteristic equation

$$\lambda^2 + (k+2+1/\epsilon)\lambda + k/\lambda = 0 \tag{9.29}$$

whose two solutions are given by

$$\lambda_{1,2} = \frac{-(k+2+1/\epsilon) \pm \sqrt{k^2+4k+4+1/\epsilon^2+4/\epsilon-2k/\epsilon}}{2}. \tag{9.30}$$

If the real parts of λ_1 and λ_2 are negative, then both SETJs can be synchronized.

For convenience in the following theoretical analysis, let us recast Eq. (9.28) into the following form:

$$\begin{aligned} \frac{d\mathbf{x}}{dt} &= \mathbf{A}^*\mathbf{x}, \quad \text{if } t \neq \tau_k, \\ \Delta r &= h_k \triangleq \begin{cases} -2\pi, & \text{if } t = \tau_k^{[1]}, \\ 2\pi, & \text{if } t = \tau_k^{[2]}, \\ 0, & \text{if } t = \tau_i^{[1]} = \tau_j^{[2]}, \end{cases} \end{aligned} \tag{9.31}$$

where $\tau_k^{[1]}$ is the kth tunneling moment caused by the first SETJ, $\tau_k^{[2]}$ is the kth tunneling moment caused by the second SETJ, and

$$\{\tau_k\} = \{\tau_k^{[1]}\} \cup \{\tau_k^{[2]}\}, \quad \mathbf{x} = \begin{pmatrix} r \\ \theta_{\text{in}} \end{pmatrix}, \quad \mathbf{A}^* = \frac{\gamma}{k\pi} \begin{pmatrix} -k-2 & 2 \\ 1/\epsilon & -1/\epsilon \end{pmatrix}.$$

Thus we know that the following conditions hold

$$0 \le \tau_0 < \tau_1 < \tau_2 \cdots < \tau_k < \tau_{k+1} \cdots, \quad \lim_{k\to\infty} \tau_k = \infty. \tag{9.32}$$

The solution of Eq. (9.31) with initial condition $\mathbf{x}(t_0^+)$ is given by

$$\begin{aligned} \mathbf{x}(t, t_0) &= \mathbf{x}(t_0^+) U_0(\tau_0, t_0) \prod_{j=1}^{k} U_j(\tau_j, \tau_{j-1}^+) U_{k+1}(t, \tau_k^+) \\ &\quad + \sum_{i=1}^{k+1} U_0(\tau_0, t_0) \prod_{j=i}^{k} U_j(\tau_j, \tau_{j-1}^+) h_{i-1} \\ &= \mathbf{x}(t_0^+) e^{\mathbf{A}^*\left[(t_0-\tau_0)+\sum_{j=1}^{k}(\tau_j-\tau_{j-1})+(t-\tau_k)\right]} \\ &\quad + \sum_{i=1}^{k+1} h_{i-1} e^{\mathbf{A}^*\left[(t_0-\tau_0)+\sum_{j=i}^{k}(\tau_j-\tau_{j-1})\right]}, \\ &\quad t_0 \le \tau_0 < \tau_k < t \le \tau_{k+1}, \end{aligned} \tag{9.33}$$

where

$$U_k(t, s) = e^{\mathbf{A}^*(t-s)}, \quad \tau_{k-1} < s < t \le \tau_k.$$

From Theorem 4.5 of [BS89] and the condition in Eq. (9.32), we immediately have the following Theorem:

Theorem 9.2 *Each solution of system (9.31) is uniformly stable (resp., asymptotically stable, uniformly asymptotically stable, unstable), if, and only if, the solution* $\mathbf{x} = 0$ *of*

$$\frac{d\mathbf{x}}{dt} = \mathbf{A}^*\mathbf{x}, \quad \text{if } t \ne \tau_k,$$

is uniformly stable (resp., asymptotically stable, uniformly asymptotically stable, unstable).

From Theorem 9.2 we know that if both eigenvalues of $\mathbf{A}^*$ are in the open left-hand side of the s-plane, then all solutions of (9.31) are asymptotically stable. However, these solutions can be a fixed point, a limit cycle or an almost periodic attractor, as will be shown later. The type of solution actually observed in the two-coupled SETJ circuits depends on the pattern of the tunneling events. In this case (both eigenvalues have negative real parts), assuming without loss of generality $\tau_0 = t_0 = 0$, we have from Eq. (9.33)

$$\mathbf{x}(\infty) = \sum_{i=1}^{\infty} h_{i-1} e^{\mathbf{A}^*\left[\sum_{j=i}^{\infty}(\tau_j - \tau_{j-1})\right]}. \tag{9.34}$$

Observe that whenever both SETJ circuits are synchronized, they will remain in synchronization forever[§] because, in this case, for any $\tau_i^{[1]}$ there is a $\tau_j^{[2]}$ such that $\tau_i^{[1]} = \tau_j^{[2]}$. When $\tau_i^{[1]} = \tau_j^{[2]}$, from Eq. (9.31) we know that $\Delta r(t) = 0$ for any time t.

9.3.2 *Almost periodic solutions of two-coupled SETJ circuits*

In this section we investigate the almost-periodic solutions of two-coupled SETJ circuits.

Definition 9.1 A function $\phi(t)$ is *almost periodic* if

(1) for any $\epsilon > 0$ there exists a positive number $\delta = \delta(\epsilon)$ such that if two point t_1 and t_2 belong to the same interval of continuity and $|t_1 - t_2| < \delta$, then $||\phi(t_1) - \phi(t_2)|| < \epsilon$;
(2) for any $\epsilon > 0$ there exists a relatively dense set Γ of ϵ-almost periods such that if $\tau \in \Gamma$, then $||\phi(t+\tau) - \phi(t)|| < \epsilon$ for all $t \in \mathbb{R}$ which satisfy the condition $|t - \tau_i| > \epsilon, i = 0, \pm 1, \pm 2, \cdots$.

Definition 9.2 Let $\tau_i^j = \tau_{i+j} - \tau_i$, then the family of the sequences $\{\tau_i^j\}$, $j = 1, 2, \cdots$, is called *equi-potentially almost periodic* if for an arbitrary $\epsilon > 0$ there exists a relatively dense set of ϵ-almost periods, that are common to all sequences $\{\tau_i^j\}$.

Theorem 9.3 *Let* $\mathbf{h}_i \triangleq \Delta\mathbf{x}|_{t=\tau_i}$. *If the largest eigenvalue* λ *of* $\frac{1}{2}(\mathbf{A}+\mathbf{A}^\top)$ *is negative* ($\lambda < 0$)*, and* $\{\tau_i^j\}$ *is equi-potentially almost periodic, then the two-coupled SETJ circuit has a unique almost periodic solution and this solution is asymptotically stable.*

[§]In many cases, the synchronization can be destroyed by an asymmetric change in these two SETJs.

Proof. Since $\mathbf{g}(t)$ is periodic, $\mathbf{h}_i$ is almost periodic, $\lambda < 0$, and $\{\tau_i^j\}$ is equi-potentially almost periodic, the conclusion follows immediately from Theorem 81 in [SP95]. □

Remark: This theorem guarantees that the attractors of two-coupled SETJ circuits can be confined to small regions; namely, in the vicinity of some periodic orbits. This phenomenon can be exploited to ensure a robust parameter tolerance for designing digital SETJ applications.

9.3.3 *Bifurcation and coexisting attractors of two-coupled SETJ circuits*

Let us now use computer simulations to verify the theoretical results presented in this section. In this section, we investigate the bifurcation process exhibited by two-coupled SETJ circuits. We fix the parameters $b = 2$, $\gamma = 1/3$, $\epsilon = 0.5$, $k = 0.1$ and let $a = a_1 = a_2$ be the bifurcation parameter. The Poincaré section is constructed at the location where the pump signal crosses zero from the negative direction. In two-coupled SETJ circuits we have found that there are two co-existing bifurcation diagrams that are partially identical to each other. This means that in some parameter regions, there are two kinds of coexisting attractors.

In Fig. 9.9(a) we show the bifurcation diagram with random initial conditions to guarantee that all coexisting attractors are uncovered. In Fig. 9.9(b) and (c) we show two bifurcation diagrams with two symmetric sets of initial conditions in order to demonstrate the symmetric property of the two kinds of co-existing attractors. Comparing Figs. 9.9(a), (b) and (c) we can see that in some parameter regions (e.g., in the vicinity of $a = 1.37$) some dots that appear in Fig. 9.9(a) do not show up in either Fig. 9.9(b) or (c). This means that there exist two kinds of coexisting attractors in these parameter regions. The examples of two co-existing waveforms will be shown in Figs. 9.10(e) and (f).

In view of the symmetric structure of the two-coupled SETJs, if one attractor is asymmetric then there should be a coexisting attractor such that the union of both attractors forms a symmetric shape with respect to the plane defined by $\theta_1 = \theta_2$ in θ_1-θ_2-θ_{in} space. This phenomenon has been observed from the bifurcation diagrams and is shown in Fig. 9.10. Figure 9.10(a) shows that when the "DC bias" a is small enough, no tunneling event occurs in either SETJ. In this case, both SETJ circuits are identically

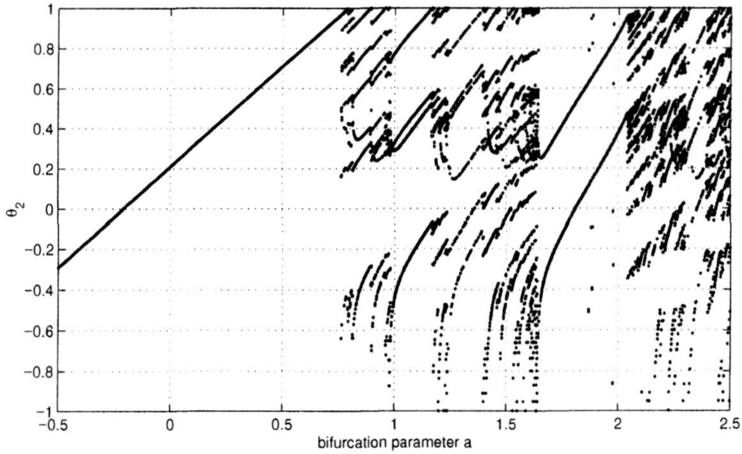

(a)

Fig. 9.9 Bifurcation diagrams of two-coupled SETJs with respect to parameter a. (a) With random initial conditions. (b) With initial condition $\theta_1(0) = -1$, $\theta_2(0) = 1$, and $\theta_{\rm in}(0) = 0$. (c) With initial condition $\theta_1(0) = 1$, $\theta_2(0) = -1$, and $\theta_{\rm in}(0) = 0$.

synchronized to each other because all eigenvalues of the matrix $\mathbf{A}$ are negative. This is predicted by the theoretical results. Figure 9.10(b) shows a symmetric limit cycle with tunneling events occurring in both SETJs. Observe that $\theta_1(t)$ and $\theta_2(t)$ are the same except for a 4π phase shift in their tunneling events. Observe also that the period for the tunnel events in each SETJ is 8π, as shown in Fig. 9.10(c). Figure 9.10(d) shows two coexisting asymmetric limit cycles with tunneling events occurring in both SETJs. The basin of attractions of both attractors are symmetric with respect to the plane $\theta_1 = \theta_2$. Observe from Fig. 9.10(e) that $\theta_1(t)$ and $\theta_2(t)$ has different shapes, and this asymmetric shape implies that there are two coexisting attractors. The waveforms of the dark and shadowed attractors in Fig. 9.10(d) are shown in Figs. 9.10(e) and (f), respectively. Observe that the dark and shadowed waveforms exchange their positions in Figs. 9.10(e) and (f). One promising application of a two-coupled SETJ array is an ultra-high-density memory which uses both attractors to represent the two binary bits. Since these two asymmetric limit cycles can be distinguished easily by monitoring the junction voltage, or by other means, a digital bit can be stored robustly.

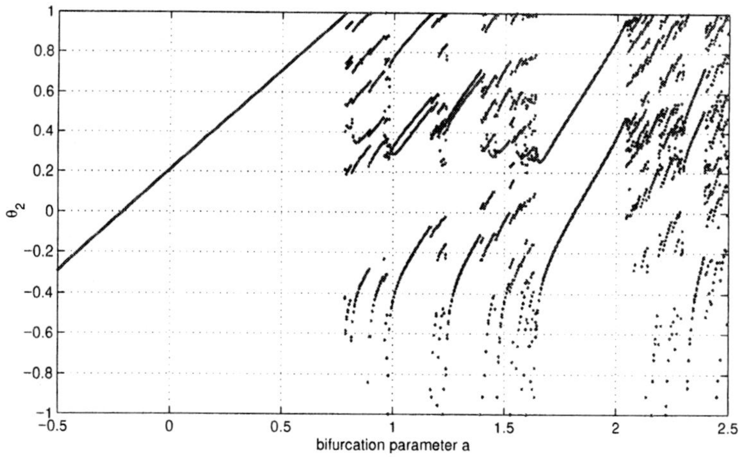

(b)

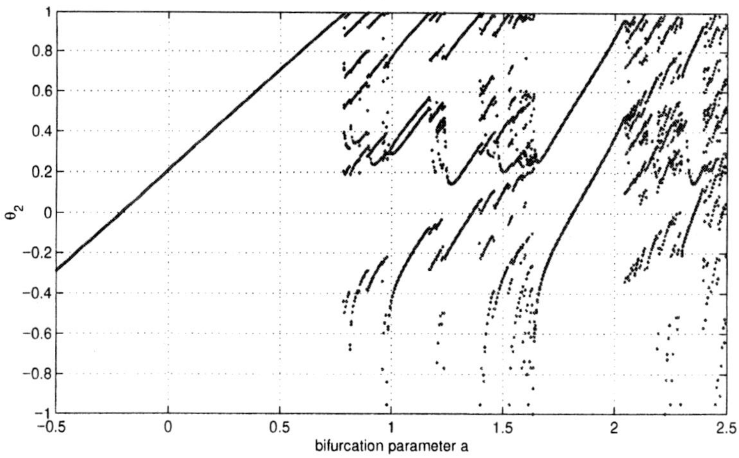

(c)

Fig. 9.9. (*Continued*)

9.3.4 *Poincaré maps of two-coupled SETJ circuits*

Figure 9.11(a) shows a bifurcation diagram when a is chosen over a large interval. In this bifurcation diagram, we choose the Poincaré section as the plane where $\theta_1(t)$ crosses zero from the negative direction. To investigate the structure of the attractors along the bifurcation route, we construct Poincaré maps at the Poincaré sections when the pump is zero. Figure

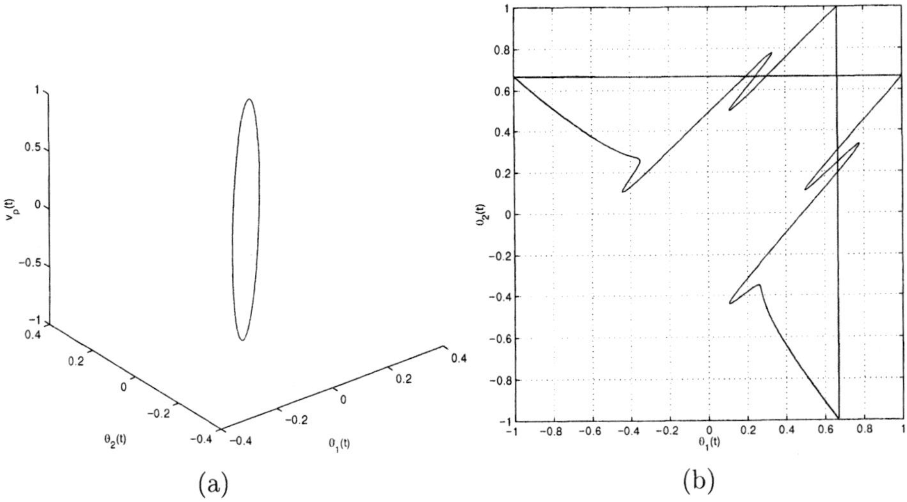

Fig. 9.10 Distinct limit cycles of two coupled SETJs. (a) $a = 0$. $\theta_1(t)$ and $\theta_2(t)$ tend to synchronize with each other asymptotically. No tunneling event occurs. (b) $a = 1$. Only one symmetric limit cycle is found. (c) The waveforms corresponding to the limit cycle in (b), where $\theta_1(t)$, $\theta_2(t)$ and $v_p(t)$ are shown in dark, shadowed and light, respectively. $(\theta_1(0), \theta_2(0), \theta_{\text{in}}(0)) = (-1, 1, 0)$. (d) $a = 0.95$. Two coexisting limit cycles are found. The initial conditions $(\theta_1(0), \theta_2(0), \theta_{\text{in}}(0))$ for the shadowed and dark limit cycles are $(1, -1, 0)$ and $(-1, 1, 0)$, respectively. (e) The waveforms corresponding to the dark attractor in (d) with initial condition $(\theta_1(0), \theta_2(0), \theta_{\text{in}}(0)) = (-1, 1, 0)$. (e) The waveforms corresponding to the shadowed attractor in (d) with initial condition $(\theta_1(0), \theta_2(0), \theta_{\text{in}}(0)) = (1, -1, 0)$.

9.11(b) shows the Poincaré map in the case of $a = 1.6$. Observe that in this case, the Poincaré map consists of isolated points. This indicates that the attractor shown in Fig 9.11(c) is periodic. Figure 9.11(d) shows the Poincaré map in the case of $a = 10$. Observe that in this case, the Poincaré map consists of two curves. This indicates that the attractor as shown in Fig 9.11(c) is quasi-periodic.

9.4 Concluding remarks

Since the phases of tunneling events in a SETJ can be used to encode and process information, it is very important to understand the dynamic behaviors of both periodic and almost periodic solutions of driven SETJs.

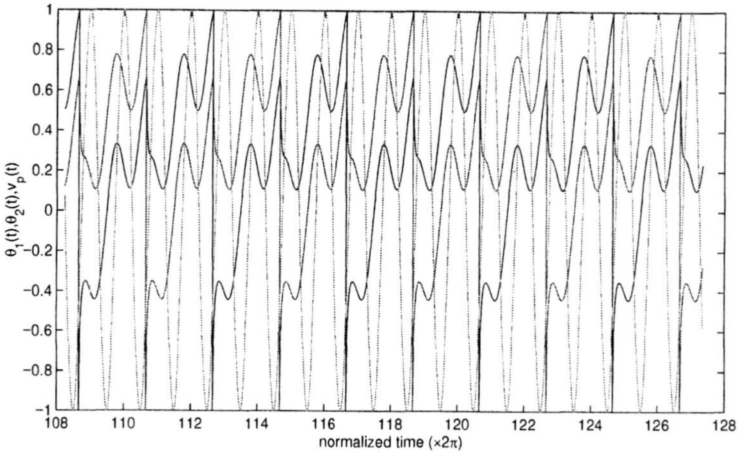

(c)

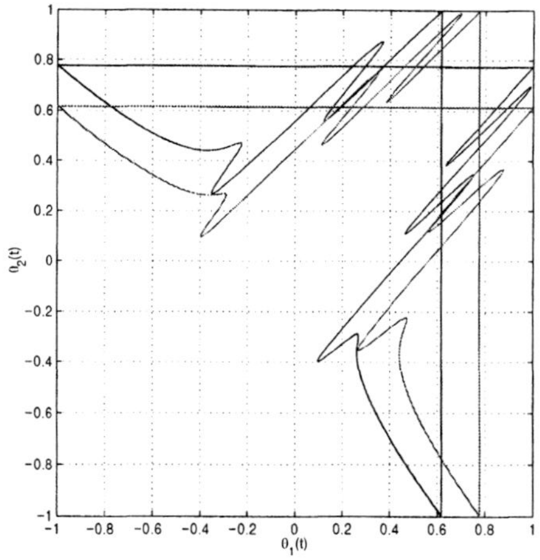

(d)

Fig. 9.10. (*Continued*)

The mathematical model of the SETJ circuits provides us with a rigorous method for exploiting the application of coupled SETJ structures based on driven SETJs.

The two distinct phase states in an isolated SETJ circuits are just two special 4π-periodic solutions that are generated by the same limit cycle. It is possible to find n phase states if we can generate a $2n\pi$-periodic trajectory. The theory guarantees that if any of these n distinct phase states is

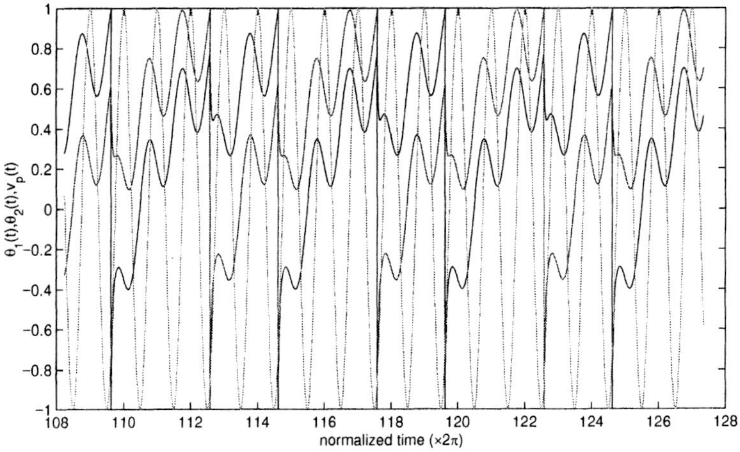

(e)

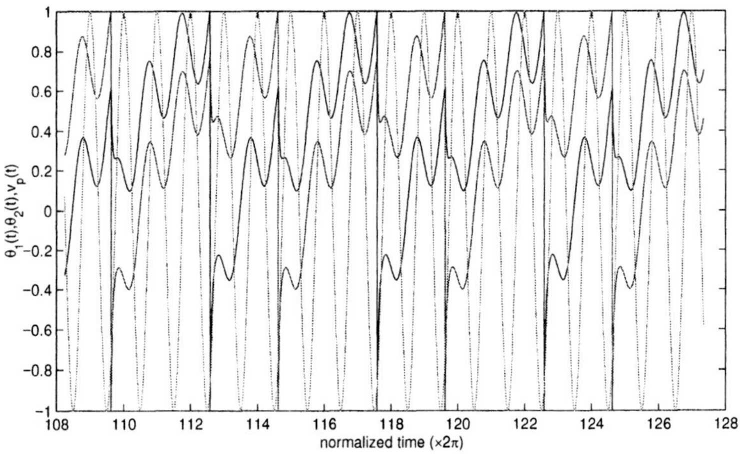

(f)

Fig. 9.10. (*Continued*)

observed, then the others $(n-1)$ distinct phase states *must* also exist. This observation can be used to design multi-valued logic operations based on driven SETJ circuits.

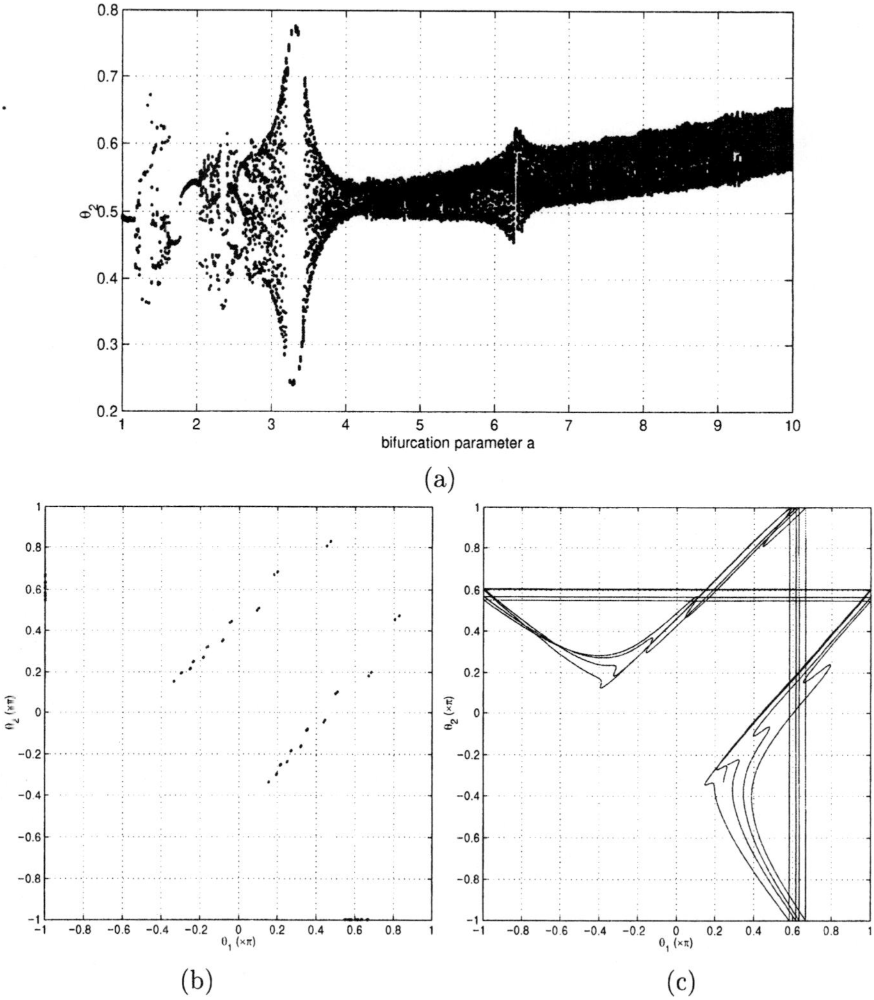

Fig. 9.11 Bifurcation diagram with respect to a and Poincaré maps. (a) Bifurcation diagram. (b) Poincaré map with $a = 1.6$. (c) Attractor with $a = 1.6$. (d) Poincaré map with $a = 10$. (e) Attractor with $a = 10$.

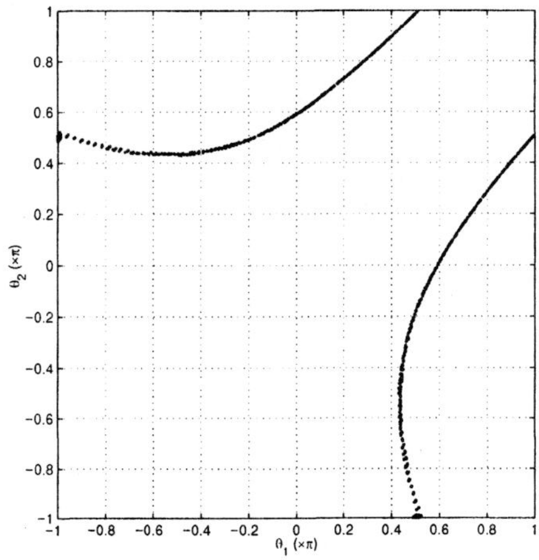

(d)

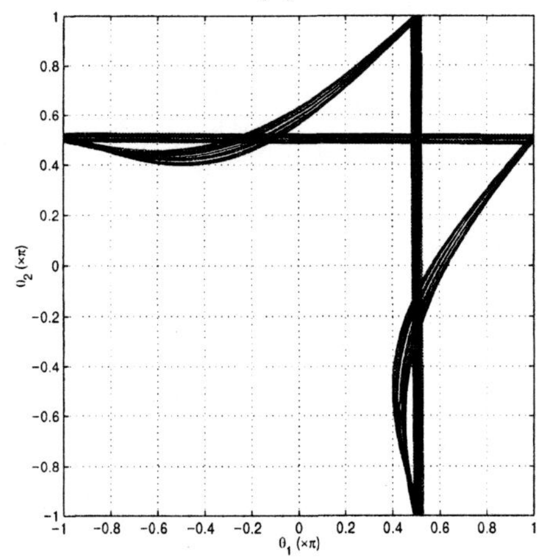

(e)

Fig. 9.11. (*Continued*)

Chapter 10

Experiment Results of Impulsive Synchronization of Chaotic Systems

Impulsive synchronization between two chaotic circuits is the key technology used in many chaotic communication schemes. Since impulsive synchronization involves changing the state variables of continuous dynamical systems rapidly, it is necessary to investigate the problems that may arise in the *practical* implementation of impulsive synchronization. In this chapter we present some experimental results on impulsive synchronization between three kinds of chaotic systems. In these experiments, only one synchronizing channel is used to transmit synchronization impulses. The robustness of impulsive synchronization with respect to variations in the frequency and the width of impulses is studied.

10.1 Experimental 1: Chaotic circuits

10.1.1 *Experimental configurations*

Chua's circuit[Chu94](shown in Fig. 10.1(a)) is chosen as the chaotic system which is defined by

$$\begin{aligned}\frac{dv_1}{dt} &= \frac{1}{C_1}\left[\frac{v_2 - v_1}{R} - f(v_1)\right] \\ \frac{dv_2}{dt} &= \frac{1}{C_2}\left[\frac{v_1 - v_2}{R} + i_3\right] \\ \frac{di_3}{dt} &= -\frac{1}{L}v_2 \qquad (10.1)\end{aligned}$$

where $f(\cdot)$(shown in Fig. 10.1(b)) is the nonlinear characteristic of Chua's diode, defined by

$$f(v_1) = G_b v_1 + \frac{1}{2}(G_a - G_b)(|v_1 + E| - |v_1 - E|) \tag{10.2}$$

and E is the breakpoint voltage of Chua's diode.

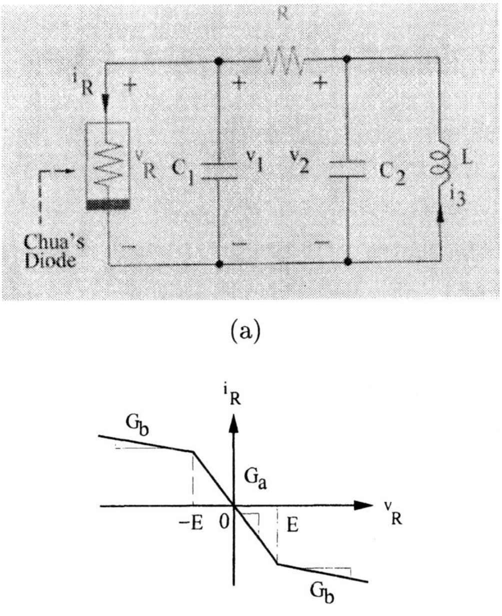

Fig. 10.1 Chua's circuit. (a) Circuit block diagram. (b) Piecewise linear characteristic of Chua's diode.

The parameters chosen for this implementation are $C_1 = 5100pF$, $C_2 = 47nF$, $L = 18mH$ and $R = 1.65k\Omega$. The parameters for implementing Chua's diode are given by $E = 1.56V$, $G_a = -0.757mS$, and $G_b = -0.459mS$. The operational amplifiers chosen in the implementation are type KR1401UD2A. With these parameters we obtain the double scroll strange attractor shown in the v_1-v_2 plane in Fig. 10.2, where the scales of the horizontal and the vertical axes are $0.2V$/div and $50mV$/div, respectively.

The block diagram of the impulsive synchronization using a single syn-

Fig. 10.2 The double scroll strange attractor observed from the Chua's circuit used in the experiments.

chronizing impulse sequence is given in Fig.10.3. Figure 10.3(a) shows the block diagram of impulsive synchronization via v_1 between two Chua's circuits. The switching signal $s(t)$ shown in Fig.10.3(c) is used to switch on/off an electronic switching device (AD6 212 AKN). Whenever $s(t)$ assumes a high voltage level, the switch is on and a synchronizing impulse is transmitted from the driving system to the driven system. We call the width of the high voltage peaks as the *impulse width* w_i. The high voltage level of $s(t)$ is chosen to be greater than $5V$. The time interval between two consecutive impulses is called the *impulse period* $1/f_i$. Whenever $s(t)$ assumes a low voltage level, the switch is turned off and the two Chua's circuits are separated electronically. The low voltage level of $s(t)$ is chosen to be $0V$. Whenever $s(t)$ assumes a high voltage level, the two voltages $v_1(t)$ and $\tilde{v}_1(t)$ are connected by two voltage buffers, which force their output voltages to be equal to their input voltages, and hence $\tilde{v}_1(t)$ must tend rapidly to $v_1(t)$.

Figure 10.3(b) shows the block diagram for impulsive synchronization via v_2 between two Chua's circuits. In this case, the voltages of capacitors C_2 in both Chua's circuits are connected impulsively by two voltage buffers and an electronic switch. It is the same structure as that in Fig.10.3(a) except that a channel gain K is introduced.

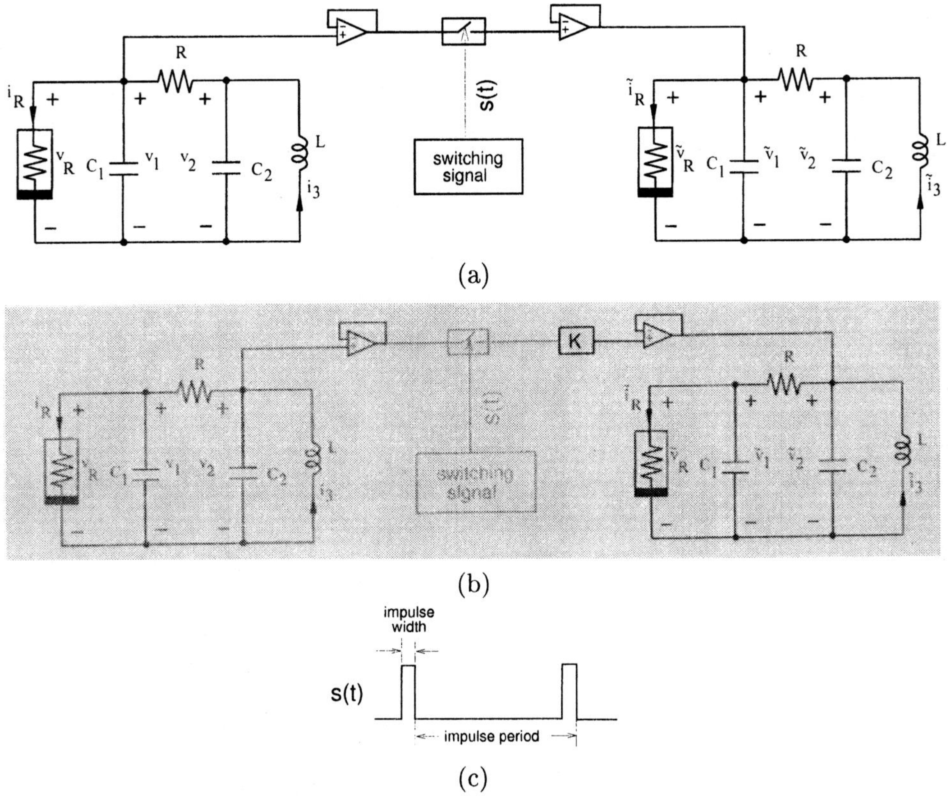

Fig. 10.3 The block diagram of impulsive synchronization between two Chua's circuits. (a) Impulsive synchronization via $v_1(t)$. (b) Impulsive synchronization via $v_2(t)$. (c) Illustration of switching signal $s(t)$.

10.1.2 *Experimental results*

In this section, the experimental results of the configuration presented in Section 10.1.1 are shown. All pictures are taken directly from the screen of an oscilloscope.

10.1.2.1 v_2 *is the transmitted signal*

Let us first study the cases when the voltage v_2 is the transmitted signal as shown in Fig.10.3(b). The experimental results are shown in Fig.10.4. Figure 10.4(a) shows in the $v_1 - \tilde{v}_1$ plane, that the two Chua's circuits are originally desynchronized if impulsive coupling is not applied. Figure 10.4(b) shows in the $v_1 - \tilde{v}_1$ plane, that the two Chua's circuits are synchronized if the impulsive coupling is applied. The parameters for the synchronization impulses are: frequency of impulses is $f_i = 18kHz$, width of impulses is $w_i = 32\mu s$. Figure 10.4(c) shows in the $v_1 - \tilde{v}_1$ plane that the two Chua's circuits become desynchronized if we decrease the width of the impulse beyond a certain limit. The parameters for the synchronization impulses are: $f_i = 18kHz$ and $w_i = 24\mu s$. Figure 10.4(d) shows in the $v_1 - \tilde{v}_1$ plane that the two Chua's circuits become desynchronized if we increase the width of the impulse beyond a certain limit. The parameters for the synchronization impulses are: $f_i = 18kHz$ and $w_i = 46\mu s$. Figure 10.4(e) shows in the $v_2 - \tilde{v}_2$ plane that impulsive synchronization between two Chua's circuits can be maintained with narrow impulses if the amplitude of the impulses is amplified by a gain $K > 1$. In this case, the parameters for the synchronization impulses are: $f_i = 18kHz$, $w_i = 12\mu s$ and $K = 2.67$.

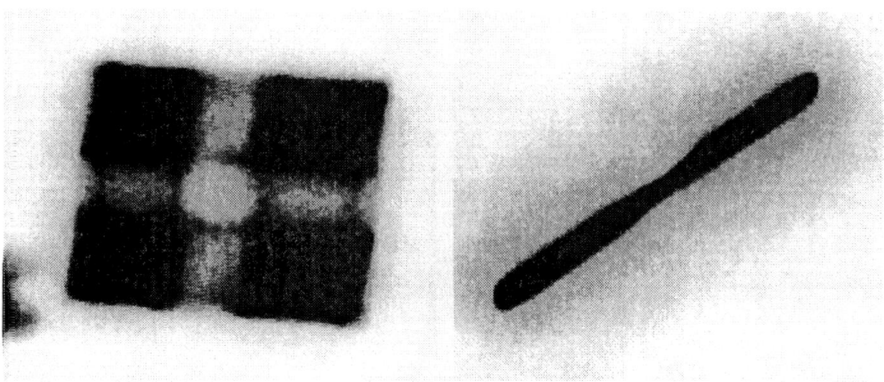

(a) (b)

Fig. 10.4 The experimental results of impulsive synchronization when v_2 is sampled as the synchronizing impulse sequence. In this figure, the scale of the oscilloscope is $0.2V/div$ for both horizontal and vertical axes. (a) The driving system and the driven system are originally desynchronized. (b) Impulsive synchronization ($K = 1$). (c) Desynchronization due to narrow impulses ($K = 1$). (d) Desynchronization due to wide impulses ($K = 1$). (e) Impulsive synchronization with $K > 1$ and narrow impulses.

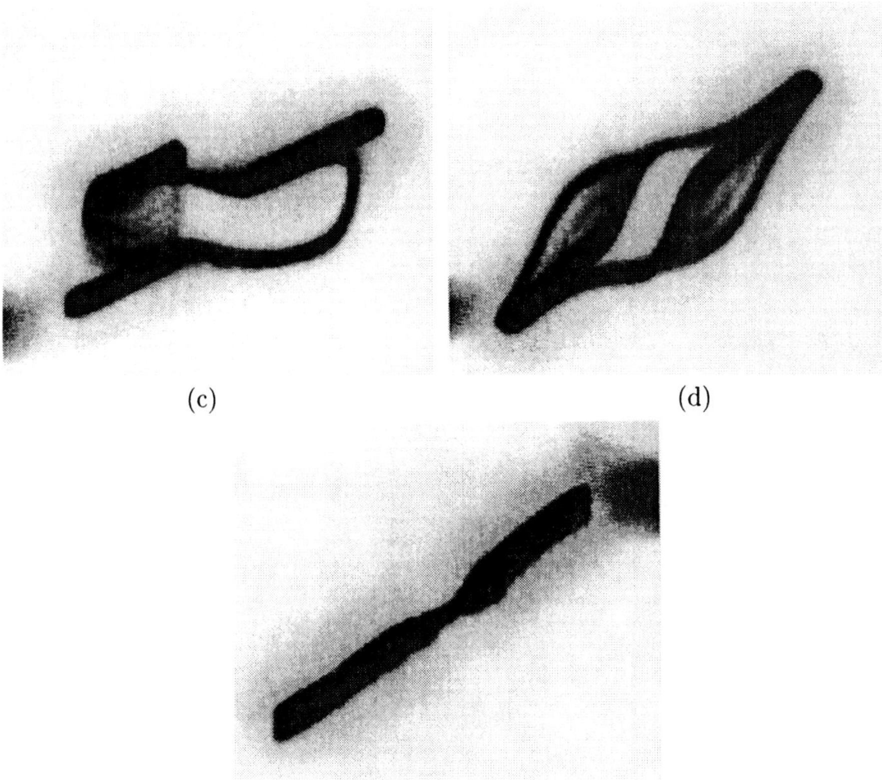

(c) (d)

(e)

Fig. 10.4 (*Continued*).

10.1.2.2 v_1 *is the transmitted signal*

Let us then study the cases when the voltage v_1 is the transmitted signal. The experimental results are shown in Fig.10.5. Figure 10.5(a) shows in the $v_2 - \tilde{v}_2$ plane that the two Chua's circuits are originally desynchronized if impulsive coupling is not applied. Figure 10.5(b) shows in the $v_2 - \tilde{v}_2$ plane that the two Chua's systems are synchronized if impulsive coupling is applied. The parameters for the synchronization impulses are: $f_i = 20kHz$ and $w_i = 32\mu s$. Figure 10.5(c) shows in the $v_2 - \tilde{v}_2$ plane that impulsive synchronization in this case is very robust using only *narrow* impulses, where the parameters of the synchronization impulses are: $f_i = 20kHz$ and $w_i = 8\mu s$. Figure 10.5(d) shows the switching signal $s(t)$ used in

Fig.10.5(c).

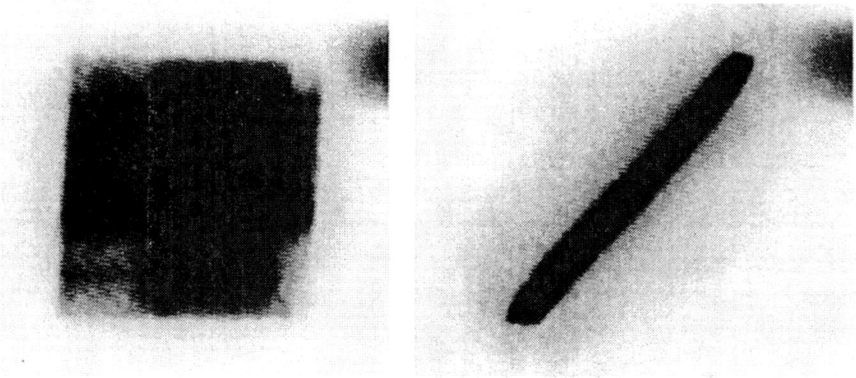

(a) (b)

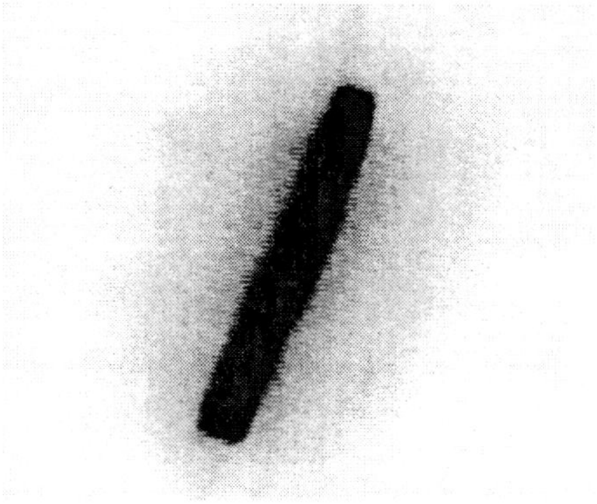

(c)

Fig. 10.5 The experimental results of impulsive synchronization when v_1 is sampled as the synchronizing impulse sequence. In this figure, the scale of the oscilloscope is $50mV/div$ for both horizontal and vertical axes. (a) The driving system and the driven system are originally desynchronized. (b) Impulsive synchronization. (c) Impulsive synchronization with narrow impulses. (d) The switching signal $s(t)$ used in (c).

10.1.3 *Remarks*

The experimental results show that the v_1-driving scheme is more robust than the v_2-driving scheme. This is because v_1 is directly connected to

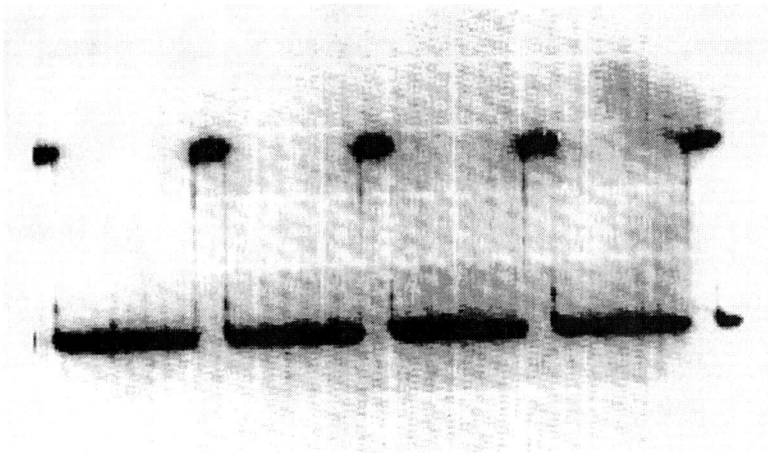

(d)
Fig. 10.5 (*Continued*).

the Chua's diode, which is the only nonlinearity in Chua's circuit. The experimental results also show that robust impulsive synchronization can be achieved under noisy conditions and a 2% parameter mismatch between the driving system and the driven system. An amplified impulse sequence with a gain greater than unity, can make the impulsive synchronization more robust.

10.2 Experimental 2: Chaotic circuits

In this section, experimental results on impulsive synchronization of a kind of chaotic circuit; namely, Chua's oscillator, are presented. To impulsively synchronize two Chua's oscillators, synchronization impulses sampled from *one* state variable of the driving circuit are transmitted to the driven circuit. Let us denote the impulse width and impulse period by Q and T, respectively.

The dynamics of Chua's oscillator, whose block diagram is shown in

Fig. 10.6, are given by the following state equation:

$$C_1 \frac{dv_1}{dt} = \frac{v_2 - v_1}{R} - f(v_1), \tag{10.3}$$

$$C_2 \frac{dv_2}{dt} = \frac{v_1 - v_2}{R} + i, \tag{10.4}$$

$$L \frac{di}{dt} = -v_2 - ri, \tag{10.5}$$

where $f(\cdot)$ is the nonlinear characteristic of the Chua's diode shown in Fig. 10.6, defined by

$$f(v_1) = G_b v_1 + \frac{1}{2}(G_a - G_b)(\mid v_1 + E \mid - \mid v_1 - E \mid), \tag{10.6}$$

where E is the breakpoint voltage of Chua's diode.

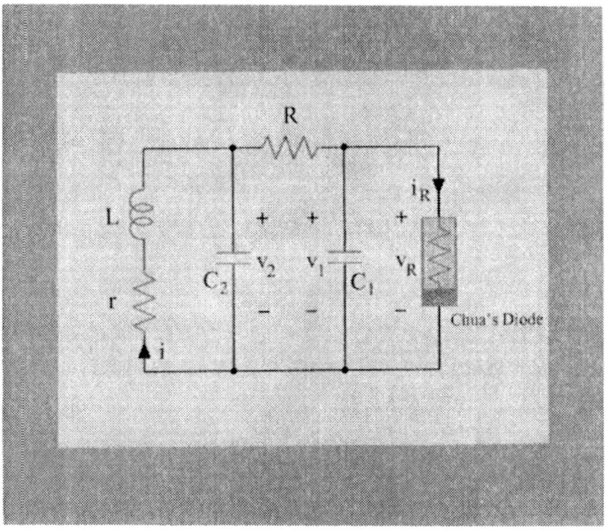

Fig. 10.6 Circuit diagram of Chua's oscillator.

The parameters for Chua's oscillators used in the experiments are given by:

$$C_1 = 10.3[nF],\ C_2 = 99.6[nF],\ L = 22.3[mH],\ R = 885[\Omega],$$
$$r = 68.9[\Omega],\ G_a = -0.875[mS],\ G_b = -0.516[mS],\ E = 0.78[V].$$

Different projections of the experimentally observed double scroll attractor are shown in Fig. 10.7.

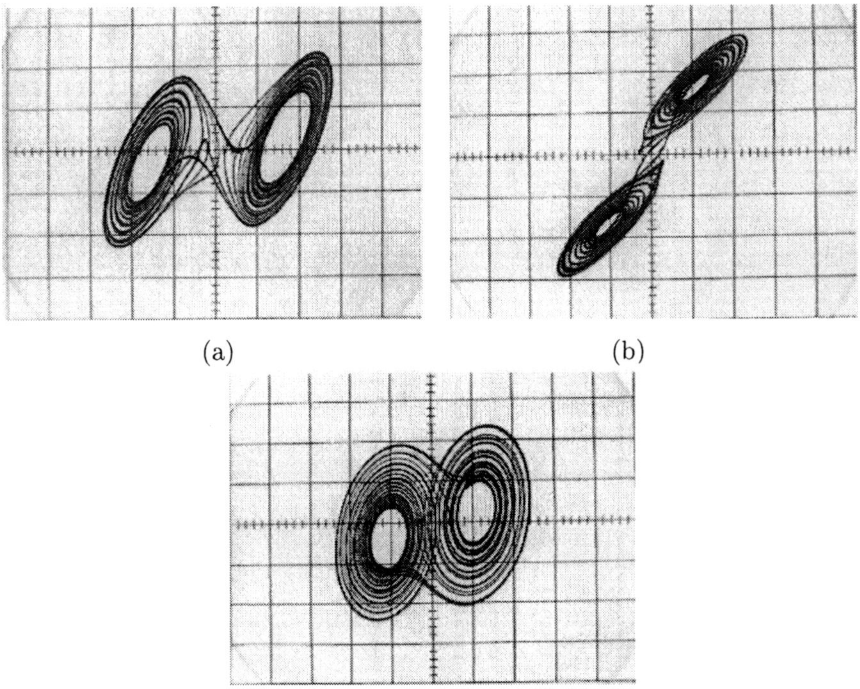

(a) (b) (c)

Fig. 10.7 Observed strange attractor of Chua's oscillator used in the experiments. (a) Projection of the double scroll attractor onto the (v_1, v_2)-plane. (b) Projection of the double scroll attractor onto the (v_1, i)-plane. (c) Projection of the double scroll attractor onto the (v_2, i)-plane.

The experimental setting for the impulsive synchronization between two Chua's oscillators is shown in Fig. 10.8. Chua's oscillators #1 and #2 are the driving circuit and the driven circuit, respectively. The voltage buffer is used to set up a uni-directional coupling from Chua's oscillator #1 to Chua's oscillator #2. The switch is implemented by a CMOS IC (4066BP) chip, which is controlled by a binary(on/off) signal called the *switching signal.* The parameter mismatch between two Chua's oscillators is within 0.5%. In the experimental setting shown in Fig. 10.8 the voltage v_1 of the Chua's oscillator #1 is chosen as the driving signal. The sequence of synchronization impulses and the corresponding switching signals are shown in Fig. 10.9. Observe that the switching signal has a "high" voltage level and a "low" voltage level. When the switching signal is in the "high" voltage level, the switch shown in Fig. 10.8 is turned on and a synchronization impulse is transmitted to the driven circuit. When the switching signal is in the "low" voltage level, the switch shown in Fig. 10.8 is turned off and the two Chua's oscillators are separated electronically.

When we observe the synchronized state from an oscilloscope, we feed one state variable of the driving circuit into the x-axis input of the oscilloscope and the corresponding state variable of the driven circuit into the y-axis input. If a 45-degree line appears on the screen, then synchronization is achieved. Figure 10.10(a)-(c) show the oscilloscope traces of the synchronized states on the v_1-v_1' plane, v_2-v_2' plane, and i-i' plane, respectively. Since a current can not directly be fed into inputs of an oscilloscope, we feed the voltages $r_1 i$ and $r_1 i'$ to display the tracing on the i-i' plane*. The experimental conditions are $Q = 0.64 \times 10^{-6}s$ and $T = 8.0 \times 10^{-6}s$. Observe that an *almost-identical synchronization*† is achieved.

We then added noise‡ to the driving signal and examined the robustness of the impulsive synchronization to additive noise. The additive noise is shown in Fig. 10.11. Observe that the amplitude range of the noise is approximately 1/10 of that of v_1. In the experiments, the noise is added after the switch.

The experimental results are shown in Fig. 10.12 with $T = 5.0 \times 10^{-6}s$. Figure 10.12(a)-(c) show the oscilloscope traces on the v_1-v_1' planes with $Q/T = 8\%$, 50% and 100%, respectively. There is no observable difference between impulsive synchronization and continuous synchronization because the thickness of 45-degree lines for impulsive synchronization and for continuous synchronization are almost the same. Thus, the impulsive synchronization is robust to small additive noise. Furthermore, more accurate

*In the experimental circuits, the resistor r is divided into two resistors $r_1 = 30.0\Omega$ and $r_2 = 38.9\Omega$.

†An almost-identical synchronization defined by the following conditions:
(1) The 45-degree line on the oscilloscope screen is sufficiently thin.
(2) No significant distortion is found along this 45-degree line.
The experimental results can depend strongly on the measuring instruments and circuit components.

‡FM receivers tuned at frequency 106MHz are used to generate noise. No commercial radio station broadcasts in the frequency band around 106MHz.

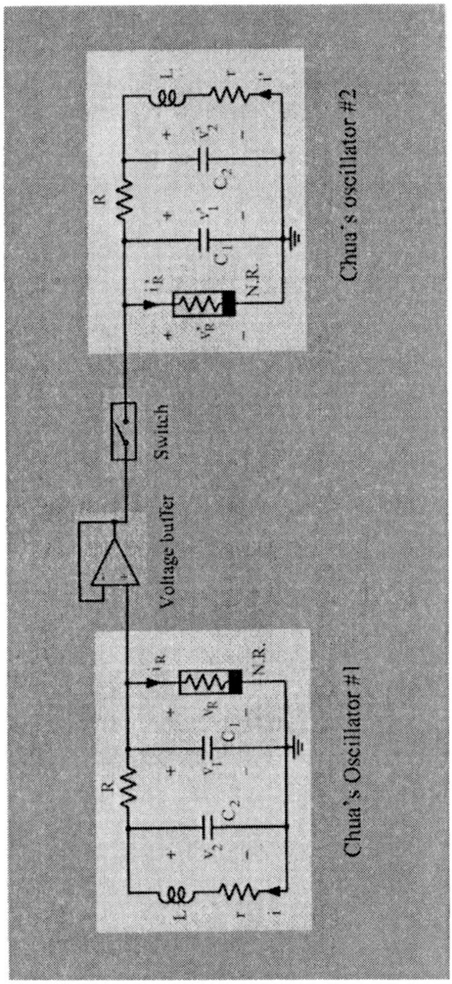

Fig. 10.8 Circuit diagram of impulsive synchronization between two Chua's oscillators.

circuit components may be chosen to make Q much shorter.

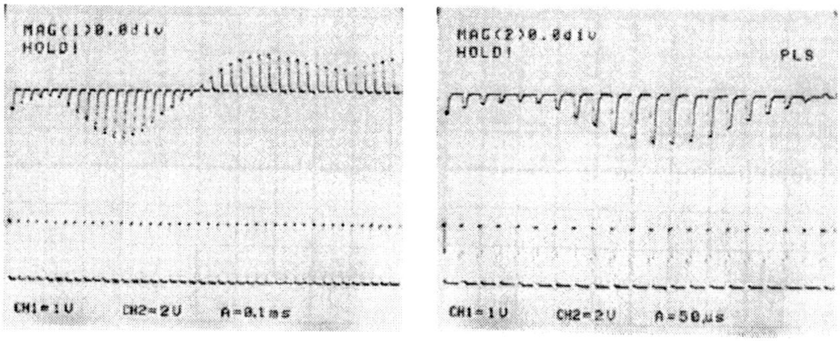

(a) (b)

Fig. 10.9 (a) The sequence of synchronization impulses (the upper trace with vertical scale 1V/div) and switching signal (the lower trace with vertical scale 2V/div). The horizontal scale is 0.1ms/div. (b) Details of the sequence of synchronization impulses (the upper trace) and switching signal (the lower trace). The vertical scales for both channels are the same as those in (a). The horizontal scale is 0.1μs/div.

10.3 Experimental 3: Hypechaotic circuits

In this section, experimental results on impulsive synchronization of a kind of hyperchaotic circuit are presented. To impulsively synchronize two hyperchaotic circuits, synchronizing impulses sampled from *two* signals of the driving circuit are sent to the driven circuit.

The block diagram of the hyperchaotic circuit used in our experiments are shown in Fig. 10.13. The state equation of this circuit is given by

$$C_1 \frac{dv_1}{dt} = g(v_2 - v_1) - i_1, \tag{10.7}$$

$$C_2 \frac{dv_2}{dt} = -g(v_2 - v_1) - i_2, \tag{10.8}$$

$$L_1 \frac{di_1}{dt} = v_1 - r_1 i_1 + R i_1, \tag{10.9}$$

$$L_2 \frac{di_2}{dt} = v_2 - r_2 i_2, \tag{10.10}$$

where $g(\cdot)$ is the v-i characteristic of Chua's diode, defined by

$$\begin{aligned} g(v_2 - v_1) &= G_a(v_2 - v_1) \\ &\quad + \frac{1}{2}(G_b - G_a)(| v_2 - v_1 - B_p | - | v_2 - v_1 + B_p |), \end{aligned}$$

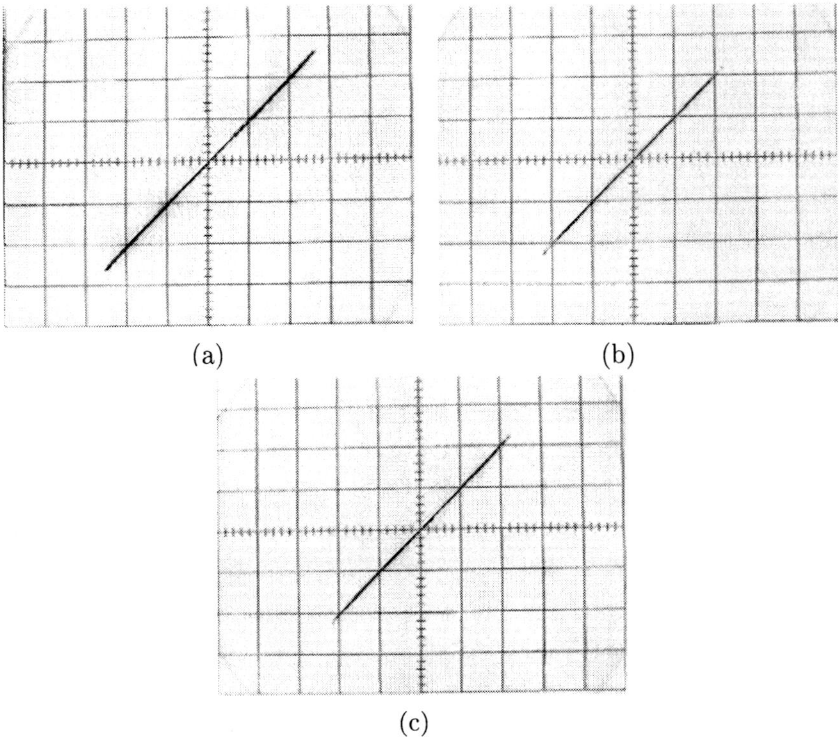

Fig. 10.10 Experimental results of the almost-identical synchronization between two Chua's oscillators with $Q = 6.4 \times 10^{-7} s$ and $T = 8.0 \times 10^{-6} s$. (a) Oscilloscope trace of v_1 versus v_1'. (b) Oscilloscope trace of v_2 versus v_2'. (c) Oscilloscope trace of i versus i'.

where B_p is the breakpoint voltage of Chua's diode. The parameters are chosen as follows:

$$\begin{aligned}
&C_1 = 102[nF],\ C_2 = 318[nF],\ L_1 = 29.9[mH],\ L_2 = 22.3[mH],\\
&R = 330[\Omega],\ r_1 = 71.2[\Omega],\ r_2 = 42.9[\Omega],\ G_a = 31.6[mS],\\
&G_b = -0.667[mS],\ B_p = 0.50[V].
\end{aligned}$$

Different projections of the observed attractor for this hyperchaotic circuit are shown in Fig. 10.14.

The circuit diagram of impulsive synchronization of the two hyperchaotic circuits is shown in Fig. 10.15. The two voltage buffers send two signals $v_2 - v_1$ and $-Ri_1$ to a switch controlled by two switching signals.

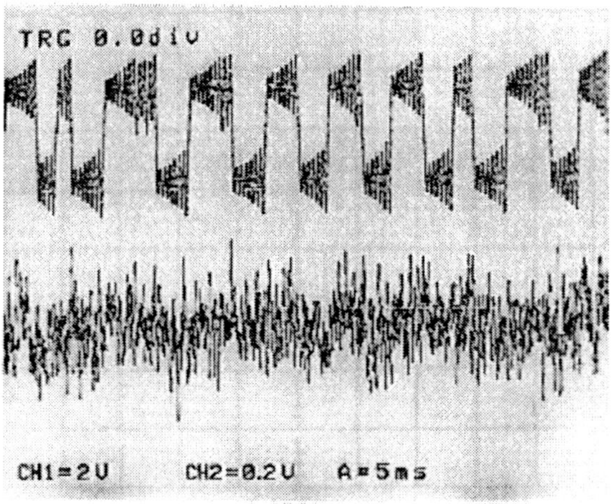

Fig. 10.11 The transmitted signal v_1 and the additive noise. Vertical scales are $2V$/div for the upper trace and $0.2V$/div for the the lower trace. Horizontal scale is $5ms$/div.

Although samples from two signals are sent to the driven circuit, we use a time-division scheme to transmit *both* impulse sequences through *only one* communication channel. In this case, each frame consists of three regions. The first two regions are synchronization regions consisting of driving impulses. The first region is used to transmit synchronization impulses sampled from signal $v_2 - v_1$, and the second region is used to transmit those sampled from signal $-Ri_1$. The third region can be used to transmit scrambled message signals. In the experiments presented in this section, we did not use the third region to transmit message signals.

Figure 10.16 shows the two switching signals used to choose and set timing of two kinds of synchronization impulses sampled from $v_2 - v_1$ and $-Ri_1$. Observe that the time shift between the high voltage levels in these two switching signals correspond to the time-division scheme for sending both impulse sequences through a single channel.

Figure 10.17(a) shows two synchronization impulse sequences sampled from $v_2 - v_1$ and $-Ri_1$. Observe that the slight time shift between these two sequences corresponds to the two switching signals shown in Fig. 10.16. Figure 10.17(b) shows the impulse sequence transmitted through the single channel. Observe that these two kinds of synchronization impulses are transmitted over different time slots in each frame.

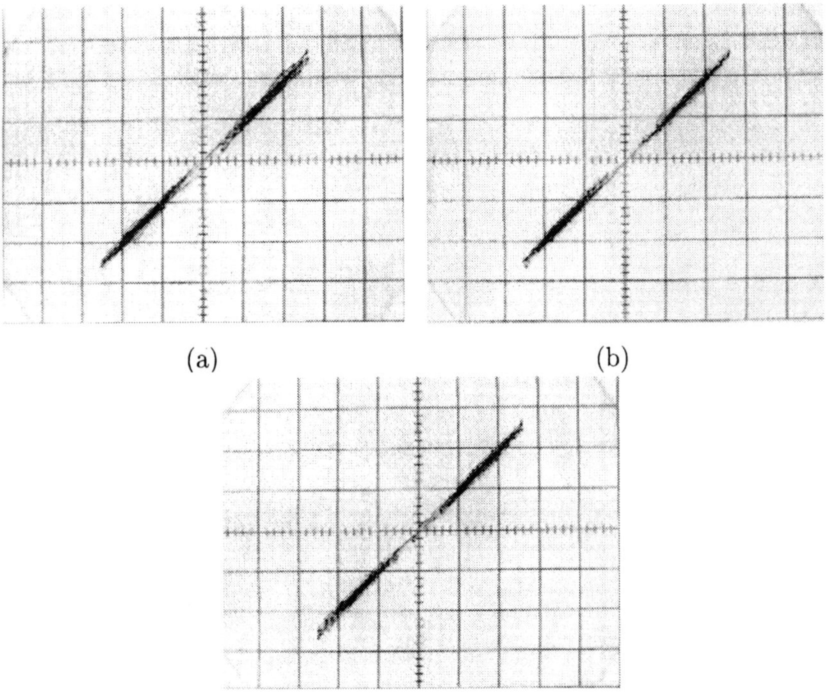

Fig. 10.12 Robustness of impulsive synchronization between Chua's oscillators to additive noise under the condition $T = 5.0 \times 10^{-6}s$. Oscilloscope traces on the v_1-v_1' planes are shown. (a) Q/T is equal to 8%. (b) Q/T is equal to 50%. (c) Q/T is equal to 100%. This condition is equivalent to continuous synchronization.

Figure 10.18(a) and (b) show the tracings in the v_1-v_1' plane and v_2-v_2' plane for $Q/T = 0.26$ and $T = 4.0 \times 10^{-6}s$, respectively. Observe that an almost-identical synchronization is achieved.

We then examine the robustness of the impulsive synchronization to small additive noise. Figure 10.19(a) shows the transmitted impulse sequence and the additive noise. To compare the amplitudes of impulses and noise, Fig. 10.19(b) shows the signal $v_2 - v_1$ and the additive noise.

The experimental results are shown in Figs. 10.20 with $T = 4.0 \times 10^{-6}s$. Figure 10.20(a)-(c) show the oscilloscope traces on the v_1-v_1' planes with $Q/T = 26\%$, 40% and 100%, respectively. There is no observable difference between impulsive synchronization and continuous synchronization because

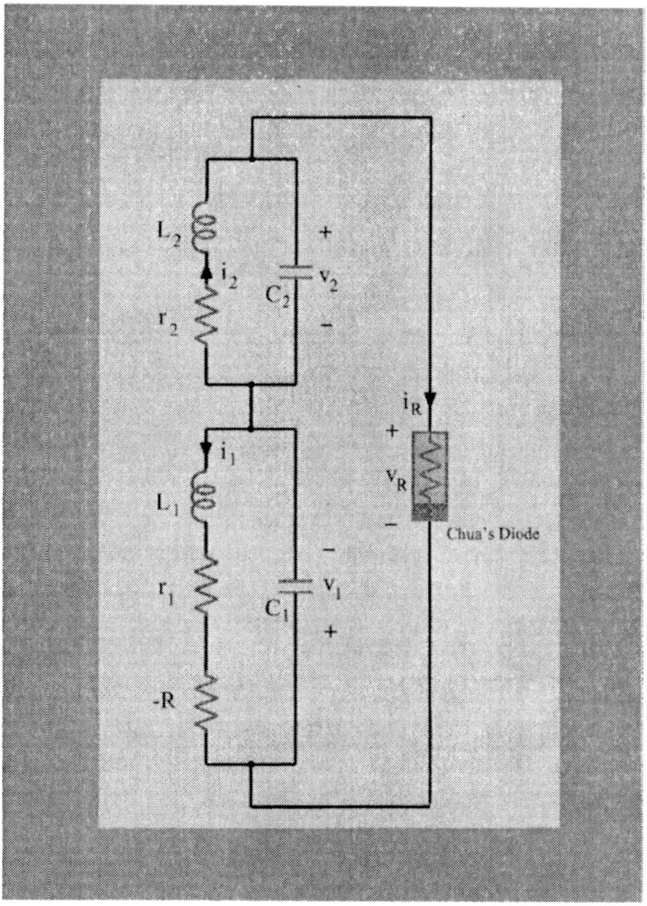

Fig. 10.13 Circuit diagram of the hyperchaotic circuit used in the experiments.

the thickness of the 45-degree lines for impulsive synchronization and for continuous synchronization is almost the same. The above observations show that impulsive synchronization of the hyperchaotic circuit in Fig. 10.13 is robust to small additive noise.

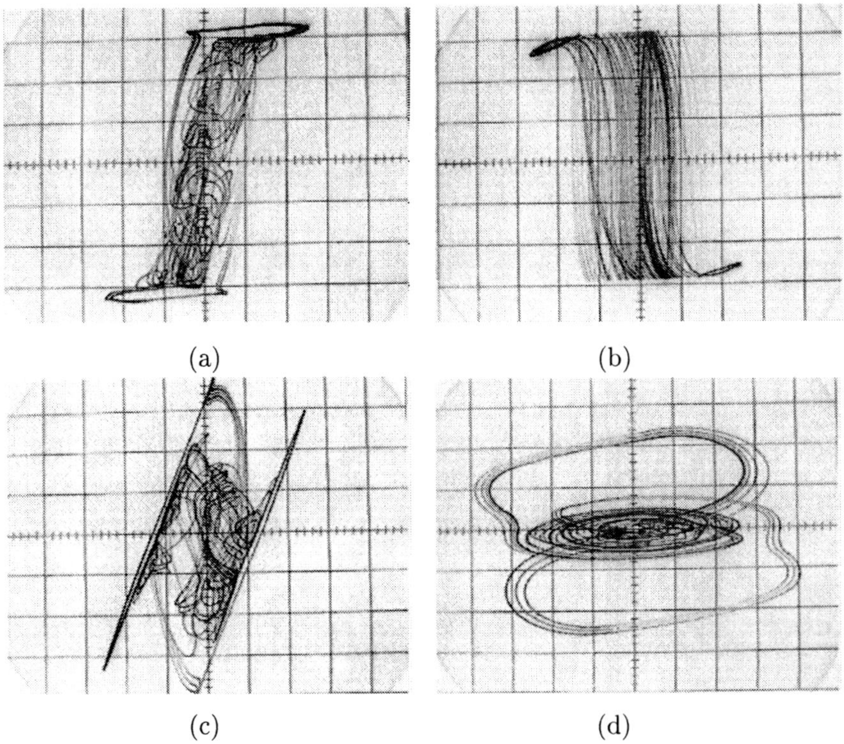

Fig. 10.14 Observed strange attractor from the hyperchaotic circuit used in the experiments. (a) Projection of the chaotic attractor onto the $(v_1, v_2 - v_1)$-plane. (b) Projection of the chaotic attractor onto the (v_1, v_2)-plane. (c) Projection of the chaotic attractor onto the (v_1, i_1)-plane. (d) Projection of the chaotic attractor onto the (v_2, i_2)-plane.

10.4 Concluding remarks

The experimental results show that the impulsive synchronization is very robust to both parameter mismatch and channel noise. We have also found that the performance of impulsive synchronization depends heavily on the frequency and the width of the synchronization impulses. The experimental results show that the accuracy of impulsive synchronization depends on both the period and the width of the impulse. The ratio between the impulse width and impulse period for "almost-identical" *synchronization* increases as the impulse period increases. The robustness of impulsive syn-

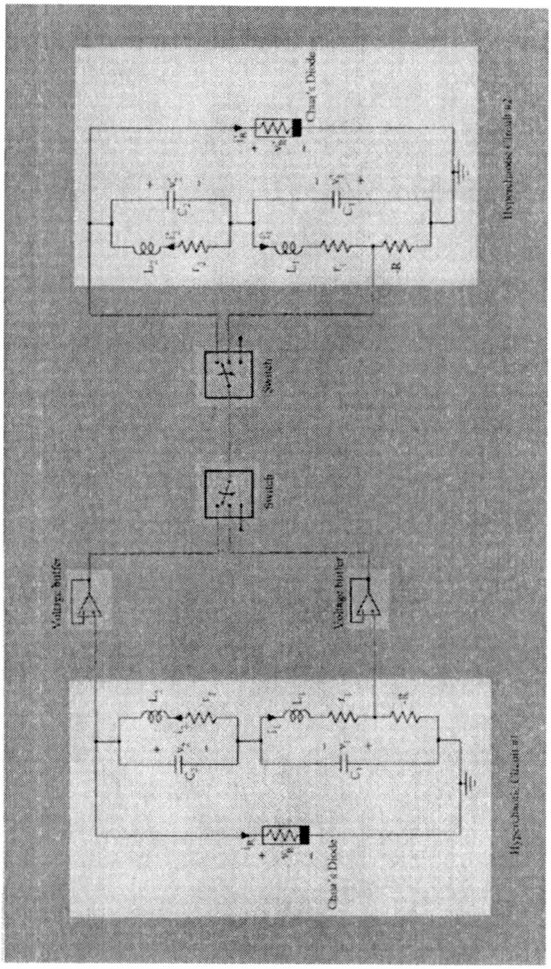

Fig. 10.15 Circuit diagram of impulsive synchronization between two hyper-chaotic circuits.

chronization to additive noise is also experimentally studied. For sufficiently short impulse periods, no significant differences are observed between impulsive and continuous synchronizations.

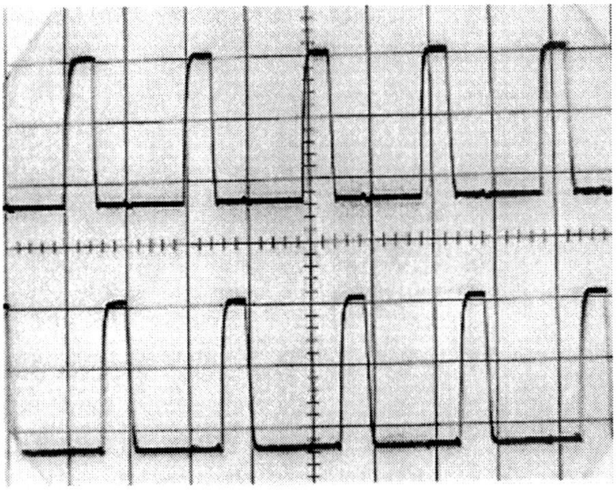

Fig. 10.16 Two switching signals are used to implement a time-division scheme for sending two impulse sequences through a single channel.

Notes

The theory of impulsive synchronization of chaotic systems can be found in [YYY97b; YL97b; YL97c; YC97; YC00a; Yan99a; SYC98; SYVC99]. The experimental results of impulsive synchronization between chaotic circuits can be found in [PYC98; IYC99].

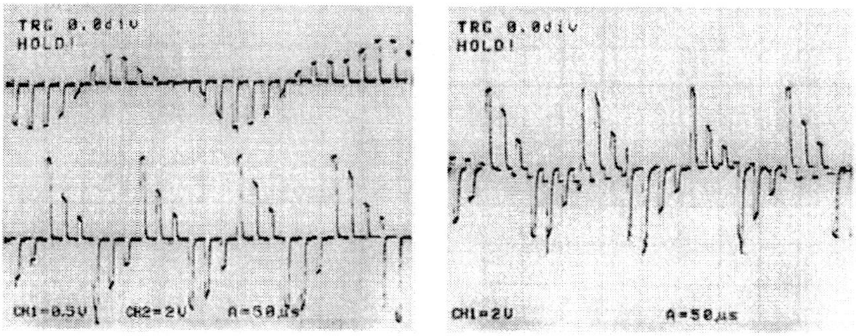

(a) (b)

Fig. 10.17 Impulse sequences for impulsive synchronization of two hyperchaotic circuits. (a) Two impulse sequences shown separately. The upper trace is the impulse sequence sampled from the voltage $v_2 - v_1$ across Chua's diode. The lower trace is the impulse sequence sampled from the voltage $-Ri$ across the negative resistor. (b) The transmitted impulse sequence through a single channel shared by the two impulse sequences shown in (a) under the time-division scheme.

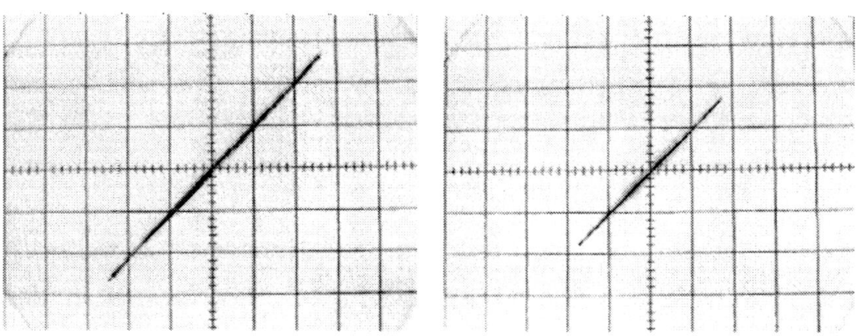

(a) (b)

Fig. 10.18 Impulsive synchronization between two hyperchaotic circuits under the conditions $\frac{Q}{T} = 0.26$ and $T = 4.0 \times 10^{-6}s$. (a) Oscilloscope trace of v_1 versus v_1'. (b) Oscilloscope trace of v_2 versus v_2'.

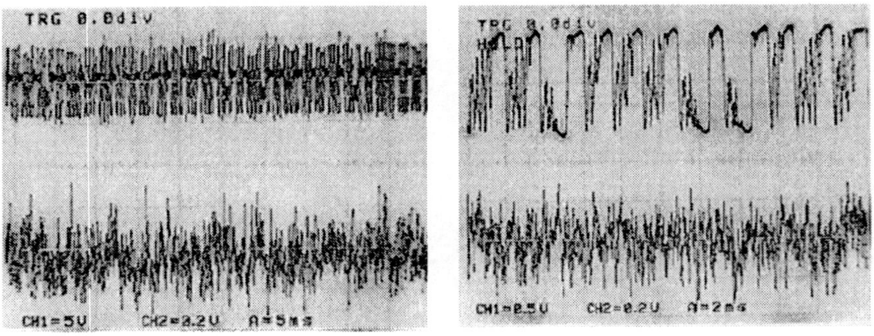

(a) (b)

Fig. 10.19 Impulse sequence and additive noise. Horizontal scale is $2ms/div$. (a) The transmitted signal(upper trace) and the additive noise(lower trace). Vertical scales are $5.0V/div$ for the upper trace and $0.2V/div$ for the lower trace, respectively. (b) The signal $v_2 - v_1$(upper trace) and the additive noise(lower trace). Vertical scales are $0.5V/div$ for the upper trace and $0.2V/div$ for the lower trace, respectively.

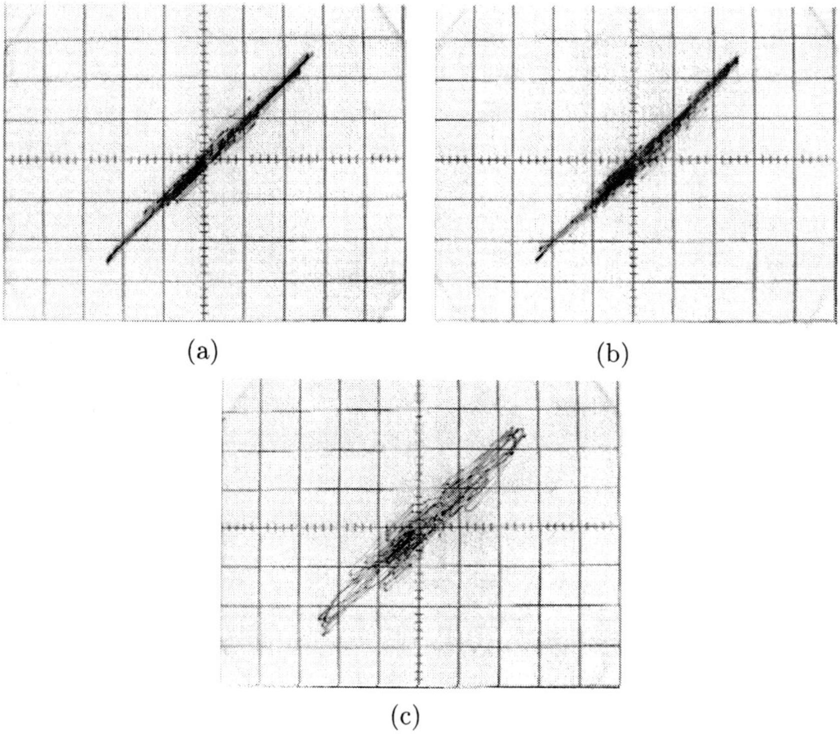

Fig. 10.20 Impulsive synchronization between two hyperchaotic circuits when the additive channel noise shown in Fig. 10.19 is added with $T = 4.0 \times 10^{-6} s$. (a) Oscilloscope trace of v_1 versus v_1' with $\frac{Q}{T} = 0.26$. (b) Oscilloscope trace of v_1 versus v_1' with $\frac{Q}{T} = 0.40$. (c) Oscilloscope trace of v_1 versus v_1' with $\frac{Q}{T} = 1.0$; namely, continuous synchronization.

Chapter 11

Applications to Chaotic Communication Systems

In this chapter we present the applications of impulsive control to chaotic communication systems with nonlinear carriers. By impulsively synchronized two RF chaotic circuits, we can design chaotic secure communication systems. Since chaotic RF carriers can function as both carrier and spreading sequence, impulsive synchronization between two chaotic systems can also be used to design chaotic digital code-division multiple access ($(CD)^2MA$) communication systems.

11.1 Secure communication using chaotic carriers

Since the publication of several chaotic cryptanalysis results in *low-dimensional chaos-based* secure communication systems[Yan95; YYY98a; YYY98d; YYY98c; YYY98b], there were concerns that such communication schemes may not be secure enough. To overcome this objection, on the one hand we can exploit hyper-chaos based secure communication systems. On the other hand, we can enhance the security of low-dimensional chaos-based secure communication schemes by combining conventional cryptographic schemes with a chaotic system[YWC97]. To overcome the low security objections against low-dimensional continuous chaos-based schemes, we should overcome the following problems:

(1) make the transmitted signal more complex,

(2) reduce the redundancy in the transmitted signal.

To achieve the first goal, it is not necessary to use hyper-chaos. We can combine a conventional cryptographic scheme with low-dimensional chaos to obtain a very complex transmitted signal. To achieve the second goal, impulsive synchronization offers a very promising approach.

In this section, we combine chaotic cryptography and impulsive synchronization to give a new chaotic secure communication scheme. The block diagram of this scheme is shown in Fig. 11.1.

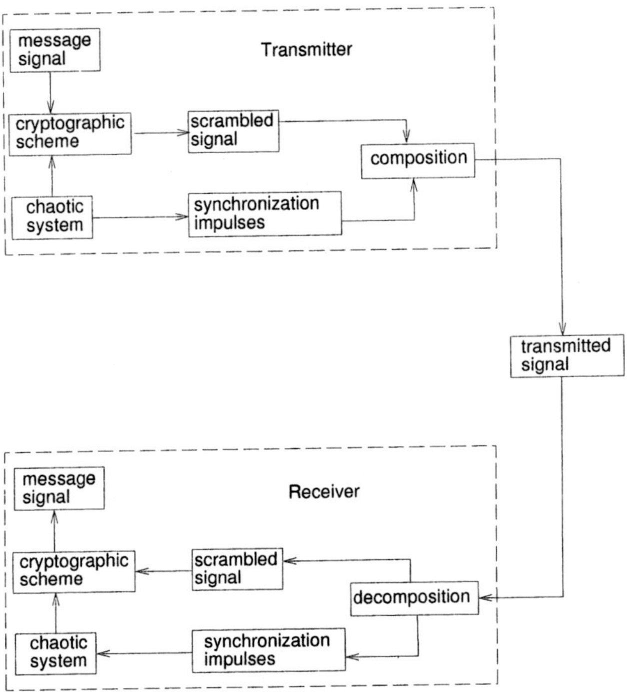

Fig. 11.1 Block diagram of the impulsive-synchronization based chaotic secure communication system.

From Fig. 11.1 we can see that this chaotic secure communication system consists of a transmitter and a receiver. In both the transmitter and the receiver, there exist two identical chaotic systems. Also, two identical conventional cryptographic schemes are embedded in both the transmitter and the receiver. Let us now consider the details of each block in Fig. 11.1. The transmitted signal consists of a sequence of time frames. Every frame has a length of T seconds and consists of two regions. In Fig. 11.2 we show the concept of a time frame and its components. The first region of the time frame is a synchronization region consisting of synchronization impulses. The synchronization impulses are used to impulsively synchronize the chaotic systems in both transmitter and receiver. The second region is the scrambled signal region where the scrambled signal is contained. Within every time frame, the synchronization region has a length of Q and the remaining time interval $T - Q$ is the scrambled signal region.

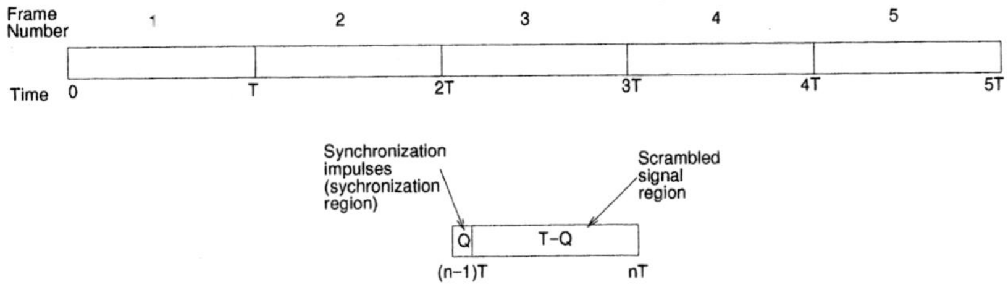

Fig. 11.2 Illustration of the concept of a time-frame and its components.

The composition block in Fig. 11.1 is used to combine the synchronization impulses and the scrambled signal into the time frame structure shown in Fig. 11.2. The simplest combination method is to substitute the beginning Q seconds of every time frame with synchronization impulses. Since Q is usually very small compared with T, the lost of time for packing a message signal is neglectable. The decomposition block is used to separate the synchronization region and the scrambled signal region within each frame at the receiver end. Then the separated synchronization impulses are used to make the chaotic system in the receiver to synchronize with that in the transmitter.

In the transmitter and the receiver, we use the same cryptographic scheme block for purposes of bi-directional communication. In a bi-directional communication scheme, every cellular phone should function both as a receiver and a transmitter. Here, the key signal is generated by the chaotic system. The cryptographic scheme is as follows:

We use a continuous n-shift cipher to encrypt the plain signal(message signal). The n-shift cipher is given by

$$e(p(t)) = \underbrace{f_1\Big(...f_1\Big(f_1}_{n}\Big(p(t), \underbrace{k(t)\Big), k(t)\Big), ..., k(t)\Big)}_{n} = y(t) \tag{11.1}$$

where h is chosen such that $p(t)$ and $k(t)$ lie within $(-h, h)$. Here, $p(t)$ and $k(t)$ denote the plain signal and the key signal, respectively, and $y(t)$ denotes the encrypted signal. The key signal $k(t)$ is chosen as a state variable of the chaotic system. The notation $f_1(\cdot,\cdot)$ denotes a scalar nonlinear function of two variables defined as follow:

$$f_1(x,k) = \begin{cases} (x+k)+2h, & -2h \le (x+k) \le -h, \\ (x+k), & -h < (x+k) < h, \\ (x+k)-2h, & h \le (x+k) \le 2h. \end{cases} \tag{11.2}$$

This function is shown in Fig. 11.3.

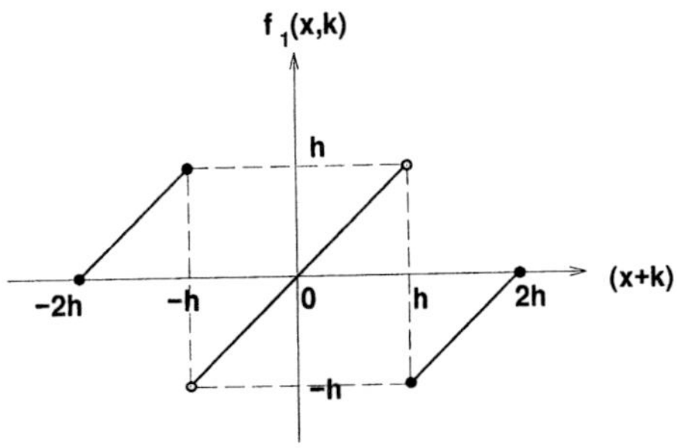

Fig. 11.3 Nonlinear function used in the continuous shift cipher.

The corresponding decryption rule is the same as the encryption rule

$$\begin{aligned} p(t) &= d(y(t)) = e(y(t)) \\ &= \underbrace{f_1\Big(\ldots f_1\Big(f_1}_{n}\Big(y(t), \underbrace{-k(t)\Big), -k(t)\Big), \ldots, -k(t)\Big)}_{n} \end{aligned} \tag{11.3}$$

To decode the encrypted signal, the same key signal should be used.

The simulation results are as follows. We use an FM scheme to modulate the synchronization impulses such that the synchronization region is located in the initial 1% of every time frame. We choose the frame length as $T = 1s$. In the synchronization region of every time frame, we transmit the impulses of the three state-variables of the Chua's oscillators. The parameter of the encrypted signal is chosen as $h = 0.4$. A continuous 10-shift cipher was used. We choose x and $\tilde{x}$ as the key signals and normalized them to fall within the amplitude range $[-0.4, 0.4]$.

Figure 11.4 shows the simulation results of the above proposed secure communication system for transmitting a speech signal. Figure 11.4(a) shows the waveforms of the sampled speech of four Chinese digits "NING"(zero)—"YI"(one)—"ER"(two)—"SANG"(three). The sampling rate is 8K. Figure 11.4(b) shows the waveforms of the received scrambled speech signal and the additive channel noise with $SNR = 16dB$. This additive noise can not change the value of the synchronization impulses which are modulated by FM. We can see that the structure of the signal in Fig. 11.4(a) was entirely covered. Figure 11.4(c) shows the waveforms of the descrambled speech signal. We can see that some noises were introduced into the recovered results due to the channel noise. But the structure of the speech signal was perfectly recovered.

11.2 Chaotic digital code-division multiple access communication systems

In this section, the structure, principle and framework of chaotic digital code-division multiple access ($(CD)^2$MA) communication systems are presented. Unlike the existing CDMA systems[Vit95; GSs97], $(CD)^2$MA systems use continuous pseudo-random time series to spread the spectrum of message signal and the spreaded signal is then directly sent through channel to the receiver. In this sense, the carrier used in $(CD)^2$MA is a continuous pseudo-random signal instead of a single tone as used in CDMA. In a $(CD)^2$MA system, every mobile station has the same structure and parameters, only different initial conditions are assigned to different mobile stations. Instead of synchronizing two binary pseudo-random sequences as in CDMA systems, we use an impulsive control scheme to synchronize two chaotic systems in $(CD)^2$MA.

In CDMA systems, pseudo-random signals are used to 1) spread the bandwidth of the modulated signal to the larger transmission bandwidth and 2) distinguish among the different user signals which are using the same transmission bandwidth in the multiple-access scheme. Ideally, these pseudo-random signals should be samples of a sequence of independent random variables, uniformly distributed on an available alphabet or range. In this case, the CDMA system is equivalent to a one-time pad used in cryptographic system requiring the highest level of security. Since the key signal in a one-time pad should be as long as the message signal, it is not feasible to use it in CDMA[SOSL94].

We must figure out a way to store/generate good pseudo-random signals

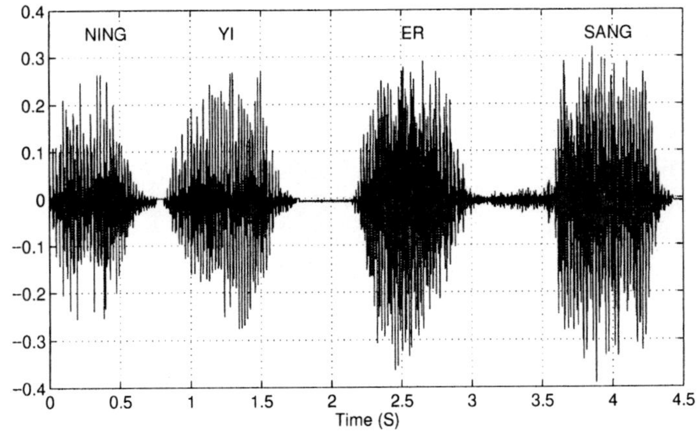

(a)

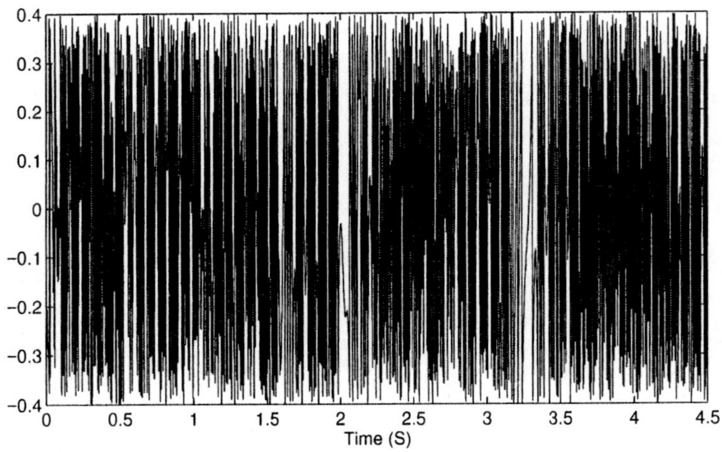

(b)

Fig. 11.4 The simulation results. (a) The time-domain waveform of the speech signal. (b) The time-domain waveform of the scrambled speech signal. (c) The time-domain waveform of the descrambled speech signal.

in both the transmitter and the receiver, despite the finite storage capacity/generating capacity of physical processing systems. It would be very expensive and energy efficient to generate such a sequence by using some chaotic circuits, e.g., Chua's circuits. In fact, some methods to generate good pseudo-random signals for cryptographic purposes by using Chua's

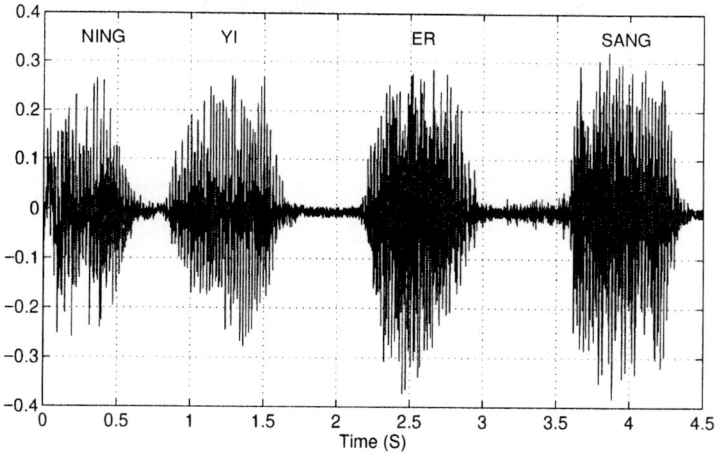

(c)

Fig. 11.4. (*Continued*)

circuits[YWC97] have already been developed.

11.2.1 *Synchronization considerations*

In a CDMA system, the heart of the receiver is its synchronization circuitry, and the heartbeats are the clock pulses which control almost every step in forming the desired output. There exist three levels of synchronization in a CDMA system: 1) correlation interval synchronization, 2) spread-spectrum generator synchronization and 3) carrier synchronization.

To correlate the Walsh codes at the receiver requires that the receiver be synchronized with the transmitter. In the forward direction, the base station can transmit a pilot signal to enable the receiver to recover synchronization. Just as the designers of the IS-665 wide-band CDMA system believed, with a wider bandwidth the base station can also recover the pilot signal sent by mobile stations. In $(CD)^2$MA, we also use this symmetric system between base station and mobile station.

In the $(CD)^2$MA system, we need to synchronize two chaotic systems. If we use the continuous synchronization scheme, we need a channel to transmit the chaotic signal. Even though we can embed a message signal into a chaotic carrier[HWIC93], we can not achieve a chaotic CDMA system. A promising method to improve this is the framework of impulsive synchronization[YL97a; YL97b]. $(CD)^2$MA system does not need the correlation interval synchronization and carrier synchronization. This make the receiver in $(CD)^2$MA simple and low-power.

The $(CD)^2$MA system can increase the capacity of a radio channel. For mobile subscribers, this increased capacity translates to better service at a lower price. On the other hand, $(CD)^2$MA is also a promising technology for low-cost implementation of cellular and PCS.

11.2.2 *Structures of CDMA and $(CD)^2MA$ systems*

Information can be modulated into the spread-spectrum signal by several methods. The most common method is to add the information into the spectrum-spreading code before it is used for modulating the carrier frequency. The corresponding CDMA system is shown in Fig.11.5. In Fig.11.5, $d(t)$ and $c(t)$ are respectively called "message signal" and "spreading signal". The signal $cos(\omega t)$ is called the carrier, and $x(t)$ is the encoded signal. At the receiver end, $I(t)$ is interference signal which consists of all channel noise, interference and/or jamming, and $r(t)$ is the recovered signal. $d(t)$ is a low frequency digital message signal with a data rate $9.6kbps$ for IS-95 CDMA system. $c(t)$ is a high frequency spreading signal with a chip rate $1.2288Mcps$ for IS-95 CDMA system. We can see that the chip rate is much higher than data rate. This is the method by which the bandwidth of the message signal is spread.

On the other hand, the message signal can also be used to modulate the carrier directly given the carrier also functions as spectrum-spreading signal. In $(CD)^2$MA system, we use this kind of modulation scheme. The corresponding $(CD)^2$MA system is shown in Fig. 11.6.

Comparing the $(CD)^2$MA system in Fig. 11.6 with the CDMA system in Fig. 11.5, we can see that both schemes use the synchronization of two identical spreading carrier generators. Instead of modulating a single tone (an RF sinusoidal signal) as in the CDMA system, a $(CD)^2$MA system transmits a pseudo-random RF spreading carrier directly. We give the details of every block in Fig.11.6. To enhance the security of the $(CD)^2$MA system, we can also use two key signals to scramble the spreading carrier. The key signal can be assigned to each transmitter and receiver pair by the base station(cell). Since each mobile station can function either as a transmitter or as a receiver, we suppose that the key signals are set by both the receiver and the transmitter and can be refreshed during conversations. The

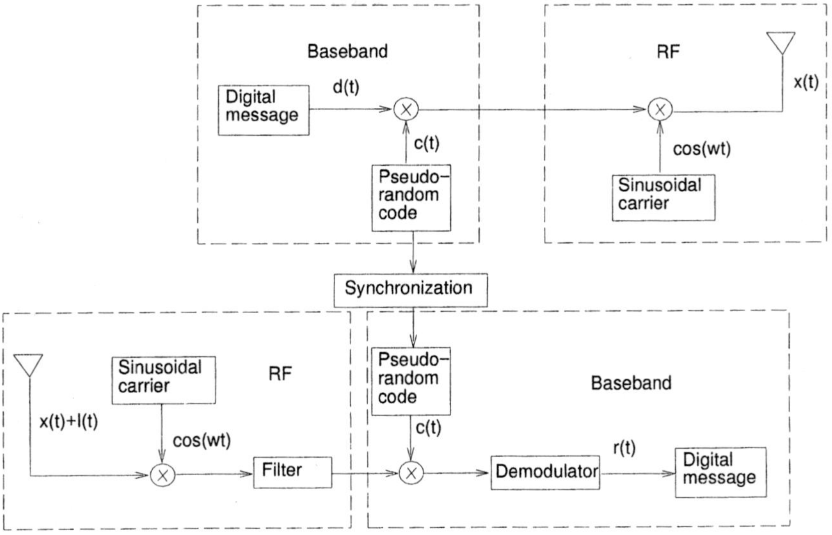

Fig. 11.5 The Block diagram of the CDMA system.

key signals may only be used in cases where a very high security should be taken into account(e.g., military applications). In commercial applications, the spreading signal along is secure enough.

In the $(CD)^2$MA paradigm, every mobile station has the same chaotic circuit, e.g., Chua's circuit. Whenever two users are connected by the base station, they are assigned the same set of initial conditions to their chaotic circuits. Then, the impulsive synchronization scheme is used to maintain the synchronization between the two chaotic circuits.

Since the chaotic circuit works in a low frequency range, we need a spreading function to spread it into bandwidths in the MHz range. In principle, there are many nonlinear functions which can be used as spreading functions. Here, the spreading function is chosen according to that used in [YWC97]. That is, we use an n-shift scheme to spread the spectrum of the chaotic signal. The chaotic systems in both transmitter and receiver are identical and the synchronization between them are achieved by *impulsive synchronization*. The demodulator consists of some low-pass filtering and thresholding blocks.

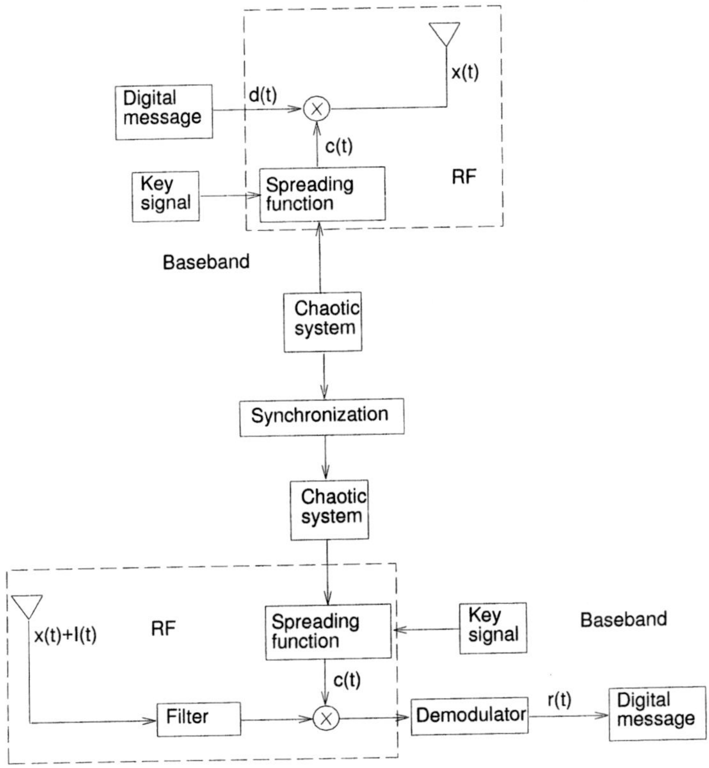

Fig. 11.6 The Block diagram of the $(CD)^2$MA system.

11.3 Experimental results of chaotic spread spectrum communication systems

The performance of chaotic spread spectrum communication systems based on impulsive synchronization is studied experimentally in this section. The experimental results show that both additive modulation and DS modulation are robust enough to channel noise. The experimental results also suggest that impulsive synchronization is a promising approach in the design of practical chaotic spread spectrum communication systems.

In chaotic spread spectrum communication systems, there are different ways to modulate the message signal $s(t)$ with a chaotic spreading carrier $c(t)$. For example, one can multiply $c(t)$ with $s(t)$ (direct sequence modulation). One can also add $s(t)$ to $c(t)$ (chaotic masking).

For relatively low security situations, we can easily implement the chaotic masking scheme by using impulsive synchronization. The synchronization region Q of each frame consists of synchronization impulses while the rest of the frame consists of the masked signal $p(t) = s(t) + c(t)$. The recovery of the message signal is performed by subtracting $\tilde{c}(t)$ from $p(t)$ within the region $T - Q$ (here, $\tilde{c}(t)$ is the chaotic carrier recovered using impulsive synchronization at the receiver end). Similarly, the direct sequence (DS) modulation can be easily implemented by impulsive synchronization.

In the spread spectrum communication systems based on impulsive synchronization presented in this section, within each frame of T seconds, the message signal takes only $T - Q$ seconds. Since Q is not small with respect to T, the recovered signal $r(t)$ is periodically sampled at every T seconds. Thus, for a reliable recovery the message signal $s(t)$ should be bandlimited to a frequency range below $\frac{1}{2T}$ in accordance with the sampling theorem.

The block diagram of the chaotic spread spectrum communication systems based on impulsive synchronization is shown in Fig. 11.7. Figure 11.7(a) shows the block diagram based on Chua's oscillators. The modulator at the transmitter end will scramble the message signal with the chaotic masks by using either addition or multiplication. The switch is switched between v_1 and the modulated message signal. Correspondingly, Fig. 11.7(b) shows the block diagram based on hyperchaotic circuits.

11.3.1 *Additive modulation*

11.3.1.1 *Chaotic circuit*

The experimental results with Chua's oscillators and additive modulation are shown in Fig. 11.8. The messages signal is a triangular waveform. Figure 11.8(a) shows the message signal and the recovered message signal under the conditions $Q/T = 10\%$ and $T = 2.0 \times 10^{-5}s$. Figure 11.8(b) shows the message signal and the recovered message signal under the conditions $Q/T = 50\%$ and $T = 2.0 \times 10^{-5}s$. Observe from Figs. 11.8(a)-(b) that the recovered signals have very little distortions. Figure 11.8(c) shows the message signal and the recovered message signal under the conditions $Q/T = 8\%$ and $T = 8.0 \times 10^{-6}s$. Significant distortions in the recovered message signal can now be observed. Figure 11.8(d) shows the message signal and the recovered message signal under the conditions $Q/T = 18\%$ and $T = 8.0 \times 10^{-6}s$. Observe from Fig. 11.8(d) that the recovered signals

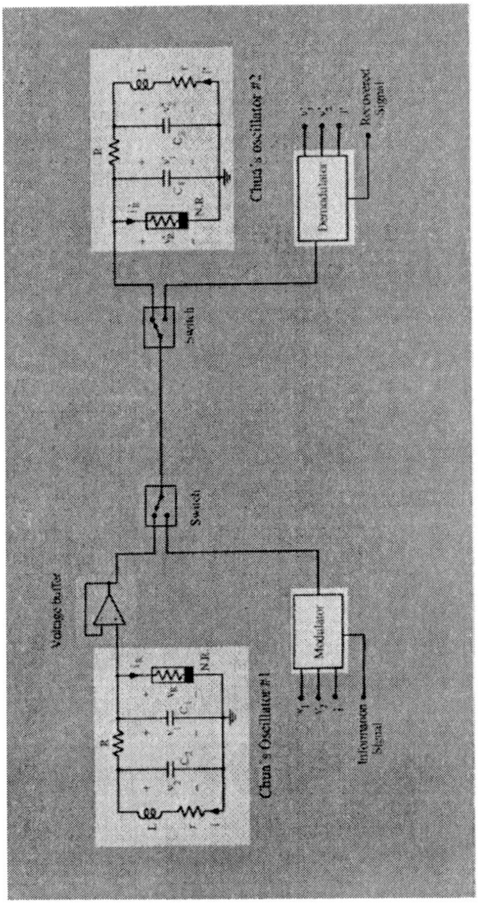

(a)

Fig. 11.7 Circuit diagrams of chaotic spread spectrum communication systems used in our experiments when additive and DS modulations are used. (a) When the Chua's oscillators are used to generate the chaotic spreading carriers. (b) When the hyperchaotic circuits are used to generate the chaotic spreading carriers.

have very little distortions.

To show what the transmitted signal in the channel looks like, in Fig. 11.9(a) the mixture of the impulse sequence and the chaotic carrier $c(t)$ is shown. Figure 11.9(b) shows the mixture of the impulse sequence and the modu-

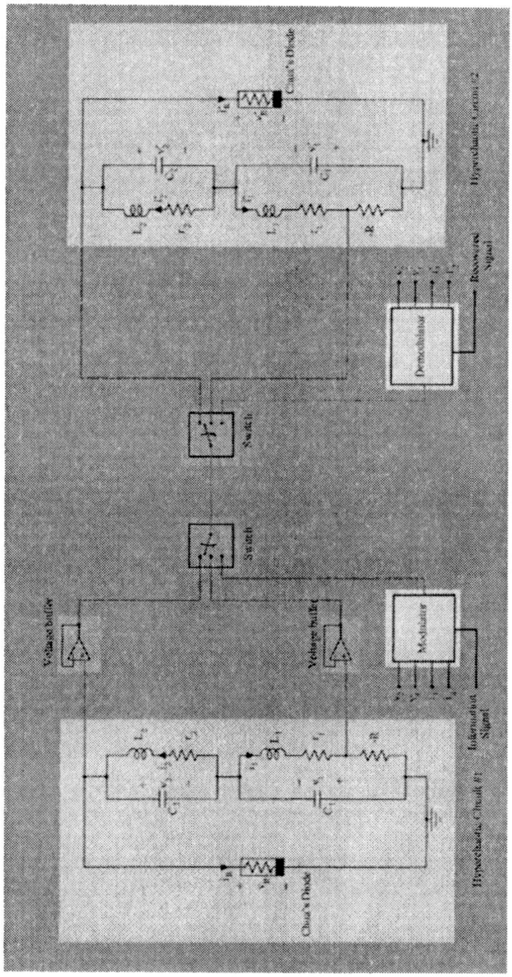

(b)

Fig. 11.7. (*Continued*)

lated message signal $c(t) + s(t)$.

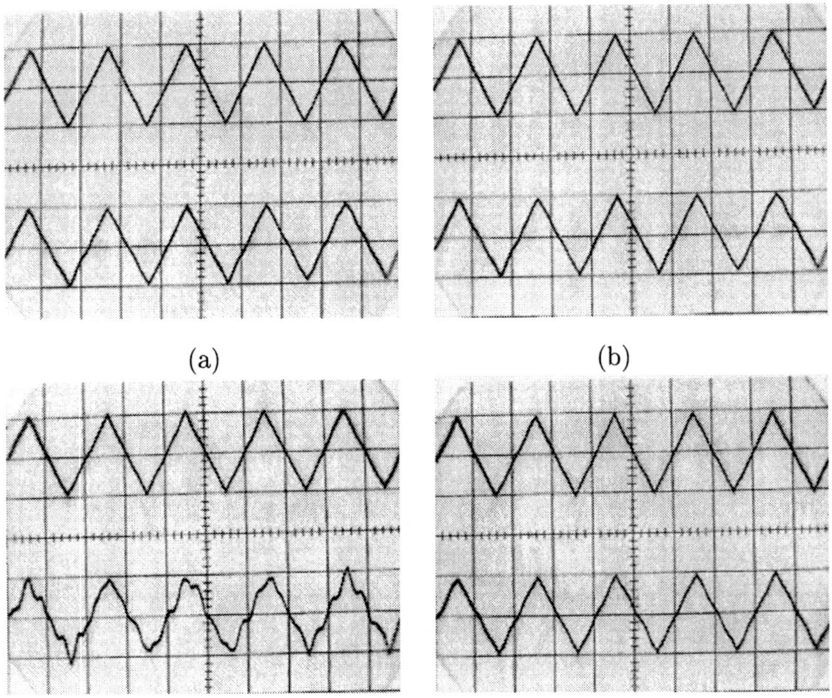

(a) (b) (c) (d)

Fig. 11.8 Message signals and recovered message signals when the systems in Fig. 11.7(a) is used with additive modulation. Message signals are shown in upper traces and the recovered massage signals are shown in lower traces. (a) $Q/T = 10\%$ and $T = 2.0 \times 10^{-5}s$. (b) $Q/T = 50\%$ and $T = 2.0 \times 10^{-5}s$. (c) $Q/T = 8\%$ and $T = 8.0 \times 10^{-6}s$. (d) $Q/T = 18\%$ and $T = 8.0 \times 10^{-6}s$.

11.3.1.2 *Hyperchaotic circuit*

The experimental results on the hyperchaotic circuits with additive modulation are shown in Fig. 11.10. The message signal is also a triangular waveform. Figure 11.10(a) shows the message signal and the recovered message signal under the conditions $Q/T = 26\%$ and $T = 5.0 \times 10^{-6}s$. Figure 11.10(b) shows the message signal and the recovered message signal under the conditions $Q/T = 26\%$ and $T = 8.0 \times 10^{-6}s$. Observe that there exists almost no distortions in the recovered signals. To show what the transmitted signal in the channel looks like, in Fig. 11.10(c) the mixture of the impulse sequence and the chaotic carrier $c(t)$ is shown. Figure 11.10(d)

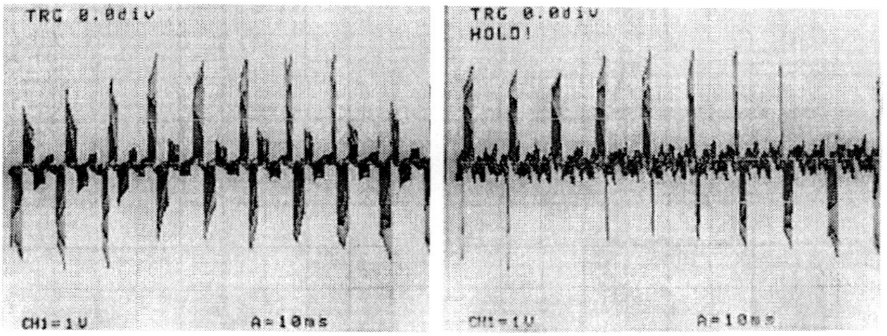

(a) (b)

Fig. 11.9 The transmitted signals with or without modulated message signals under the conditions $Q/T = 10\%$ and $T = 2.0 \times 10^{-5}s$. The chaotic systems are Chua's oscillators. (a) Transmitted signal without the modulated message signal. (b) Transmitted signal with the modulated message signal when additive modulation is used.

shows the mixture of the impulse sequence and the modulated message signal $c(t) + s(t)$.

In the DS modulation, the message signal is multiplied by the chaotic spreading carrier at the transmitter and the transmitted signal is divided by the recovered chaotic spreading carrier at the receiver end. In our experiments, we use AD533 IC chips as multipliers and dividers. In the AD533 chip, the division operations are not performed in the negative regions of the input signals, thereby giving rise to interruptions in the recovering process. Furthermore, the recovered signals are saturated in the negative regions. If AD532 chips are used instead, then the division is performed well in the negative regions and better results can be expected.

11.3.2 *DS modulation*

11.3.2.1 *Chaotic circuit*

Figure 11.11 shows the results when the message signal $s(t)$ is modulated by $v_1(t)$; namely, the modulated message signal is $v_1(t)s(t)$. Conditions are given by: $Q/T = 10\%$ and $T = 2.0 \times 10^{-5}s$. Figure 11.11(a) shows the triangular message signal and the recovered message signal. Figure 11.11(b) shows the binary message signal and the recovered message signal. Observe that whenever the recovered chaotic carrier $\tilde{v}_1(t)$ is in the negative region,

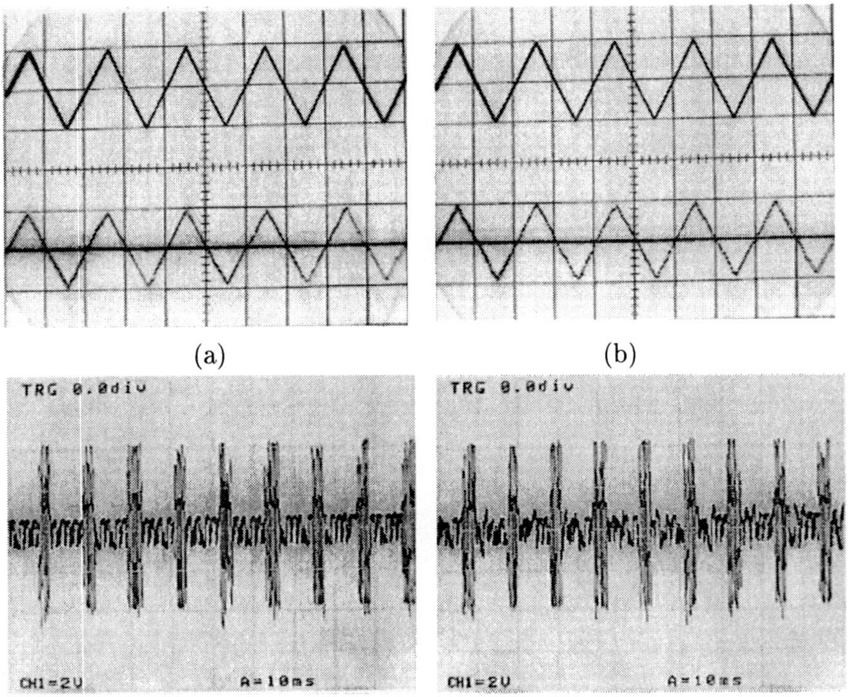

(a) (b) (c) (d)

Fig. 11.10 Different kinds of signals observed in experiments when hyperchaotic circuits and additive modulation are used. (a) Message signal (the upper trace) and recovered message signal (the lower trace). $Q/T = 26\%$ and $T = 5.0 \times 10^{-6}s$. (b) Message signal (the upper trace) and recovered message signal (the lower trace). $Q/T = 26\%$ and $T = 8.0 \times 10^{-6}s$. (c) Transmitted signal without the modulated message signal. (d) Transmitted signal with the modulated message signal.

the recovered result jumps to a saturated value. But the message signal is recovered when $\tilde{v}_1(t)$ is positive.

Figure 11.12 shows the results when the message signal $s(t)$ is modulated by $v_2(t)$; namely, the modulated message signal is $v_2(t)s_1(t)$. Conditions are given by: $Q/T = 10\%$ and $T = 2.0 \times 10^{-5}s$. Figure 11.12(a) shows the triangular message signal and the recovered message signal. Figure 11.12(b) shows the binary message signal and the recovered message signal. Since the recovered carrier $\tilde{v}_2(t)$ switches between positive and negative regions at a rather high frequency, the recovered signals are modulated by short

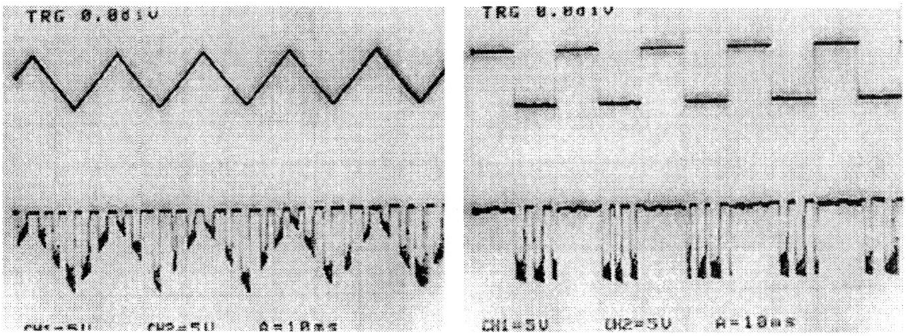

(a) (b)

Fig. 11.11 Experimental results when DS modulation and Chua's oscillators are used. v_1 is chosen as the chaotic spreading carrier. $Q/T = 10\%$ and $T = 2.0 \times 10^{-5}s$. (a) The triangular message signal $s(t)$ (the upper trace) and the recovered message signal $r(t)$ (the lower trace). (b) The binary message signal $s(t)$ (the upper trace) and the recovered message signal $r(t)$ (the lower trace).

impulses. By using low-pass filters, we can easily improve the recovered message signals.

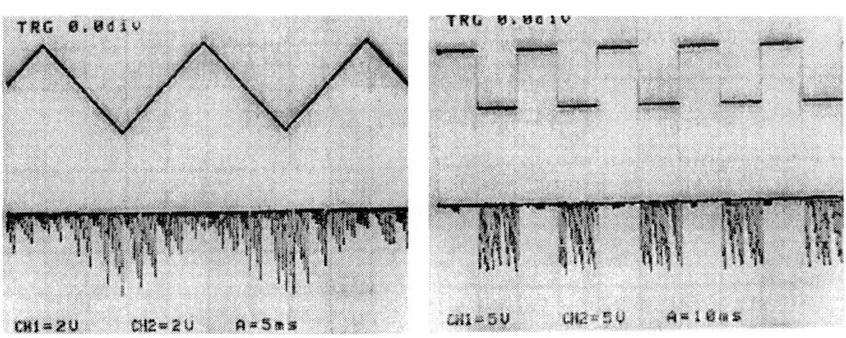

(a) (b)

Fig. 11.12 Experimental results when DS modulation and Chua's oscillators are used. v_2 is chosen as the chaotic spreading carrier. $Q/T = 10\%$ and $T = 2.0 \times 10^{-5}s$. (a) The triangular message signal $s(t)$ (the upper trace) and the recovered message signal $r(t)$ (the lower trace). (b) The binary message signal $s(t)$ (the upper trace) and the recovered message signal $r(t)$ (the lower trace).

11.3.2.2 *Hyperchaotic circuit*

The experimental results for the DS modulation when hyperchaotic circuits are used are shown in Fig. 11.13. The message signal $s(t)$ is modulated by $(v_2(t) - v_1(t))$; namely, the modulated message signal is $(v_2(t) - v_1(t))s_1(t)$. Conditions are given by: $Q/T = 34\%$ and $T = 2.5 \times 10^{-6}s$. Figure 11.13(a) shows the triangular message signal and the recovered message signal. Figure 11.13(b) shows the binary message signal and the recovered message signal. Figure 11.13(c) shows the transmitted signal, which consists of both synchronization impulses and the modulated message signal.

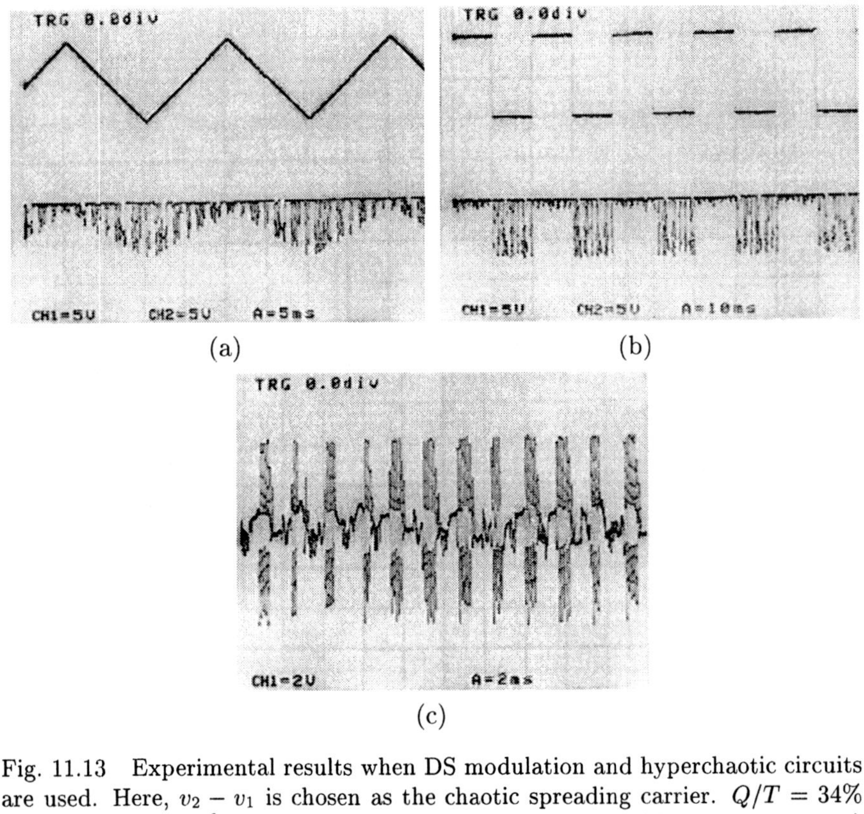

Fig. 11.13 Experimental results when DS modulation and hyperchaotic circuits are used. Here, $v_2 - v_1$ is chosen as the chaotic spreading carrier. $Q/T = 34\%$ and $T = 2.5 \times 10^{-6}s$. (a) The triangular message signal $s(t)$ (the upper trace) and the recovered message signal $r(t)$ (the lower trace). (b) The binary message signal $s(t)$ (the upper trace) and the recovered message signal $r(t)$ (the lower trace). (c) Transmitted signal.

From the above experimental results we conclude that both chaotic spread spectrum communication systems work well. We also conclude that impulsive synchronization is useful for synchronizing chaotic systems in the transmitter and the receiver even though Q is not small.

11.3.3 *Effects of channel noise*

The robustness of chaotic spread spectrum communication systems based on impulsive synchronization is studied by adding noise into the communication channel. Only cases when the message signals are binary are studied in our experiments. Figure 11.14 shows experimental results when Chua's oscillators and DS modulations are used. Figure 11.14(a) shows the binary message signal and the recovered message signal. Observe that the result can be easily improved by low-pass filtering and thresholding. The channel noise is shown in Fig. 11.14(b). Observe that the amplitude of the additive channel noise is around 1/10 of that of the transmitted signal. Figure 11.14(c) and (d) show the oscilloscope traces on the v_1-v_1' and v_2-v_2' planes, respectively. Observe that the synchronization error is small.

Figure 11.15 shows experimental results when the hyperchaotic circuits and DS modulations are used. The chaotic spreading carrier is $v_2(t)-v_1(t)$. Figure 11.15(a) shows the binary message signal and the recovered message signal. We can easily improve the recovered results by low-pass filtering and thresholding. The channel noise is shown in Fig. 11.15(b). In this case, the amplitude of the noise is almost 1/2 of that of the transmitted signal. Figure 11.15(c) and (d) show the oscilloscope traces on the v_1-v_1' and v_2-v_2' planes, respectively. Observe that the synchronization error is small.

From our experiments we found that the binary message signal can be completely recovered if the SNR is around 12 dB. In this case, the 45-degree lines become very thick. This means that we can recover the binary information signal even if the synchronization is far from the desired almost-identical synchronization. Furthermore, if we use some error-correction coding techniques, the binary message signal can be completely recovered under an even low SNR.

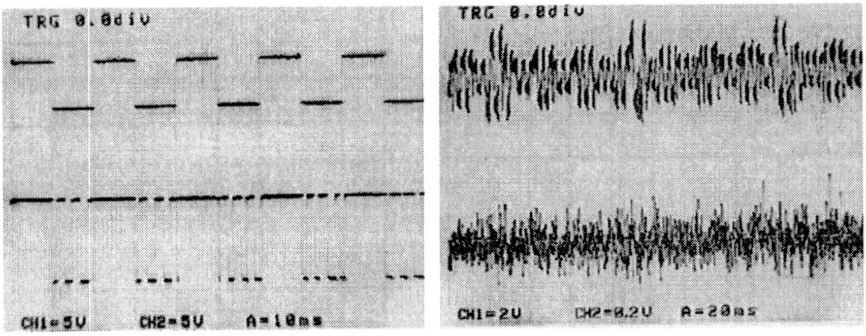

(a) (b)

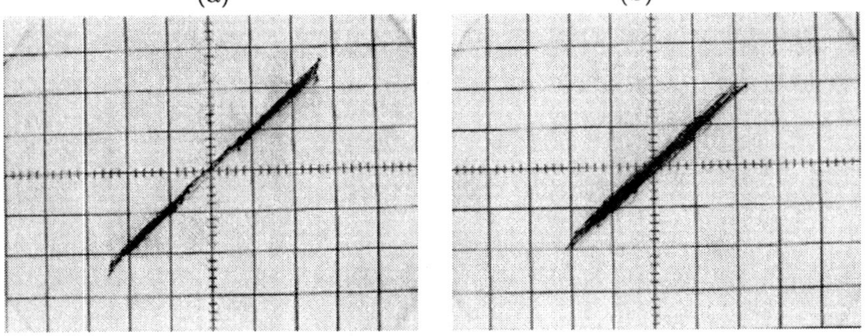

(c)

Fig. 11.14 Experimental results on robustness of spread spectrum communication systems to additive channel noise. The voltage $v_1(t)$ of Chua's oscillator is chosen as the chaotic spreading carrier and DS modulation is used. $Q/T = 10\%$, $T = 20\mu s$. (a) The binary message signal $s(t)$ (the upper trace) and the recovered message signal $r(t)$ (the lower trace). (b) Transmitted signal (the upper trace) and additive channel noise (the lower trace). (c) Oscilloscope trace of v_1-v_1' plane. (d) Oscilloscope trace of v_2-v_2' plane.

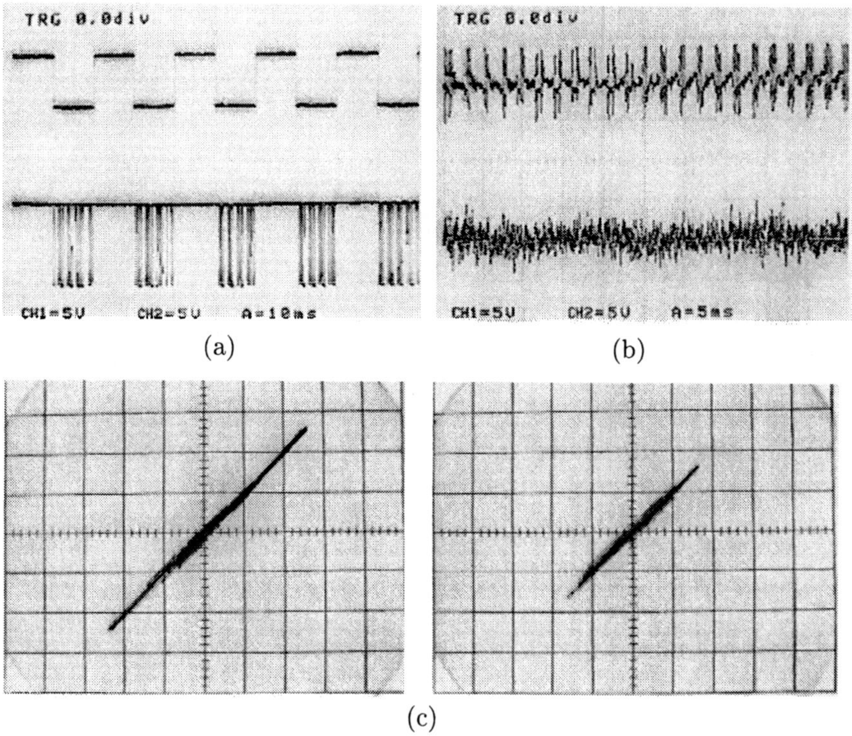

Fig. 11.15 Experimental results on robustness of spread spectrum communication systems to additive channel noise. The voltage $v_2(t) - v_1(t)$ of the hyperchaotic circuit is chosen as the chaotic spreading carrier and DS modulation is used. $Q/T = 34\%$, $T = 2.5\mu s$. (a) The binary message signal $s(t)$ (the upper trace) and the recovered message signal $r(t)$ (the lower trace). (b) Transmitted signal (the upper trace) and additive channel noise (the lower trace). (c) Oscilloscope trace of v_1-v_1' plane. (d) Oscilloscope trace of v_2-v_2' plane.

Bibliography

D. Bainov and V. Covachev. *Impulsive Differential Equations with a Small Parameter*, volume 24 of *Series on Advances in Mathematics for Applied Sciences*. World Scientific, Singapore, 1994.

D.D. Bainov and P.S. Simeonov. *Systems With Impulsive Effect: Stability, Theory and Applications*. Ellis Horwood Limited, Chichester, 1989.

D.D. Bainov and P.S. Simeonov. *Impulsive Differential Equations: Periodic Solutions and Applications*. Longman Group UK Limited, 1993.

T.E. Carter. Optimal impulsive space trajectories based on linear equations. *Journal of Optimization Theory and Applications*, 70(2):277–297, Aug. 1991.

L.O. Chua. Chua's circuit—An overview ten years later. *Journal of Circuit, Systems, and Computers*, 4(2):117–159, Jan. 1994.

V.K. Garg, K.F. Smolik, and J.E. sWilkes. *Applications of CDMA in Wireless/Personal Communications. Feher/Prentice Hall Digital and Wireless Communication Series*. Prentice Hall PTR, Upper Saddle River, NJ, 1997.

E. Huller. Trial/awareness advertising decisions: A control problem with phase diagrams with non-stationary boundaries. *Journal of Economic Dynamics and Control*, 6:333–350, 1983.

K.S Halle, C. W. Wu, M. Itoh, and L.O. Chua. Spread spectrum communication through modulation of chaos. *International Journal of Bifurcation and Chaos in Applied Sciences and Engineering*, 3(2):469–77, April 1993.

M. Itoh, T. Yang, and L. O. Chua. Experimental study of impulsive synchronization of chaotic and hyperchaotic circuits. *International Jour-*

nal of Bifurcation and Chaos, 9(7):1393–1424, Jul. 1999.

Journal. *Dynamics of Continuous, Discrete and Impulsive Systems.*

R.A. Kiehl and T. Ohshima. Bistable locking of single-electron tunneling elements for digital circuitry. *Applied Physics Letters*, 67(17):2494–2496, 23 Oct. 1995.

F.Y. Liu, F.-T. An, and R.A. Kiehl. Ternary single electron tunneling phase logic element. *Applied Physics Letters*, 74(26):4040–4042, 28 June 1999.

V. Lakshmikantham, D.D. Bainov, and P.S. Simeonov. *Theory of Impulsive Differential Equations.* World Scientific, Singapore, 1989.

X.-X. Liao. *Absolute Stability of Nonlinear Control Systems.* Kluwer Academic Publisher and Science Press, Boston, 1993.

X.-Z. Liu. Practical stabilization of control systems with impulse effects. *Journal of Mathematical Analysis and Applications*, 116(2):563–76, 15 May 1992.

X.-Z. Liu. Impulsive control and optimization. *Applied Mathematics and Computation*, 73(1):77–98, Nov. 1995.

V. Lakshmikantham and X.-Z. Liu. Stability for impulsive differential systems in terms of two measures. *Applied Mathematics and Computation*, 29(1, pt.1):89–98, Jan. 1989.

V. Lakshmikantham and X.-Z. Liu. *Stability Analysis in Terms of Two Measures.* World Scientific, Singapore, 1993.

E.N. Lorenz. Deterministic nonperiodic flow. *J. Atmos. Sci.*, 20:130–141, 1963.

F. A. McRae. Practical stability of impulsive control systems. *J. of Mathematical Analysis and Applications*, 181:656–672, 1994.

K. Murali, M. Lakshmanan, and L.O. Chua. Bifurcations and chaos in the simplest dissipative non-autonomous circuit. *International Journal of Bifurcation and Chaos in Applied Sciences and Engineering*, 4(6):1151–1524, 1994.

T. Ohshima and R.A. Kiehl. Operation of bistable phase-locked single-electron tunneling logic elements. *Journal of Applied Physics*, 80(2):912–923, 15 July 1996.

J.M. Ortega. *Matrix Theory—A Second Course.* Plenum Press, New York, 1987.

T. Pavlidis. Stability of systems described by differential equations containing impulses. *IEEE Trans. on Automatic Control*, AC-12(1):43–45,

Feb. 1967.

A. I. Panas, T. Yang, and L. O. Chua. Experimental results of impulsive synchronization between two Chua's oscillators. *International Journal of Bifurcation and Chaos*, 8(3):639–644, Mar. 1998.

M.K. Simon, J.K. Omura, R.A. Scholtz, and B.K. Levitt. *Spread Spectrum Communications Handbook.* McGraw-Hill, New York, 1994.

A.M. Samoilenko and N.A. Perestyuk. *Impulsive Differential Equations.* World Scientific, Singapore, 1995.

J.A.K. Suykens, T. Yang, and L. O. Chua. Impulsive synchronization of chaotic Lur'e systems by measurement feedback. *International Journal of Bifurcation and Chaos*, 8(6):1371–1381, Jun. 1998.

J.A.K. Suykens, T. Yang, J. Vandewalle, and L.O. Chua. *Impulsive Control and Synchronization of Chaos*, in G.R. Chen(Ed.), Controlling Chaos and Bifurcations in Engineering Systems, pp.275-297, CRC Press: Boca Raton, 1999.

A.J. Viterbi. *CDMA: Principles of Spread Spectrum Communication.* Addison-Wesley wireless communications series. Addison-Wesley Pub. Co., Reading, Mass., 1995.

T. Yang. Recovery of digital signals from chaotic switching. *International Journal of Circuit Theory and Applications*, 23(6):611–615, Nov.-Dec. 1995.

T. Yang. Chaotic secure communication systems: History and new results. *Telecommunications Review*, 9(4):597–631, Jul.-Aug. 1999.

T. Yang. Impulsive control. *IEEE Trans. on Automatic Control*, 44(5):1081–1083, May 1999.

T. Yang. *Advances in Computational Verb Systems.* Nova Science Publishers, Inc., Huntington, NY, 2000. in press.

T. Yang and L.O. Chua. Chaotic digital code-division multiple access (CDMA) systems. *International Journal of Bifurcation and Chaos*, 7(12):2789–2805, Dec. 1997.

T. Yang and L. O. Chua. Practical stability of impulsive synchronization between two nonautonomous chaotic systems. *International Journal of Bifurcation and Chaos*, 10(4):859–867, Apr. 2000.

T. Yang and L.O. Chua. Single-electron tunneling junction cellular nonlinear networks. Technical Report Memorandum No. UCB/ERL M00/10, Electronics Research Laboratory, College of Engineering, University of California, Berkeley, CA 94720, 9 Mar. 2000. page 1-20.

T. Yang and L.O.Chua. Impulsive control and synchronization of chaotic systems and secure communication. Memorandum UCB/ERL M97/12, Electronics Research Laboratory, College of Engineering, University of California, Berkeley, CA 94720, 29 Jan. 1997.

T. Yang and L.O.Chua. Impulsive control and synchronization of nonlinear dynamical systems and application to secure communication. *International Journal of Bifurcation and Chaos*, 7(3):645–664, Mar 1997.

T. Yang and L.O.Chua. Impulsive stabilization for control and synchronization of chaotic systems: Theory and application to secure communication. *IEEE Transactions on Circuits and Systems-I*, 44(10):976–988, Oct. 1997.

H. Ye, A.N. Michel, and L. Hou. Stability theory for hybrid dynamical systems. *IEEE Trans. on Automatic Control*, AC-43(4):461–474, Apr. 1998.

T. Yang, J.A.K. Suykens, and L. O. Chua. Impulsive control of nonautonomous chaotic systems using practical stabilization. *International Journal of Bifurcation and Chaos*, 8(7):1557–1564, Jul. 1998.

T. Yang, C.W. Wu, and L.O. Chua. Cryptography based on chaotic systems. *IEEE Transaction on Circuits and Systems—I: fundamental theory and applications*, 44(5):469–472, May 1997.

T. Yang and L.B. Yang. The global stability of fuzzy cellular neural network. *IEEE Transactions on Circuits and Systems–I: Fundamental Theory and Applications*, 43(10):880–883, Oct. 1996.

L.B. Yang and T. Yang. Impulsive synchronization of nonautonomous chaotic systems. *Acta Physica Sinica*, 49(1):33–37, Jan. 2000.

T. Yang, L.-B. Yang, and C.-M. Yang. Impulsive control of Lorenz system. *Physica D*, 110(1-2):18–24, 1 Dec. 1997.

T. Yang, L.-B. Yang, and C.-M. Yang. Impulsive synchronization of Lorenz systems. *Physics Letters A*, 226(6):349–354, Mar. 3 1997.

T. Yang, C.M. Yang, and L.B. Yang. Break chaotic switching using generalized synchronization: Examples. *IEEE Transaction on Circuits and Systems-I*, 45(10):1062–1067, Oct. 1998.

T. Yang, L.B. Yang, and C.M. Yang. Application of neural networks to unmasking chaotic secure communication. *Physica D*, 124(1-3):248–257, 1 Dec. 1998.

T. Yang, L.B. Yang, and C.M. Yang. Breaking chaotic secure communication using a spectrogram. *Physics Letters A*, 247(1-2):105–111, 5 Oct. 1998.

T. Yang, L.B. Yang, and C.M. Yang. Cryptanalyzing chaotic secure communications using return maps. *Physics Letters A*, 245(6):495–510, 31 Aug. 1998.

INDEX

D

E

F

G

H

I

J

L

M

N

U

V

W